MATH/STAT
LIBRARY

RANDOM CURVES

Neal Koblitz

RANDOM CURVES

**Journeys
of a Mathematician**

 Springer

Neal Koblitz
Department of Mathematics
University of Washington
Seattle, WA 98195
USA
koblitz@math.washington.edu

ISBN 978-3-540-74077-3   e-ISBN 978-3-540-74078-0

DOI 10.1007/978-3-540-74078-0

Cover photo: The author at Machu Picchu in 1993.

Library of Congress Control Number: 2007934828

Mathematics Subject Classification (2000): 01A70, 01A60, 01A65, 94-03, 97C60

This work is subject to copyright. All rights are reserved, whether the whole or part of the material is concerned, specifically the rights of translation, reprinting, reuse of illustrations, recitation, broadcasting, reproduction on microfilm or in any other way, and storage in data banks. Duplication of this publication or parts thereof is permitted only under the provisions of the German Copyright Law of September 9, 1965, in its current version, and permission for use must always be obtained from Springer. Violations are liable for prosecution under the German Copyright Law.

© 2008 Springer-Verlag Berlin Heidelberg

The use of general descriptive names, registered names, trademarks, etc. in this publication does not imply, even in the absence of a specific statement, that such names are exempt from the relevant protective laws and regulations and therefore free for general use.

Typesetting by the author and EDV-Beratung Frank Herweg using a Springer LATEX macro package

Cover Design: WMXDesign GmbH, Heidelberg

Printed on acid-free paper

9 8 7 6 5 4 3 2 1

springer.com

To the memory of my mother

Minnie Weinberg Koblitz

November 17, 1923 – April 9, 2007

ACKNOWLEDGMENTS

I would like to thank Scott Vanstone and Ian McKinnon for suggesting that I write these memoirs. This happened at Crypto 2006 after a dinner and several beers; I'm not sure whether they would have made the suggestion had it not been for the latter. I also want to thank the following people for reading all or part of an earlier version and making helpful comments: L. Timmel Duchamp, Tom Duchamp, Darrel Hankerson, Ann Hibner Koblitz, Ellen Koblitz, Robert J. Koblitz, Alfred Menezes, Leon Pintsov, Luther M. Ragin, Jr., Bruce Richmond, and Scott Vanstone. Needless to say, any errors that remain are my sole responsibility, as are all of the opinions expressed on various subjects.

Finally, I wish to thank the people at Springer-Verlag, in particular Ruth Allewelt, Martin Peters, Angela Schulze-Thomin, and Ute Bujard, for working closely with me and putting great effort into ensuring a high quality of production.

Neal Koblitz
Seattle, October 2007

CONTENTS

CHAPTER 1	**EARLY YEARS**	1
CHAPTER 2	**HARVARD**	9
CHAPTER 3	**SDS**	21
CHAPTER 4	**THE ARMY**	47
CHAPTER 5	**SPRING OF 1972**	63
CHAPTER 6	**ACADEMICS**	85
CHAPTER 7	**THE SOVIET UNION**	95
CHAPTER 8	**RACISM AND APARTHEID**	123
	PHOTO SECTION	143
CHAPTER 9	**VIETNAM PART I**	163
CHAPTER 10	**VIETNAM PART II**	195
CHAPTER 11	**NICARAGUA AND CUBA**	227
CHAPTER 12	**EL SALVADOR AND PERU**	253
CHAPTER 13	**TWO CULTURES**	277
CHAPTER 14	**CRYPTOGRAPHY**	297
CHAPTER 15	**EDUCATION**	331
CHAPTER 16	**ARIZONA**	353
	INDEX	375

CHAPTER 1 EARLY YEARS

My academic career got off to a very inauspicious beginning: I was held back in kindergarten. After my first year in school, my family moved to a different district, which had a higher age requirement for children entering the first grade. My parents tried to argue that an exception should be made for me, since I had already had one year of kindergarten. However, the school authorities pointed to the undeniable signs of my developmental immaturity — most obviously, that I stuttered — and I was forced to repeat the year.

Many years later, when my wife Ann felt that I was being too harsh on my students — holding them to what she considered unrealistically high standards — she would threaten to "out" me as a hypocrite. "Mr. High Academic Standards here," she was going to tell my students, "flunked *kindergarten.*"

My prospects improved when my family spent the academic year 1955-1956 in India. This was only ten years after the end of World War II and eight years after Indian independence from British rule. It was also just a few years after the U.S. government established the Fulbright program, and my father (Robert J. Koblitz) had a Fulbright fellowship to teach American government at the University of Baroda, located in Gujarat province roughly halfway between Bombay (now Mumbai) and Delhi. He went by plane in the spring in order to arrive in time for the start of the Indian semester. My mother (Minnie Weinberg Koblitz) waited until June so that I could finish school — my second year of kindergarten — and she then took me and my younger brother and sister on a month-long boat trip to India. To children of ages two, four, and six a month on an ocean-liner seemed a long time — my recollection is that during most of the voyage we had some combination of seasickness, heat exhaustion, and/or terminal boredom — and to our mother the voyage undoubtedly seemed even longer.

In India not only did I stop stuttering, but, more importantly, I was allowed to skip the first grade. I attended a Catholic school, because that was the only one available where the teaching was in English rather than Hindi or Gujarati. The math curriculum was relatively advanced compared to that of American schools. In the second grade we learned how to multiply multi-digit numbers, and we memorized the times-tables through 16 times 16.

This turned out to be crucial. When I returned to the U.S. and entered the third grade, I showed my teacher the arithmetic I had learned how to do, at which point she mistakenly believed that I had a special gift for mathematics. Like other erroneous notions that teachers get into their heads, this sort of belief has a way of becoming a self-fulfilling prophecy. My teacher sent me upstairs to a fifth-grade classroom to borrow a textbook for me to work through — this was a heady experience for a little kid. As a result of all the encouragement I received after my year in India, I was on my way toward a career as a mathematician.

I am indebted to India not only for pointing me in the direction of mathematics, but also for deeply influencing my worldview in another way: I trace my progressive politics to that year. As a six-year-old, I became acutely conscious of the horrible injustice of poverty. Everywhere we went, I saw destitute children my own age who were begging. I wanted to give them something, but my parents tried to explain that that wouldn't solve anything. I daydreamed about becoming rich some day and returning to India to pass out money to all the beggars.

If I had been a few years older or a few years younger, I probably would not have felt such an impact. When we grow up, we learn how to rationalize and accept extreme differences in material well-being. But while a six-year-old can understand and think about what he is seeing, I was not yet old enough to have developed an explanation that would have made it easier to accept.

Two decades later, when I was at Princeton, I met a student whose father was the director of UNICEF and whose family had lived in India for a while. He told me the story of a party his father had given for friends and colleagues, all of whom were foreigners or well-to-do Indians. During the evening his little brother went to the coatroom, picked the jacket pockets of all the guests, took the money out to the streets of Delhi, and gave it to the beggars. Even though I would not have had the initiative to do what this boy did, I fully understood the impulse that caused him to do it.

Obviously, this is not what progressive politics means — although our right-wing critics might accuse us of advocating picking the pockets of the rich in order to redistribute the wealth. Nevertheless, I would guess that many political activists can trace their views to some type of early childhood exposure to extreme social injustice.

In 1959 my family moved to Scarsdale, an affluent suburb of New York City known for its good schools. My mother taught fourth grade there, and my father commuted on a weekly basis to Bard College, located about eighty miles (130 kilometers) to the north, where he taught political science.

Soon after we came to Scarsdale, I skipped the sixth grade. I was considerably younger and less mature than my classmates, some of whom started calling me "Baby Univac" after the new generation of computers that were in the news in the late 1950's and very early 1960's. This was a friendly nickname — in those years even preteens had a real respect for anyone who was good at math and science. The anti-intellectualism that later became so prevalent in America was largely absent then, at least in the better schools.

My junior and senior high school years coincided with the post-Sputnik era, which saw a national obsession with the "space race" and an atmosphere of public enthusiasm for science and technology. In 1963, after the tenth grade, I spent the summer at a program in mathematics at San Diego State College (now University) that was sponsored by the National Science Foundation. Until then I had studied some higher math with the help of tutors my parents hired or on my own, and I had just finished reading my high school's Advanced Placement Calculus textbook. The director of the summer program, a former student of R. L. Moore named Dr. Deaton, advised me that a much better book to read would be Courant's two-volume calculus text. I started studying that immediately.

A few months later I had some mathematical questions that none of my high school teachers could answer. Through friends of friends of my parents, I contacted the mathematician Mark Kac, who lived in Scarsdale about a 20-minute walk from our house. A member of the generation of brilliant East European refugees who fled the Nazis in the 1930's, Kac was not only a world-renowned mathematician and a leader in the field of probabilities and statistical mechanics, but also a terrific teacher and generous mentor.

One Sunday morning in December 1963 he invited me to his house to talk with him about mathematics. I told him that I had had some trouble understanding what was really going on in Courant's chapter on Fourier series, and Kac said he'd give me an alternative explanation that I would probably find more natural. To test my mathematical level, he also gave me a series of challenging calculus problems that he assured me "only one in 800 Cornell freshmen would be able to solve." (He had taught for many years at Cornell before moving to Rockefeller University in New York City.)

I solved those problems without difficulty. After that, Kac met with me for a couple of hours almost every Sunday morning until I graduated from high school in 1965. He also helped me in many personal ways. When I applied to college, he wrote letters to his friends in the math departments at Harvard, Princeton, and Cornell, so that I would be admitted to those universities. And much later, when I was incarcerated in a U.S. Army prison, he wrote to the Commandant urging him to release me as soon as possible so that I could "contribute to society." Shortly before he died in 1984, Mark Kac wrote a wonderful autobiography titled *Enigmas of Chance*.

My most positive memories of Scarsdale High School are of my English classes. In that subject I had excellent teachers, and they had high standards for student writing. My first long paper, on the Romantic poet Percy Bysshe Shelley, was for my tenth grade teacher Doris Breslow. Once in the 1990's, when my wife Ann was commenting on the difficulties her students were having with their writing, out of curiosity I went to the basement and fished my essay on Shelley out of a box of old papers and gave it to her to read. She said that it was better written than anything she had received in the previous ten years from her college students.

In high school I was not, however, considered a particularly good English student; for instance, I didn't make the cut into the Advanced Placement class. When I entered Harvard, I assumed that my level in most of the non-sciences would lag behind that of other students, many of whom had been valedictorians or editors of their high school newspapers. But to my surprise, after taking the placement exam I was admitted to an Honors freshman writing class. At that point I started to appreciate how fortunate I had been to have gone to a high school that took great pains to teach us how to write well.

Postscripts

From 1997 to 2002 my mother belonged to a memoir-writing group on Cape Cod that consisted of women of the generation that had lived through the Great Depression and World War II. She wrote about eighty pages of reminiscences. With her permission, I am including four vignettes in an edited form. The first describes her meeting in 1955 with Eleanor Roosevelt, who was Ambassador to the United Nations and widow of President Franklin D. Roosevelt; the second is a sketch of my family's living arrangements in Baroda, India in 1955-1956; the third tells about our lunch with the U.S. Ambassador to India; and the fourth recounts a stop we made in Denmark in the course of a four-month camping trip through Europe on our way home from India in the summer of 1956.

Bard College, where my husband Bob taught, had some illustrious neighbors in the Hudson Valley of New York. One of them was Eleanor Roosevelt, to whom I wrote a letter in early 1955 telling her how much I enjoyed her book on India. We were leaving for India in June with our three children, ages two, four, and six.

Mrs. Roosevelt's secretary phoned to invite me for tea at her home, Valkill, which was located about 18 miles (30 kilometers) from us in Hyde Park. When I asked if my husband was included, the secretary asked Mrs. Roosevelt and assured me that he was.

We were pretty excited to have such an invitation, and we assumed there would be many other guests. Imagine our surprise when we drove into the driveway of Valkill to see only one other car, a station wagon belonging to Mrs. Roosevelt.

We came in the back door, and a casually dressed, large man, probably a bodyguard, answered. He told us we were expected, and called upstairs, "Mrs. Roosevelt, your guests are here." Her voice, familiar to us from radio broadcasts, came down, "I'll be right there."

We were ushered into the living room, and she joined us bringing a pot of tea. The three of us sat around and talked about India for an hour-and-a-half. Mrs. Roosevelt told us of her recent visit and was very enthusiastic about the people she had met. She said that the Indians would be pleased to have our children among them, and added that we had courage to take our small children to India because it was such a poor country. She told us to be sure to sleep under mosquito netting, and commented, "You will see women squatting down and sweeping and doing all kinds of menial tasks."

When Mrs. Roosevelt heard that we would be at the University of Baroda, she told us that that was quite a coincidence. Her close friend on the United Nations Human Rights Commission was the Indian representative, Hansa Mehta. Mrs. Mehta was Vice-Chancellor (what we would call the president) of the University of Baroda!

What a marvelous afternoon! We were enthralled with this gracious woman and felt completely comfortable talking with her, perhaps partly because of the lack of pretentiousness in our surroundings. The living room was plain with a large piano covered with family photographs. The furniture was slip-covered and looked well used. The house was inviting and informal, a contrast to the formal Hyde Park home of Franklin D. Roosevelt across the road.

The day after I arrived in India I received an invitation to tea from Mrs. Hansa Mehta. There were about fifteen women present, and I was introduced to all of them as "Mrs. Roosevelt's friend." Mrs. Mehta told me that she had received a letter from Mrs. Roosevelt advising her that I was coming. During the year Bob and I were invited to the Vice-Chancellor's house on several occasions.

I wrote to Mrs. Roosevelt a few times, thanked her, and sent my impressions of India. I still have her replies among my cherished possessions.

When we were traveling in India, we met a woman who told a story about Mrs. Roosevelt that I believe is typical. The woman had been staying in a Quaker guest house in Madras, when she was asked if she'd mind sharing her room for a night. To her amazement her companion was Eleanor Roosevelt. She was even more astounded when Mrs. Roosevelt apologized for intruding and thanked her for sharing her room!

For ten months we lived in a house on the edge of the University of Baroda. In 1955-1956 Baroda had about 400,000 residents. I knew little about it before we lived there, but I had read a novel called *The Rains Came* by Louis Bromfield which takes place in Baroda.

Our house was called Shevanti Bagh. I surmised that "bagh" meant "dwelling" in Hindi or Gujarati, but I didn't know what "shevanti" meant.

We had a large lawn cared for by the gardener or *mali*, who spent a lot of time chasing away the cows or throwing stones at the monkeys. The monkeys loved to jump on our metal roof at 5 a.m., making a tremendous clatter and waking everyone up.

It was a two-family house, and we lived downstairs. We had a large living-dining room, and a typical day began with breakfast with the two big doors wide open. Our children loved to see the birds come swooping in the doors and eat their cereal as they sat watching and attempting to get a few bites. They also liked to look at the small chameleons crawling on the inside walls catching mosquitos.

Off the hall was a large semi-tiled shower-room with small drains on the side of the floor. We were warned to check the room every day during monsoon season. Small frogs came up the drains, and occasionally snakes followed. We also had to check our shoes in the morning to snatch any frogs lounging inside.

Our house was the only one in the city with screens on the windows. Bob had purchased them and put them in himself before we arrived. People had been amused to see the odd American professor nailing screens over the windows. "Keeps out the fresh air," our colleagues told us.

There was a good-sized garage attached to the house. Living in that one room for many years were our servant Motilal, his wife Monabai, and their three children. He came with the house, and was our cook, helper, and interpreter for all occasions. Monabai was our *ayah* or babysitter.

One of the highlights of our stay in India was a two-week visit to Delhi sponsored by the Fulbright office. During this time we received an invitation to lunch from Ambassador and Mrs. John Sherman Cooper. Bob and Ambassador Cooper had been in the Army together during World War II and had struck up a friendship. In his civilian career Cooper had been a judge, then ambassador, and later a U.S. senator.

Neal, not quite seven, was excited at the prospect of having lunch at the American Embassy in New Delhi. We had probably created inflated expectations by putting his sister Ellen in her best dress and leaving his two-year-old brother Donnie in the care of the Fulbright staff.

When we arrived at the Embassy, in the British tradition the children were separated from us and led to a special table set up for them in the garden. There was an *ayah* to watch them while the grownups had lunch in the dining room. The four adults had a lively time, with Bob and the Ambassador discussing their Army experiences and India. I noticed at the other end of the garden that Neal looked unhappy and agitated.

As we got up to leave after a long, leisurely meal, Neal was insistent that he had to say something very important to the Ambassador. He said, "There are three reasons why I hate the British. First, they fought against America in the Revolutionary War; second, they forced India to be their colony; and third, they don't eat with their children." We all howled.

The Scandinavian portion of our 1956 camping trip had a special destination. We were headed for Ejsberg, Denmark to visit a family that Bob's parents had corresponded with for ten years and had never met. The father was a fisherman who had brought up a wallet in his fishing net that he saw had belonged to an American serviceman lost at sea. He found the parents' address inside and mailed the wallet back to them. It was all that was ever recovered from Bob's younger brother Alvin, who was shot down over the North Atlantic during World War II.

When we arrived at the home of the fisherman in our Volkswagen microbus, we parked across the street. Bob knocked, but there was no reply, although I could see movement upstairs from behind a window-curtain. Based on our earlier experience in Scandinavia, we decided that our German car must have caused this reaction. It was ten years after the War, and Germans were still disliked here. So we parked around the corner, and Bob went back to the house. Sure enough, they welcomed him with open arms.

An English-speaking relative was sent for, and our host and hostess explained that the whole extended family was coming over to meet us and have a day of celebration. They also wanted to thank Bob's parents for the gifts of food they had sent immediately after the War, when they had had severe hardships.

What a day! We felt like official American representatives as everybody who came told us how much they appreciated the American troops and their victory over the Nazis. Our children got lots of attention, and even though they hadn't expected us, food and wine appeared in great quantities.

It was a special pleasure to visit the family church. There was an official church in Denmark, but this family and their friends didn't belong. They had built their own dissident church, and it was beautiful. Made in a simple style out of wood, it was resplendent with wrought iron carvings. It turned out that the fisherman and his family were also ironmongers. The altar and the stairways were decorated with beautiful handwork. We had never seen anything like this. After a tour of the town and lots of farewells we went back to our campsite for the night.

CHAPTER 2 **HARVARD**

I was an undergraduate at Harvard from 1965 to 1969. During that time Harvard changed from a rather socially conservative place that expected students to wear a suit jacket to meals, to a hotbed of leftist radicalism and the setting for some of the most tumultuous anti-war protests in the country.

When I entered college, I was already strongly against the Vietnam War. I had some angry and unproductive arguments with my two roommates, one of whom thought that the government knew best and the other of whom was undecided. At this time public opinion, even among liberal Harvard students, was still generally pro-war. Most Democrats supported President Lyndon Johnson and his vision for a "Great Society," and they were tremendously relieved that the "warmonger" Barry Goldwater had lost the election in a landslide the year before. It was during my freshman and sophomore years that opposition to the War started to pick up force, especially among students.

I recall the first time the Vietnam War entered my consciousness. My parents and I were watching President Kennedy's news conference in March 1961. He had a map of Southeast Asia on an easel and was pointing to Laos and Vietnam and explaining that his administration was alarmed at the growth of communist influence and that the U.S. had a vital security interest in the area. I remember that my father exclaimed something about how foolish that was.

But I knew little about the War until 1965, when my family spent the summer in Ithaca, New York. My father attended a special workshop on Southeast Asia that was taught by Cornell's leading specialists George McT. Kahin and John Lewis. He gave me a long paper they had written in the *Bulletin of the Atomic Scientists*.

The article by Kahin and Lewis put the Vietnam War in a historical and cultural context. It explained, for example, that during the war against French colonialism the Việt Minh under Hồ Chí Minh had been able to win overwhelming popular support through its agrarian reforms. For the first time I learned why the democratic elections to unify Vietnam, promised for 1956 in the 1954 Geneva Accords, never took place: in his memoirs Eisenhower said that U.S. intelligence estimates had Hồ Chí Minh winning about 80% of the nationwide popular vote.

But most people relied on the newspapers and television, not articles by Southeast Asia experts, for information about the Vietnam War. And until 1968 the mainstream media faithfully repeated the Johnson administration's version of events and gave poor coverage to the anti-war movement.

The other major social and political issue of the 1960's was, of course, civil rights. Harvard had recently started to admit a significant number of black students. During my six years in the Scarsdale school system I had never had a black classmate, and I was glad to see that Harvard was more integrated than my high school had been. But I was afraid that my social awkwardness might come off badly with the black students. I followed the national news about civil rights, but I didn't know if it was okay to specifically initiate a conversation about that with a black student who was sitting at the table. Would it seem strained and patronizing? Was it better to talk about other things?

I asked my mother about this. She recalled an analogous situation a generation before. In the 1940's she had been one of a small number of Jewish students at Ohio State University. This was a time when anti-semitism and the Nazi Holocaust were very much on people's minds. Sometimes other students would seek out Jewish classmates to talk about it. Yes, that would occasionally be awkward; but she understood that their intentions were good.

The summer after my freshman year I decided to spend a few weeks as a voter-registration volunteer in the South. For years I had been hearing about the civil rights movement secondhand, and now I felt that I was old enough to be a small part of it. Shortly before, the Student Non-violent Coordinating Committee (SNCC) had kicked out the white volunteers. However, Martin Luther King's group, the Southern Christian Leadership Conference (SCLC), was still happy to welcome students from the North. In August 1966, I went to Grenada, Mississippi to work with SCLC. A black family in the town generously provided me with lodging, and I participated in a wide range of SCLC activities. I stayed for only twelve days, not the several weeks I had planned to, but I learned a tremendous amount during that time.

It was an intense experience. My impressions are preserved in the form of a long letter I wrote immediately afterwards to a friend of my family named Vera Rony. She worked for a progressive organization called the Workers Defense League, and at the time she was the most committed political activist I knew.

As I wrote Vera, "To the Negroes, local and SCLC alike, I was a curiosity, not only as the youngest of the only half dozen or so white civil rightniks. I must have seemed from a different world with my briefcase, my Northern accent... and my habit of snacking on cabbage. Because of my relatively formidable vocabulary... and my Harvard identification (even the leadership among the Negroes had very poor formal education), they treated me as if I had some mystical power which could help them." At a mass meeting soon after I arrived that filled the local church, the black leader Leon Hall mentioned, as if it would reassure his listeners, that "one of the volunteers who just came down to help us is from Harvard University."

However, as I explained to Vera, it was never clear to me exactly what I could do. When Hosea Williams came to town (he was one of Martin Luther King's lieutenants and was Leon Hall's superior in SCLC), he decided that as a Harvard math student, I could best be utilized helping a group of local business and professional people set up a black-owned store in Grenada. I tagged along with a delegation that went to Selma, Alabama, where a successful project of that sort was in operation. But I knew that my year at Harvard studying advanced calculus had not equipped me with any relevant knowledge or insights. It was largely because of my "feeling of guilt and frustration... about sponging off the good will and good cooking of the local Negroes without helping them in any tangible way" that I decided to leave Grenada earlier than I had planned.

I couldn't help feeling like a fish out of water. When all of us congregated for the nightly march around the town square, I was unable to immerse myself in the singing and clapping and dancing. The other marchers thought that I was scared and intimidated by the local whites, who were looking on and jeering. In reality, I would have been equally awkward if there had been no hostile spectators. It was social inhibition, not fear, that was my problem.

On the other hand, I enjoyed going with the other Movement people to integrate the beach at Grenada dam. As soon as we showed up, the local whites would leave the water. I wrote:

> Once, a couple young white kids didn't leave when the others did and had to be fetched by an older kid to be brought to the grownups to receive their first lesson in Mississippi racism. That same day I got some crushed ice thrown at me as we left a refreshment stand where we had been debating (I had thought with little bitterness) with some local whites.

Although SCLC had not been influenced by the emerging Black Power ideology to the extent that SNCC had been, there was tension between the white volunteers and the local black leaders who had to rely on them to perform a lot of the bureaucratic tasks of the Movement. I wrote that when Hosea Williams came to town, "Although he was in a terrific hurry that evening, I waited over an hour to go with him to the business and professional meeting while he argued with Leon over some trivial point of keeping the accounts. Simple bookkeeping... evidently is a major problem... and necessitates a reluctant reliance on whites for most of the paperwork."

I went on to describe an incident that showed me how much tension lay beneath the surface:

> Mel Gissen, a law student, had arranged through his father in Mount Vernon, N.Y. for a shoe company to send down some models of shoes they would let us order from them for the local Negroes at wholesale plus transportation cost. The idea was that a local boy and I would set ourselves up in a room in the church and record the orders as the local Negroes came in with the money. When Mel asked me to announce the plan at the mass meeting that night, I went to get permission from Leon, who informed me, to my surprise, that nothing had been settled about the shoes and I would have to wait. I reported this to Mel, who told me I shouldn't have gone to Leon, who, in his words, is unintelligent, incompetent, and only interested in getting credit for himself. Since Reverend Cunningham had O.K.'d the use of the church, none of this need involve SCLC, he told me, or Leon's approval. I then asked Leon why he didn't go along with the idea. First, he explained that the Movement had to be careful about whose ideas, especially those involving money, it could support, since in spite of the likelihood that the Gissens have the best of intentions, there is always the chance that someone is trying to exploit the local Negroes and the Movement. When I asked about Cunningham's approval, he told me that he was a "white man's Negro," who could be duped by any white who seemed to want to help the Negroes.... Leon explained that the key to SCLC is a chain of command which Mel "and other white volunteers" are constantly trying to bypass.

I concluded, "This may give a clue as to how the Negroes in other organizations, such as SNCC, turned anti-white. They must have decided that the roles the whites could play to help the Movement were not sufficiently important to justify the political threat." Although politically inexperienced, I was already beginning to understand the problems and pitfalls that arise when people from relatively privileged backgrounds volunteer to go help the less fortunate. I could see that some whites in the civil rights movement were out for power and control, and I recognized the mistrust that this engendered. In high school and college I was infatuated with Dostoevsky's novels and stories, and my awareness that activists might have mixed motives came in part from his writings, which explored the extreme contradictions in the psychology of his characters.

When Hosea Williams was in Grenada, I listened in on a discussion he was having with some much younger black activists. He was explaining to them why non-violence is preferable to armed struggle. His argument was purely tactical — if you try to use violence, you'll find that the white racists have you outgunned. He spoke derisively of the blacks who were talking of arming themselves — as soon as they see the weapons that the other side has, they'll be hightailing it out of there as fast as they can run! What was interesting to me was that Williams made no reference at all to the moral and philosophical arguments of Gandhi and Martin Luther King.

Although most of the report on my trip to Grenada still reads well four decades later, there are a few passages that make me cringe. After reading them, my wife Ann said that she was glad she hadn't known me then. On occasion my comments were snide and arrogant. I complained about the "middle class Babbittry" of the black professionals I met, sneered at the "tasteless artificial flowers [and] a record of songs from 'Hello Dolly'" that were in the home of the minister whose church was being used by the SCLC, and pompously asked, "Is it too idealistic to hope for loftier objectives?" Ann said she felt like going back in a time machine and smacking my seventeen-year-old face.

During my four years at Harvard I was what would later come to be called a "nerd" or "geek." In Harvard student jargon of the time the term was "wonk" ("know" spelled backwards). I was socially awkward, self-consciously intellectual, extremely serious about my courses, and oblivious to my generation's style of dress. Because of inertia and a dislike of buying clothes, I continued to wear a suit jacket around campus long after the rule requiring formal dress to meals had receded into the dustbin of history. My nerdiness continued even after I became active in anti-war politics in my senior year. The April 25, 1969 issue of *Life* magazine ran a large picture showing about thirty radical students meeting in Harvard's University Hall after it had been seized by protesters and the deans had been forcibly evicted. Most look more or less like what you'd expect students of the 1960's to look like. But one of the students appears completely out of place, with short hair, a chubby face, and a coat and tie. That was me.

Like other major research universities, Harvard has often been accused of neglecting undergraduate education, and the faculty has been described as aloof and inaccessible. That was not my experience. As freshmen, my roommate Rich Ellis and I would regularly go to the office hours of our physics professor, Edward M. Purcell. He kept them without fail and always had time for us. It was a heady experience having one's physics questions answered by a Nobel prizewinner.

In the math department, when I became interested in number theory, John Tate suggested that I read the newly-translated Russian textbook by Borevich and Shafarevich, and then he met with me several times to explain concepts that I was having difficulty with. Even though I never took a course from him, he was always supportive and generous with his time. I have equally warm memories of my senior thesis adviser, Raoul Bott, who was a true humanitarian. I'll come back to him when I discuss faculty reaction to the student protests and police bust of 1969.

Outside of math and physics, my favorite course during the four years was "Dostoevsky, Camus, and Faulkner." Taught by a visiting professor from Chicago named Edward Wasiolek, it was a popular course, with an enrollment of over 150. In his lectures Wasiolek somehow captured not only the life and times of those three authors, but also the spirit of our generation as well. His true passion (and mine) was Dostoevsky. At one point, while lecturing on *The Brothers Karamazov*, he asked the class to vote. Which of the three brothers did we most identify with: Dmitri, the hedonist and party animal; Alyosha, the religious mystic; or Ivan, the intellectual, atheist, and radical? The students voted overwhelmingly for Ivan. I often remember that when I think of later generations of American students, who would undoubtedly vote for Dmitri first, closely followed by Alyosha, with Ivan a distant third.

My most serious interest outside of math was Russian literature. I had started learning Russian in high school in 1962 when, with a lot of fanfare, Scarsdale High School began offering it. At the start about forty students came to the class, which at first was given after school. By the time I was a senior we were down to about a half dozen, and I was the only one who was putting much effort into the course.

In the years following Sputnik and the Cuban missile crisis, the Soviet Union was on everyone's mind. The general philosophy behind expanding Russian studies was the "know your enemy" principle. Indeed, my high school Russian teacher was the wife of the director of Radio Liberty, a virulently anti-communist radio station broadcasting into Eastern Europe that made Voice of America look tame by comparison.

But I was never anti-Soviet. My motives for learning Russian were different, and they were, strangely enough, careerist. Even in high school I realized that, good as I was at math, I would never be at the top. However, if I could go easily between the two mathematical superpowers — the U.S. and the U.S.S.R. — then I'd have a unique role to play. There'd be people who were better mathematicians than I, and there'd be people who knew the Russian language and culture better than I, but there'd be very few Americans who were in both categories. At age thirteen, I was already fashioning myself as some sort of future mathematical ambassador who would help bridge the gap caused by the Cold War. A peculiar and somewhat pretentious notion, perhaps, but it turned out to be very close to what actually occurred.

Starting in high school and continuing through college, I studied Russian on my own as well as in school. I bought recordings of short stories, poems, and book excerpts read aloud by native speakers. I would listen to them dozens of times, following along with the printed text, until I almost memorized whole sections. The writers ranged from Gogol, Tolstoy, and Chekhov to Pasternak, Tiutchev, and Zoshchenko.

Although I eventually took some Russian literature classes at Harvard — one-semester courses on Pushkin and Gogol, and a reading course on the medieval classic *The Lay of Igor's Campaign* — I never took a language course for credit. I did, however, receive special permission to sit in on the Russian conversation sections that were given in conjunction with language courses. Since I wasn't an enrolled student, I was told that I could listen but not speak. I chose the 8 a.m. sections, which were sparsely attended by the enrolled students, so before long the Russian emigrés running the classes forgot that they weren't supposed to let me participate fully.

At that time Harvard strictly limited the number of courses a student could take (without paying extra tuition) to four per semester. The only exception was the small-group tutorials given to a department's majors. I decided to declare a major in Slavic Languages and Literatures rather than in Mathematics (which didn't have any special tutorial courses for its majors) so that I could get a free fifth course and learn more Russian literature than would otherwise have been possible.

After I declared my major, I had an excellent sophomore tutorial with Katherine O'Connor. We read and wrote short papers about some of the less famous works of 19th-century Russian literature. She led us to a much deeper appreciation of the nuances of the novels and short stories. Sometimes our discussions would take us well past the scheduled end of class.

I didn't think much of the graduate student who led our junior tutorial, and so decided to switch my major back to Mathematics a year earlier than I had originally planned. But I continued to take and audit classes in Russian literature, including a graduate course on 18th-century poetry taught in Russian by Kiril Taranovski.

I was the only undergraduate sitting in on Taranovski's lectures. He was an advocate of statistical methodology in the analysis of poetry; in that way he thought he could document the influence of early folk poetry on Pushkin, for example. During one lecture I noticed that he had made a simple arithmetic error in a table, and I told him about it after class. The following week he spent several minutes explaining to everyone what I had pointed out. He thanked me profusely, as if I had supplied a valuable insight.

Another scholar who sometimes used quantitative methods was Roman Jakobson. He was undoubtedly the most famous Slavicist then at Harvard, a living link to the avant-garde era of early Soviet culture, having been part of the same literary circles as the great poets Mayakovsky and Khlebnikov. I met Jakobson once in 1968, just before my second visit to the Soviet Union. Knowing of his interest in mathematical methods, I asked him to suggest mathematicians I could see in Moscow or Leningrad. He gave me the name of R. L. Dobrushin, a leading specialist in probability theory. I did stop in to see Dobrushin in Moscow, but probability theory was far from my own mathematical interests, so I merely conveyed Jakobson's greetings and talked briefly with him.

I spent the summer after my sophomore year in Indiana University's Russian program, which consisted of five weeks of intensive language study followed by five weeks in the Soviet Union. That was my first trip overseas since my family's year in India over a decade earlier.

Our group of over a hundred students traveled to Moscow, Leningrad, Kiev, Pyatigorsk (a resort town in the Caucasus), and Volgograd. In 1967 American students were a novelty in the Soviet Union. Exchange programs and tours were relatively new — they had been made possible by the "thaw" in the Cold War in the late 1950's and early 1960's.

I was thrilled with the opportunities to speak Russian, and I was struck by the young people's eagerness for information about the West — especially current literature and popular music. But the experience, while exciting, was superficial. We spent a week in each city, went on excursions, and heard formal lectures. That was not the way to meet ordinary Russians, not even ordinary young people. Some of those who approached us as "friends" later turned out to be interested in black-market transactions.

It was clear to me that if I wanted a more meaningful visit, I would have to establish ties first through mathematical connections. During this early visit I did make two such contacts. In Leningrad I went to see Z. I. Borevich (a coauthor of the number theory book that I had studied) and Ju. V. Linnik (a close colleague of Mark Kac), both of whom introduced me to some students.

I very much wanted to return for a more serious visit to the Soviet Union. But there were no opportunities available for undergraduate study in the U.S.S.R., and an independent trip of several weeks to Moscow and Leningrad would be expensive.

My break came the following summer, when I learned that the American Mathematical Society had embarked on a major project to translate Soviet journals, including several years of back issues, and they badly needed translators. They paid $6/page, even for articles in analytic number theory that consisted mainly of formulas. In the summer of 1968 I feverishly translated as much as I could, going back and forth by bus to the AMS headquarters in Providence so that I could get new articles to translate without having to wait for them to be mailed. I earned $8,000 in three months, which was a princely sum for a student. (By way of comparison, the cost of a year of tuition, room, board, and expenses at Harvard was $3,400.)

In the autumn of 1968 I went to a travel agent and asked her to arrange an independent trip of three weeks in Leningrad (where I had already met some students) and two in Moscow. I planned my class schedule for the autumn semester of my senior year so that I wouldn't have any exams and could be absent during January. That took some doing, since Harvard had few courses (except for independent study) that had no final exam. But faculty members in both math and Slavic were very cooperative, and I left for the Soviet Union right after classes ended in December.

My second trip to the U.S.S.R. made a much deeper impression on me than the first. In Moscow I met the famous number theorists Igor R. Shafarevich (the other coauthor of the textbook I had read, and later a prominent anti-communist dissident) and Yuri I. Manin (under whose supervision I would later spend a year as a post-doc). In Leningrad most of the time I hung around with a circle of graduate and undergraduate math students.

Even the undergraduates were far more advanced than I was. This came as somewhat of a shock. I was one of the five or six best math majors in my year at Harvard, and yet in Leningrad I was out of my league. Many of the Soviet students had already published original research. They would ask me, for example, what John Tate was currently working on. I knew

Tate only as the professor who had answered the questions that I had after reading the introductory textbook by Borevich and Shafarevich. I had no idea what his research was about, let alone his current interests.

From a mathematical standpoint I was a disappointment to my peers in Leningrad, and I felt a real sense of inferiority. At the time I didn't realize that the Soviet system of scientific education and research, impressive as it was, wasn't perfect. Of the students whose knowledge of advanced mathematics dwarfed my own, many burnt out at a young age or got academic positions in the provinces, where the research environment was far worse than in Moscow and Leningrad. Ultimately, that generation produced some top scientists and mathematicians, but probably no more than came out of the American university system.

Despite my feeling of inadequacy as a mathematical ambassador, at least I was able to satisfy my peers' hunger for Western popular music. From my earlier trip I knew that Soviet customs allowed visitors to come with a maximum of ten records. I also knew that most of the albums brought in by tourists ended up in Moscow, where young people passed them from hand to hand, copying them onto tapes and recopying them. But other cities, including Leningrad, had far less access to American youth culture.

In 1968 I was still not interested in popular music. When I studied I liked to have Mozart or Bach on my stereo. I had a mild interest in some of the folk singers who were heroes of the early anti-war movement — especially Pete Seeger and Bob Dylan — but I didn't have their records with me in college. In any case, I knew that those weren't what the Soviet young people would want.

In the autumn of 1968 in preparation for my trip I went for help to Tom Hopkins, an English major who lived in my dormitory and was known as a sort of one-man cultural avant-garde of Winthrop House. (Later that year he singlehandedly organized a Warhol Film Festival in the dorm.) I presented him with an interesting problem: Give me a list of the ten most cutting-edge popular music albums. He consulted with his friend John Newmeyer, and together they came up with a list. I don't remember all of them, but I know that they included the most recent albums by the Beatles, the Rolling Stones, Janis Joplin, Jimi Hendrix, the Steve Miller Band, and Cream. At the time I paid little attention to the names. It was only a few months later — largely under the influence of Tom Hopkins — that I abandoned classical music and joined the rest of my generation to become a fan of rock music.

It was soon after my return from the Soviet Union in early 1969 that I became heavily involved in anti-war politics. Until that time I had been on the periphery of Students for a Democratic Society (SDS) and had admired what radical students were doing. But I hated the thought of going to meetings, and did not think that political organizing was something I was cut out to do.

I had had an experience the previous year that reinforced my doubts about how effective I could be as an organizer. In April 1968, I was invited to join Phi Beta Kappa, but I turned it down. As I wrote to the chapter president, Phi Beta Kappa has "too many of the features of fraternities," promotes "an unjustified attitude of superiority and condescension toward those not in the group," and "aggravates what is anyway one of Harvard's worst problems — complacency and self-glorification." I added that "grades make a ludicrous criterion of intellectual worth" and that "if anyone at Harvard needs more recognition, it is the militant political activists."

In high school I had not been in the honor society, and I was extremely pleased when my friend Alan Boles turned down membership in it. That was in 1964, and Alan had, in turn, been inspired by Jean-Paul Sartre's recent refusal of the Nobel Prize for Literature. To many in my generation, turning down an honor was much cooler than accepting it.

When *The Harvard Crimson* ran a front-page article on my refusal of membership in Phi Beta Kappa, many students came up and congratulated me on my stand. Even some members of the organization told me that they basically agreed with what I had said.

This all led me, in my naiveté, to think that my individual act could be turned into something more. For several days I tried to organize a mass resignation from Phi Beta Kappa, starting with the members whom I knew to be political leftists. I imagined that with the help of Alan Boles, who was then the editor of Yale's student newspaper, the attack on Phi Beta Kappa would spread to other campuses.

But I ended up bitterly disillusioned. None of the current members wanted to resign — not even the political leftists. Although they generally shared my view of the organization as antiquated and reactionary, they didn't want to anger their parents, and they didn't want to risk any adverse effect on their future careers if they gave up their Phi Beta Kappa key. So my first attempt to organize the masses fizzled miserably.

CHAPTER 3 **SDS**

In the spring of 1968, as I was trying to stir up trouble for Phi Beta Kappa, much more important things were taking place on other campuses. Columbia University's chapter of SDS had taken over several buildings in protest against Columbia's complicity with the Vietnam War and its real estate policies as it expanded into neighboring Harlem. The student uprising lasted several days and was widely covered (with harsh disapproval) in the newspapers. I wrote to a high school friend who was studying at Columbia, congratulating him on his classmates' militancy and complaining that Harvard students were such a pampered and complacent lot that they would never do anything similar. History would prove me wrong about that.

Many American students were also following events in France with great interest. That spring a massive revolt by students and workers almost brought down the De Gaulle government. A few months later I passed briefly through Paris on the way back from Moscow. I had intended to return via Spain to see my sister Ellen, who was spending the year there. However, the Soviet airline Aeroflot canceled my flight with no explanation, and I had to be rerouted with a day in Paris. When I got there, I telephoned Professor Tate, who was spending the year in France. (How I got his phone number I don't remember.) It was an indication both of my *chutzpah* as a Harvard student and Tate's easygoing, friendly manner that I thought nothing of calling him out of the blue and imposing myself on him and his wife.

They invited me over and found an inexpensive hotel for me for the night. I told Tate about the amazing students I had met in Leningrad and about their interest in his work. He and his wife described to me what it was like to be in Paris during the spring 1968 uprising, when protesters had paralyzed the city. I was fascinated, and I was envious of the French leftists.

Throughout 1967 and 1968 many students around the U.S. were becoming angry and bitter about what our government was doing in Vietnam. A special issue of the liberal Catholic magazine *Ramparts* had recently been devoted to the victims of napalm and other chemicals. The photographs of Vietnamese children burned by napalm were horrific. People started to speak of "genocide" and talk of Americans who failed to protest as being akin to the "good Germans" who acquiesced in the Nazi Holocaust.

Harvard was a center of complicity with the War. Some of the top war planners in the Kennedy and Johnson administrations — men like McGeorge Bundy, Samuel Huntington, and Henry Kissinger — had come from Harvard. Napalm, as Harvard-Radcliffe SDS never tired of pointing out, had been tested on the Harvard soccer field. And the Harvard chapter of the Reserve Officer Training Corps (ROTC, which we pronounced *rot-sea*) was preparing students to command troops in Vietnam.

It seemed to many of us that Harvard students had done little to protest. We were impatient, and in fact that feeling was a little unfair. In spring of 1967 a large number had signed a national "We Won't Go" statement. That spring also saw a major protest when a recruiter from Dow Chemical, the makers of napalm, came to campus. Anti-war students blockaded the room where the recruiter was interviewing, and wouldn't let anyone in or out. When university officials asked the protesters for their student ID's so that they could proceed with punishment, about a hundred others who were watching but not taking part in the sit-in — including me — handed in our ID's in solidarity, so as to make it impossible for the officials to determine whom to punish for the blockade.

A few months before, an incident had occurred that illustrated the depth of anger at the Vietnam War among Harvard students. Defense Secretary Robert McNamara visited the campus to speak to small groups of faculty and students. When he refused SDS demands to debate a representative of the anti-war movement, students organized a demonstration that almost turned violent. As McNamara got into his car to leave, protesters surrounded it and refused to budge. I recall the alarmed looks on the faces of his security detail as waves of angry students surged against the car, rocking it and keeping him hostage. Later, campus conservatives organized a letter of apology to the Defense Secretary, and I remember being furious that over a thousand Harvard students signed it.

Until February of my senior year I was only on the periphery of SDS. I knew several people in the organization with whom I'd talk on a regular basis. I had met Pete Bilazarian in my freshman year when he came to clean the toilets in my dorm — that was his part-time job to work his way through college. Harvard was truly admitting a more diverse class than ever before in the sense of race, ethnicity, and class, and Pete was an example of that.

Among the SDS'ers I knew best were the "two Carls" in the math department. Carl Pomerance was a graduate student who would later become a leading expert in computational and analytic number theory. Carl Offner was also in the graduate program; in 1969 he would be the only math student to be completely expelled from Harvard (not just suspended) for his role in the building takeover. A little later I met and learned a lot from talking with a graduate student in economics named Alan Gilbert. Alan was a member of the Progressive Labor Party (PLP), a Maoist breakaway group from the American Communist Party which had formed in the early 1960's at the time of the Sino-Soviet split.

Part of the student lifestyle in those days was to have long "bull sessions" hashing out political and philosophical issues. The student organizers, such as Bilazarian, Pomerance, Offner, and Gilbert, would spend hours at a time discussing politics with people like me who were sympathetic but not yet involved. In the jargon that PLP made popular, they were "building a base."

I was a particularly difficult case for their base-building efforts. Fundamentally I was still a liberal, not a radical. As late as autumn of 1968, if I had been of voting age (which then was twenty-one) I would have voted for Hubert Humphrey as the "lesser of the evils." Many anti-war people had vowed not to vote for Humphrey because he was still supporting the War and because they were disgusted by the violence against demonstrators by Mayor Daley's police at the Democratic National Convention in Chicago the previous summer. However, I believed that Humphrey would be more likely than Nixon to end the War.

I was put off by much of the anti-imperialist rhetoric of my friends in SDS and PLP. I used to taunt them by advising them not to try to sell me any of their publications, since I was much more sympathetic to their cause without reading them than I would be if I actually read what they had to say. And I had no time to go to their meetings.

But what was clear to me — and to many, many others on campus — was that if any militant protests against the War were to occur at Harvard, then they were going to be organized by SDS. And I felt a special responsibility to take part in such an action when the time came. Not only did I have a personal obligation to oppose racism and genocide — I knew that some of my mother's relatives had perished in the Holocaust and that my father's younger brother had been killed in action in World War II — but I also felt confident that my career as a mathematician would not be adversely affected by university discipline, arrests, or any notoriety I might get from being part of an SDS action. I had read about the history of mathematics and was aware of the long tradition of tolerance of eccentricity and political dissidence among mathematicians. I knew of several prominent leftist mathematicians, such as Steve Smale, who had denounced the War in Vietnam during the International Congress of Mathematicians in Moscow in 1966, and Serge Lang, who in 1968 had spoken out at Columbia University in support of SDS and in opposition to secret military research on campus. I fully expected that the older mathematicians who later would judge me for various purposes — admission to graduate school, hiring and promotion — would do so fairly, without any bias against me for political reasons. So I would not really be sacrificing anything by incurring university punishment or arrest. As it turned out, I was correct in my assumption — in all my years as a mathematician, I have never suffered any bias or reprisals because of leftist activities.

In contrast, many of my classmates were acutely conscious of the possibility that the actions they took would come back to haunt them later in life. A few of them also had to worry about angry parents who threatened to cut them off financially if they got into trouble. I had neither of these worries, so I had no excuse to hold back from participating in militant protests. It was obvious to me that at some point I was going to get much more actively involved in SDS.

In December 1968, the Harvard faculty met in Paine Hall to discuss, among other things, the status of ROTC on campus. SDS demanded to be allowed to speak and present the case for abolishing ROTC. The faculty refused; it was not their policy to allow outsiders to address the meeting. In response, SDS organized a sit-in at Paine Hall. University officials were furious, and moved to discipline the protesters.

I had missed the protest, and regretted that I had. It was a time when SDS needed more bodies on the line, and I wasn't there. Moreover, there had been a vote at the sit-in about whether or not to stay and face university punishment. A substantial minority of the SDS'ers had voted to leave. This perplexed and annoyed me. How could so many of them have voted that way? What was the point of being in SDS if you were going to call off a demonstration because of the threats of some university bureaucrats?

I shared these sentiments with the SDS'ers I knew, and they explained to me that there were, in fact, two factions within SDS. My friends all belonged to the so-called Worker-Student Alliance (WSA) caucus, which was created and led by the PLP. (In the anti-communist jargon of an earlier era, they were "fellow-travelers.") This faction was in favor of militancy, and they had all voted to stay at Paine Hall and face the consequences.

The other faction didn't really have a name. It was sometimes called the New Left (as opposed to the "Old Left," meaning the communists, Trotskyites, and socialists of various stripes), but more descriptively could be termed the "Youth Culture" faction. On the Harvard campus they had a reputation for being more moderate and less militant.

The WSA caucus differed not only in level of activism. They also pushed to include in the SDS agenda other issues besides the War that they believed would bring the student movement closer to disgruntled segments of the U.S. working class. Several members of the WSA caucus of Harvard-Radcliffe SDS had gotten arrested at demonstrations in support of welfare mothers, and others had taken an active part in the Cambridge Rent-Control Campaign, one of the main purposes of which was to stop Harvard's and MIT's expansion into Cambridge and the consequent displacement of working-class renters.

In early February 1969, when I joined SDS and attended my first meeting, I was not much interested in these other issues. It was the annual election meeting, and my WSA friends had organized their supporters to come to be sure that the Youth Culture faction was not voted into any leadership positions. I was a firm supporter of WSA for just one reason — I wanted to see SDS take action. All the speakers, both WSA and non-WSA, sounded sensible to me — I couldn't even tell which was which — so I sat next to my friend Pete Bilazarian so he could tell me how to vote.

The factional rift in Harvard-Radcliffe SDS mirrored a similar division at the national level. Many young people of the 1960's had little interest in the traditional Marxist approach to radical politics. We were impressed by the "end of ideology" viewpoint of Daniel Bell and especially the "new working class" philosophy of Herbert Marcuse. We were also influenced by the pop-psychology of journalists, who loved to write about the "generation gap" in America. There was a certain appeal in the notion that the new class that would foment revolution was defined not along economic, but rather generational lines. Indeed, the anti-WSA faction of SDS later chose to name itself the "Revolutionary Youth Movement" (RYM).

My classmates and I generally accepted the stereotype of the American worker as ignorant and hopelessly reactionary. Certainly the main confederation of American labor unions, the AFL-CIO, with its deep ties to the pro-war wing of the Democratic Party, was one of the most powerful institutions supporting the War. To many young people — including some like me who backed the Worker-Student Alliance caucus for other reasons — it seemed as if PLP and their fellow-travelers were living in a time-warp if they thought that anything good would come out of the American working class.

In February 1969 I was starting to become more optimistic about the prospects for meaningful protest at Harvard, and I began devoting much more time to meetings and political discussions. The previous autumn I had "gotten my feet wet" by helping Carl Pomerance canvass in the dorms for an anti-ROTC petition. I mostly just listened as he rebutted the arguments that many of the students made in defense of ROTC: that abolishing it would interfere with students' rights to choose their extracurricular activities, that ROTC helped the less affluent students by providing scholarships, and that Harvard's participation would help liberalize and humanize the officer corps. I was impressed with how patient and persuasive Carl was. My attitude would have been far less tolerant — I would've thought that anyone who made an argument in favor of ROTC was a fool who wasn't worth dealing with. But Carl and the other SDS organizers — especially those in WSA — believed that it was just a matter of time and some lengthy discussions before most of the middle-of-the-road Harvard students would come around to our position.

I was gradually developing more respect for the political analyses of SDS and even PLP, although you'd never have guessed it from the sarcastic remarks I made at times. In reality, despite my general identification with the liberal wing of the Democratic Party, the Vietnam War was moving me to the left. Ever since 1965, when I read the article by Kahin and Lewis in the *Bulletin of the Atomic Scientists*, I firmly believed that the Vietnamese communist guerrillas were fighting for the interests of their people, and the U.S. was totally in the wrong. Moreover, just as I thought that the U.S. would not have used nuclear weapons on Hiroshima and Nagasaki if the civilian populations of those cities had been white, I also believed that the Vietnam War was being tolerated by Americans in large part because of racism. And I did not disagree when SDS'ers used terms like "genocide" to characterize American conduct of the War.

February and March were times of constant debates, meetings, and leaflets. I was keeping up with my courses and writing a senior honors thesis on Clifford algebras under Raoul Bott's direction. But by then I was probably spending more time on politics than on academics. I read things that in those days would never have been included in a university course, such as *Labor's Untold Story* by Richard Boyer and Herbert Morais and *Fanshen: A Documentary of Revolution in a Chinese Village* by William Hinton. I read a long article in *PL Magazine* analyzing the 1936-1937 auto workers' strike in Flint, Michigan; several booklets and essays by Lenin and Mao; and an excellently researched PLP pamphlet called "Free Speech at Harvard: An Answer to Various Deans" that described Harvard's purge of suspected communists from the faculty in the 1950's.

Later in the spring I did my own research, using the archives of *The Harvard Crimson*, into what the university's role had been during the McCarthy period. I learned details about the people who had been fired or hounded out. One of those dismissed because of suspected communist sympathies (and because he refused to name names) was a junior faculty member by the name of Leon Kamin. Ironically, three years later I would encounter him at Princeton, where, as chair of the psychology department, he was on the opposite side from us in a dispute about inviting a right-wing psychologist. But that story had an amazing turn-around, which I'll describe in Chapter 5.

As the spring wore on, anti-war students were getting more and more impatient with the lack of progress in convincing the faculty and administration to abolish ROTC. The time was approaching for a major action. SDS couldn't wait too long, or we'd be in "reading period" (two weeks without classes before exams) and then exam period.

But would all the talk ever lead to action? SDS meetings were well-attended — my recollection is that between 100 and 150 students had been at the election meeting in February — even though they often went on for hours and hours of seemingly interminable debates. The evening of Tuesday, April 8, SDS met amid rumors that a militant action was likely. The WSA caucus had been meeting separately and strategizing about how to get enough students behind a direct action. I attended the April 8 meeting, which finalized a set of six demands related to the abolition of ROTC and the curtailment of Harvard's expansion into Cambridge. The meeting did not end with a clear decision about tactics and timing — a proposal to do something that night was voted down. But a consensus was reached that militant action should be taken very soon, with the details left vague. And a rally in Harvard Yard was called for noon the next day. After the meeting, well past midnight, SDS'ers marched noisily through campus.

I was not in the WSA caucus and had not even stayed to the end of the meeting the night before. I still had some doubt that a militant action would take place, and I wasn't sure that I could come to the rally at noon. Two high school seniors — my second cousin Larry Kominz and a family friend named Jim Levenson — had come to tour the campus and be interviewed by the admissions office, and I had agreed to show them around.

Between classes I ran into Pete Bilazarian, who told me that I definitely had to be at the rally, that this was the day. I explained to Larry that he'd have to tour campus and visit classes on his own. Jim Levenson, on the other hand, wanted to see the Harvard student movement first-hand, and said he'd come with me to the rally.

At noon Jim and I went to Harvard Yard, where hundreds of students were milling around. A handful of student leaders — I recognized PLP members Alan Gilbert and John Berg — were standing on the steps to University Hall and calmly announcing that SDS was taking over the building and issuing a set of six non-negotiable demands directed against Harvard's complicity in the War and its expansion into Cambridge. Then some guy from the Divinity School who had a thick beard, looked like some sort of New Age shaman, and most likely had no connection with SDS or PLP, started speaking into the microphone about how this was a non-violent building takeover in the spirit of Mahatma Gandhi and Martin Luther King.

Only a trickle of people were following the call to enter the building, and some of the students in the Yard were overtly hostile. At first it didn't look good. Perhaps some protesters were going into the building at other entrances, but I didn't know. I turned to Jim and said that I was sorry to leave him, but I really had to join in the takeover. To my surprise, he replied that he wished he could come with me, but it wouldn't look good if he missed his admissions interview because he was with SDS in the building! I then joined the other SDS'ers in University Hall.

At first there seemed to be a lot of confusion. Most of the staff were leaving voluntarily, but in another part of the building Carl Offner was in a group that was forcibly evicting Deans Epps and Watson. That scene later made the pages of *Life* magazine, and it resulted in Carl being expelled from the university and sent to prison for three months.

Gradually the occupiers started to congregate in the Faculty Room on the second floor of University Hall. Soon a meeting started to discuss and vote on all sorts of issues connected with the takeover. Drugs were banned, and no vandalism of the building would be allowed. Representatives of the administration would not be admitted to the building, but everyone else was free to come and go.

The question of what to do when the police came — as we knew they would, probably sooner rather than later — was controversial. PLP wanted people to fight back. They argued that students should follow the lead of striking workers who fight to keep scabs and the cops who protect them from crossing the picket lines. They also reminded us of the uprisings that had occurred in the black ghettos, which were anything but non-violent. However, most students were not persuaded, and PLP lost the vote. We decided to meet the police with chants and linked arms, but we wouldn't fight back.

By mid-afternoon there were still only about 150 people inside the building. At 4:15 p.m. the administration gave us an ultimatum to leave within fifteen minutes or be arrested for criminal trespass. Some people who had been in similar situations (at Columbia, at the Democratic Convention in Chicago, or at the Pentagon) gave instructions. The phone number of the Legal Aid Society was posted. Reporters and tutors started leaving hurriedly, and some *Harvard Crimson* reporters rushed into another room to confer. I was scared and sat quietly in the corner.

But by 5 p.m. it was clear that the police wouldn't be called in while Harvard Yard was full of spectators. The tension subsided as we saw that the number of students in the building was steadily increasing. In addition, the orderly, friendly debates and discussions in the Faculty Room inspired confidence in what we were doing. The meeting was chaired by Richard Hyland, a low-key, soft-spoken *Crimson* reporter who was the personification of calm in the middle of a storm. Some students who had initially come into the building out of curiosity and who had not been at all convinced of the wisdom of a building takeover, later decided to stay, in part because of the atmosphere of comradery, mutual respect, and political commitment.

Not everyone was in the Faculty Room. Many formed small groups with friends, fanning out to other rooms. By evening there was an incredibly varied group of students in the building. Some came for only a short time to see what was happening or talk with friends who were in the occupation. A few faculty members entered the building in order to try to talk students out of staying.

One professor who came with friendly intentions was Valentine Boss, whose course on Soviet history I had audited the previous semester. Besides me, several other campus radicals had been sitting in on the course, presumably so that they could better participate in the internal ideological debates that were going on within SDS. Boss was the son of a British diplomat who had been in the Soviet Union during World War II. He once described to me the train trip evacuating him from Moscow to Helsinki; as a 10-year-old boy, he saw mounds of dead bodies near the tracks from Leningrad to Finland. Perhaps because of his background, he had a sardonic and sometimes cynical attitude toward leftist politics. When he saw me in University Hall, he asked, "Where is your Lenin? Where is your Trotsky?" Boss was a tutor in Kirkland House and knew several of the students in University Hall quite well. He was kind and supportive toward the radicals despite his jaundiced view of the left.

Late at night the temporary visitors had gone, and the students remaining were those who when the police came intended either to stay in University Hall or else stand near the building in support of their friends inside. We knew that after the building takeover at Columbia the university officials decided that they had erred in letting it continue for several days in the hope that it would peter out. We expected that the Harvard administration would not wait long before calling the police. Most likely, the cops would come in the early morning when everyone would be asleep except for the University Hall occupiers.

Some of the protesters slept, but I could doze off only for a few minutes. I was twenty years old and, like most of the other students, had never been arrested.

The impressions of that night and morning remained vivid in my mind long after. With time, of course, I forgot many details. Fortunately, the following day I wrote a long letter to my parents which they preserved; it is my main source for this account.

Between 3 and 4 a.m., I took a walk around the inside perimeter of Harvard Yard. The early morning air was unseasonably chilly. The campus police had locked the gates to the Yard to keep out students from the Houses (Harvard's term for the dorms where sophomores, juniors, and seniors lived). When I returned to University Hall, there seemed to be a larger crowd and more activity than before. A few SDS'ers had looked through the locked gates and seen police cars, marked with the names of communities from all over the greater Boston area, cruising slowly down the empty streets. There was a rumor that a group of police had been spotted forming by Memorial Hall. When the hundreds of students who were sitting on the steps in solidarity with the protesters inside started singing "We shall not, We shall not be moved," I went back into the building. Just before someone pulled shut and locked the large wooden doors and chained the inner glass doors, in the blue half-brightness before sunrise I caught a glimpse of a big gray bus pulling into the Yard.

Although most of the student living areas were cut off from Harvard Yard, the freshman dormitories were located in or contiguous to the Yard. SDS'ers went to these dorms and rang the fire alarms. Hundreds of freshmen poured out of their rooms just before over 400 police in riot gear — helmets, clubs, and gas masks — entered Harvard Yard.

The police quickly cleared away several hundred students — now including many of the recently awakened freshmen — who surrounded the building. Some of them were beaten badly by police clubs. Then, led by state troopers, the police mounted the steps of University Hall with a crowbar and pincers to break through the chains that SDS had placed on the doors. Inside the building the protesters had linked arms and were chanting "Smash ROTC! No expansion!" The double-doors sprang open, and the police approached the glass doors. They fumbled with the pincers, finally broke the big red chain, and came at us.

One of the cops who had a darker uniform than the rest jumped over the banister onto the stairs and started directing the others — pressing his fists together as if to signal "crush." I was lucky to be in the third row, since the police started viciously clubbing people in the first two rows. Some protesters put their hands in front of their faces or heads to protect them. Then the police started releasing tear gas. Some people started screaming, "What are you doing?" "Pigs!" "We can't get out!" "Let us out!" The doors on the opposite side of the narrow building opened, but the police held there, and the crush continued. After some more beating, suddenly the police in back of us stepped back, and we started tripping backwards down the stairs. They dragged us one by one down the steps into the waiting paddy wagons.

The police handled me gently — I was not clubbed. But because I got a fair amount of tear gas in my eyes, I had only a blurred view of what was happening. Most likely, the reason for the police restraint in handling me was that with my coat and tie and short hair I appeared very different from their image of a student radical. They must've thought that I was a reporter, or perhaps someone who was there by mistake (at 5 a.m.!).

It was an eerie feeling to ride to jail through the deserted early-morning streets of Cambridge. I had been the last one in the paddy wagon, and through a slit in the back I looked at the dimly-lit streets with my still stinging eyes. I had studied the Russian Revolution, and I had read about the American communist labor organizers of the 1920's and 1930's. Those earlier revolutionaries were the ones who had inspired the leaders of our building takeover. But would our actions have any real effect in stopping the War and strengthening the left in America?

The reaction on campus and around the country was greater than I could have imagined. In Harvard Yard after we were taken away, police contingents remained to deal with the freshmen who were taunting them. (A few had gotten caught up in the mayhem and thrown into paddy wagons; eventually, a half dozen or so students were acquitted of criminal trespass when they produced witnesses who testified that they had not been in the building.) The police clubbed many of the onlookers, including a student in a wheelchair. Some students shouted "Pusey's pigs!" (Nathan Pusey was president of Harvard, and he had had the final responsibility for calling in the police.) Others started shouting "On strike! Shut it down!"

A key point that Pusey and his advisers had failed to understand was that SDS was not an isolated extremist organization. As I had learned when I accompanied Carl Pomerance in the anti-ROTC petition drive the previous autumn, SDS activists had been patiently "building a base" among the student body for a long time. Moreover, I wasn't the only one who was moving politically to the left as a result of the continuing U.S. barbarity in Vietnam. Many of the students who did not initially agree with the building seizure or did not participate for personal reasons were nevertheless convinced that strong action to try to end the Vietnam War was justified. The War was a central moral issue of our generation, and thousands of Harvard students felt that this was not the time for "business as usual."

Thus, all but the most conservative students believed that, even if they disapproved of SDS tactics, the police bust ordered by the administration had been far worse. Some wanted to make police brutality the issue of the day. SDS discouraged this, however, because we did not want to divert attention from the War, Harvard's anti-working-class policies in Cambridge, and our six demands. And WSA people were quick to point out that the level of police violence against us had been low compared to what striking workers or blacks in the ghettos were accustomed to.

Because this all happened at Harvard, reputed to be the most elite university in the country, the building takeover and police bust made national news. There was a front-page article in *The New York Times* and several pages of photos in *Life*. The tone of the coverage was, of course, disparaging.

The morning of Thursday, April 10, jammed into ridiculously overcrowded cells in the Cambridge jail, we were not thinking about what the national reaction would be; we had more immediate concerns. It took until early afternoon for all 184 of us to get processed, arraigned, and released. That was none too soon for the SDS leaders, who were visibly nervous about the vacuum of leadership as thousands of students joined the call for a strike. They were especially concerned that a group of opportunistic student-government types (who called themselves "moderates" but were labeled "sellouts" by the radicals) would seize the moment in their absence and try to direct students away from the SDS demands and toward what we regarded as phony issues. The main phony issue that the leftists were worried about was reform of the university's decision-making process (getting more students on committees, securing student representation on the Board of Overseers, and so on).

That night SDS met in Lowell Lecture Hall, which had 700 seats. It was full to the rafters, with hundreds standing up. The number of students who identified with SDS had mushroomed. To my surprise, my friends in WSA had mixed feelings about this. They knew that at the meeting someone would propose adding "student power" planks to the list of SDS demands. To newly radicalized but inexperienced students at an elitist university like Harvard, the appeal of student power might be very great.

The meeting, as usual, went on for hours. At one point a tutor in one of the Houses got up to address the crowd. He was popular with the students and a good speaker. He started out by saying, "I've been looking over the six demands, and I've got to say — they're not very radical! What we really have to do is effect fundamental and permanent change, not just some momentary tactical concessions, and the way to do this is to restructure the university." I was standing with John Berg of the PLP, and he looked worried.

Then the chair recognized Greg Pilkington, who was one of the few black students who worked mainly with SDS rather than with the black student organization. Greg started out, "I recently read an article in *The Wall Street Journal* about how to deal with student rebellion. They advised a two-pronged attack: first, bust early. Next, divert the issue to student power." The meeting erupted into thunderous applause, followed by a standing ovation. Later the student power demands were resoundingly voted down. It was a major victory for WSA and a striking example of the rapidly shifting dynamics of student politics in those days.

Outside of SDS, however, the radicals could not control events. A few days after the bust the "moderates" organized a gigantic meeting in the Harvard Stadium. After much debate the 10,000 people assembled there voted to continue the strike — this was a victory of sorts for us — but did not vote to endorse the SDS demands (which by then numbered eight with the addition of amnesty and the creation of a black studies program).

Powerful forces on campus were trying to marginalize the radicals. Foremost among them was the liberal caucus of the faculty, led by political scientist Stanley Hoffmann. Three days after the bust he was quoted in *The Boston Globe* saying that "discipline will have to be exerted strictly and severely." Economist Wassily Leontief, whose motion at a faculty meeting established the quasi-judicial procedure that would determine how the university would discipline the protesters, said, "There is no doubt in my mind that the action of the student group, particularly of its leaders... should be punished and... it should be the maximum punishment. They should be expelled from the university."

I was disturbed by the liberals' extreme animosity toward SDS. After all, until a few months earlier I would have looked up to those professors as leading representatives of an enlightened outlook on politics. But it now seemed to me that they were far more upset about protests disrupting their comfortable ivory tower than they were about the atrocities that were being committed in Vietnam.

I had a long talk with Valentine Boss, who explained what was going on in a more dispassionate way. Even though the liberal professors disliked the student radicals, that was not what was foremost in their minds. Their principal objective, Boss explained, was to wrest power from the administration and also outmaneuver the conservative caucus. By criticizing Pusey's ineptitude and at the same time moving aggressively to punish the students and bring the campus back to normalcy, they would be exerting leadership and showing that they deserved more decision-making authority on campus. Kicking out the radicals would merely be a means toward that end.

To many of us the most dramatic illustration of the rift between faculty and student opinion occurred at the faculty meeting that was held shortly after the bust. Because of the unusual circumstances, the faculty broke with its tradition of secrecy and allowed the meeting to be broadcast on the Harvard radio station.

Most of the professors expressed hostility toward the student radicals. The culmination of the meeting was a speech by Professor of Economics Alexander Gerschenkron. He started out by saying that during the darkest hours of Nazi Germany there was a story by Hans Christian Andersen called "The Most Incredible Thing" that brought comfort to people. There was a king who announced that he would give his daughter in marriage to the man who could do the most unbelievable thing. On the day of judging it seemed that a certain handsome young man's creation would win. It was a clock, a most amazing clock, with the figures of the leading intellectuals carved into it. But suddenly a new competitor arrived, a "lowbrow fellow" (Gerschenkron took some liberties with the actual Hans Christian Andersen story), and he smashed the clock with a sledge hammer. Everyone had to agree that that was the most unbelievable thing, and so the lowbrow fellow won.

In the story, Gerschenkron continued, at the end the figures of the great intellectuals rose up and destroyed the lowbrow fellow, and the princess married the handsome young man. Here at Harvard he didn't know if it would end so happily, but he hoped that this faculty would put an end to "all the criminal nonsense" that's happening on college campuses.

The faculty gave Gerschenkron a thunderous ovation. I was dumbfounded, as were the students listening with me. Aside from the pomposity of Gerschenkron's presentation, we were shocked that he would compare student protesters — rather than those who were committing atrocities in Vietnam — to the Nazis. The professors who were applauding Gerschenkron's hysterical hyperbole had done nothing — at least nothing effective — to oppose the War. Some Harvard professors had even been among the architects of U.S. policy in Vietnam. It seemed incredible to us that the only thing that could arouse the moral indignation of the Harvard faculty was our disruption of their cozy sanctuary.

The contrast with student opinion could not have been more glaring. In the days after the bust the vast majority of students were favorably disposed toward SDS. A petition supporting all of the SDS demands was signed by well over half of the graduating Radcliffe seniors. Among students one would have had to strain to hear an unkind word expressed toward the protesters.

Not every faculty member, of course, was hostile to us. One professor in the philosophy department, Hilary Putnam, at the time was a well-known campus radical and spokesman for WSA politics. A junior faculty member named Jack Stauder had been arrested at University Hall, and a few others (such as Boss) were at least quietly supportive.

The gesture of sympathy that especially touched me — even though it was motivated more by humanitarian concern than by political agreement — came from my senior thesis adviser Raoul Bott. He invited me to lunch shortly after the police bust. He wanted to hear a first-hand account of what had occurred, and he was particularly disturbed to learn that mace and tear gas had been used against Harvard students. A secretary in the math department reported that the day of the bust he had come in exclaiming, "Either Pusey goes, or I go."

As the committee set up by the faculty (called the "Committee of Fifteen") proceeded to hold hearings and decide on punishment for the occupiers of University Hall, I embarked on my first leaflet-writing project. The administration had started a newsletter called *Rap-Up*, the purpose of which was to show that everything was getting back to normal and the crisis was coming to an end ("wrapping up"). In response, I wrote a series of three leaflets called *Crap-Up*, in which I analyzed and ridiculed the Committee of Fifteen and compared its procedures to Harvard's actions against suspected communists in the 1950's.

My friends in SDS helped me mimeograph and distribute the leaflets, even though the focus was very different from what SDS wanted to emphasize. Questions of unfair procedures against Harvard students and issues of civil liberties were not what SDS considered important. But some radical students liked my leaflets. Many years later my wife Ann ran into a woman named Ellen Messing who had been in SDS at the time. She commented that she remembered me because she had enjoyed the *Crap-Ups*.

My leaflets were carefully researched and well-written, but in the late 1960's and early 1970's there was nothing unusual about high-quality student writing. A few years ago while cleaning out his office before retirement, one of my colleagues at the University of Washington came upon an old UW-SDS leaflet from about 1970. Knowing of my background, he gave it to me. More than anything else my reaction to it, reading it with the eyes of a teacher rather than those of a student, was "Wow, students in those days could sure write well!"

The Committee of Fifteen announced its punishments in early June, about a week before graduation. A small number of protesters — including Carl Offner — were expelled from the university in a strong sense, namely, they would never be readmitted to finish their degree program unless two-thirds of the faculty so voted. Some other students were required to withdraw for a semester or year. Most of those who had no prior disciplinary record were put on probation. Because I had been reprimanded for misuse of my student ID card at the Dow Chemical sit-in, my punishment was slightly more severe — but in a way that only a bureaucrat could appreciate. I was suspended for a year, but the sentence was suspended. If I engaged in any further "misconduct" — obviously, this could only have meant disrupting graduation — my degree would be held up for a year. I never did figure out the difference between "suspended suspension" and probation.

I knew that the Princeton math department, which had admitted me for graduate studies, would not care if my diploma were held back for a year. So I saw no compelling reason not to participate in a disruption, if there was one, at Harvard's Commencement.

Under normal circumstances I wouldn't have bothered to attend graduation. I hated long ceremonies and pompous speeches. Since my parents were out of the country that year, there wouldn't be anyone I'd disappoint by skipping it. However, these were not normal circumstances. At the very least SDS was planning a walk-out, and I wouldn't be able to walk out in protest if I didn't attend, so I made plans to go to Commencement.

SDS demanded to have a speaker at graduation. At first, this was refused. SDS threatened to seize the microphone if necessary. Several SDS'ers, including me, were well situated to do this because our grade point averages were near the top of our class. By tradition, the students graduating *summa cum laude* were seated in the front rows near the stage.

As time for the ceremony approached, I was getting very nervous. I had never taken part in anything as physically confrontational as seizing a mike. Fortunately, I didn't have to: the student master of ceremonies prevailed upon Pusey to give in to the SDS demand so that graduation could proceed without disruption.

At the beginning of the ceremonies SDS was informed that a representative would be permitted to speak briefly. Bruce Allen, who had been chosen to give the speech, delivered a lengthy denunciation of Harvard and the capitalist class, specifically pointing to David Rockefeller, who was on the stage a short distance away. The vast majority of the crowd — in which alumni, parents, faculty, administrators, and visitors far outnumbered the graduating seniors — reacted with fury, booing and calling for the mike to be shut off. Bruce's words could barely be heard over the boos.

I was later amused to find that on the TV news SDS dominated coverage of the Harvard Commencement. For technical reasons having to do with the placement of the audio equipment, the crowd reaction was not picked up, and the excerpts of Bruce's speech that made it onto the news sounded much clearer on TV — no boos — than they had in Harvard Yard.

After fifteen minutes, Bruce was coaxed off the podium, and SDS staged a walk-out. We went to another part of campus for an "alternative graduation" at which Hilary Putnam spoke.

In the spring semester of 1969 large numbers of Harvard students felt caught up in an incredible rush of events and ideas. In retrospect, a vast amount seems to have been compressed into a very short time. Like many of my classmates, I viewed my last few months as a Harvard senior as transformative. My outlook on politics and society changed in more fundamental ways from January to June of 1969 than it has in the nearly four decades since then.

After the intensely emotional experience of the building takeover and the police bust, I felt closer to the other students in SDS. I no longer made snide remarks about their long meetings or the pointlessness of trying to rouse smug and complacent Harvard students. As I wrote my parents, "Sitting in that hot, miserable... jail cell... I had time to realize that for the first time I didn't feel like knocking Harvard students or being ashamed of being a Harvard student..."

I even started to take seriously what the WSA had been saying about the working class. During the weeks after the bust, WSA organized students to leaflet factory gates at the morning shift change. The leaflets explained the student uprising at Harvard and tried to tie in the SDS demands with the interests of workers. I was impressed that my peers were waking up at 6 a.m. to make it to the factories before shift change. I was always an early riser — I had never missed an 8 a.m. Russian conversation section that I was auditing — but I knew that getting up at such an hour was anathema to the typical student.

I didn't like to pass out leaflets, and I didn't go with them to the factory gates. However, on several occasions I went with my WSA friends on what was a central part of their strategy — strike support. We would learn of a local union-sanctioned or wildcat strike and come with coffee, donuts, and leaflets. The workers would invariably appreciate the unexpected support, and the students would benefit from the interactions even more than the workers.

My most memorable experience at a strike was at the General Dynamics shipyards in Quincy, Massachusetts. If any group of workers was going to be hopelessly tied to the War, it would surely be these guys — most of their company's business and most of their jobs came from military contracts. But the workers had rejected all "patriotic" appeals not to strike during the War.

I went with Alan Gilbert and other SDS'ers, and spent much of the day standing around listening to the conversations with the workers — none of whom turned out to be pro-war and none of whom had any problem welcoming support from student radicals. And there was no noticeable difference between the responses of the older and younger workers; there was no "generation gap."

What impressed me more than anything else, though, occurred midway through the morning. The company tried to get some cars of replacement workers — scabs — through the picket lines. The scabs were accompanied by 50 or 100 cops. At first the police pushed aside strikers and escorted the scabs toward the gates. There must have been a couple of thousand workers gathered around the plant that day, and suddenly about 500 of them converged on the scab convoy, threatening violence. The police were terrified. They looked around, saw what was happening, and immediately retreated. The scabs did not get in, the police left, and no one was arrested for their threats against the replacement workers and their police escort.

No PLP rhetoric about the "power of the working class" could have had the effect of that visit to the strike lines. What I had just witnessed contrasted dramatically with the scene in Harvard Yard only a couple of months before, when students had stood helplessly and been beaten and arrested by the police. For me and other SDS'ers it was amazing to watch the striking General Dynamics workers overwhelm and intimidate the police. The WSA had found the perfect way to convince us that a worker-student alliance strategy was in the interest of the anti-war movement.

As a result of the SDS building takeover on April 9, 1969, the police bust the next morning, and the student strike that followed, the Harvard campus was transformed into a center of protest and experimentation in a broad sense. Even before those events shook up the campus, many people had been trying to change what they saw as the university's stultifying approach to education.

The previous year, over the objections of conservatives, the social relations (sociology) department had introduced a course on "Social Change in America" that served as a focal point for reading and debating radical literature and discussing tactics for the student movement. In 1968-1969 it was one of the most popular courses at Harvard. But after the events in the spring, the course was canceled.

A major issue that came to the fore in April was black studies. For many months the efforts of Afro, as the black student organization was known, had been stalled. Most of the faculty were notoriously conservative on curricular matters and believed that establishing such a department would amount to an erosion of academic standards. But after the SDS actions, the administration and the faculty's liberal caucus were extremely worried that Afro would form an alliance with SDS to disrupt the campus if their black studies demand was rejected. At a tense and carefully orchestrated faculty meeting, the professors voted to establish a black studies department.

But some of what was happening at Harvard was much less confrontational. During the strike several people tried to set up an "alternative university" consisting of off-beat informal courses. John Newmeyer (who the previous autumn had helped Tom Hopkins make up a list of the ten best rock music albums for me to take to Leningrad) taught a course on yoga and Eastern mysticism.

For a number of reasons the hard-core leftists in SDS were opposed to what they called "counter-culture" movements. They viewed them as a diversion from issues of war, racism, and class injustice, and also saw them as fitting into media stereotypes of "hippies" that most working-class Americans found objectionable. In addition, those movements often encouraged the use of marijuana and psychedelic substances. The Worker-Student Alliance caucus of SDS took an uncompromising anti-drug stand.

The counter-culture people, in turn, ridiculed the WSA activists for being so "straight." (In the 1960's the word "straight" referred to physical appearance, musical tastes and attitude toward drugs, not sexual orientation.) A famous expression of this disdainful attitude toward the left by the icons of youth culture was the Beatles' 1968 lyrics: "You better free your mind instead/ But if you go carrying pictures of Chairman Mao/ You ain't gonna make it with anyone anyhow."

Although I fit the "straight" stereotype even more than most of my WSA friends, I had a more positive view of "cultural radicals" than they had. This was part of the reason why I didn't join WSA for several months, even though I always voted with them within SDS. The best example I knew of a cultural radical was Tom Hopkins, who had a major influence on me and many other of his classmates, especially in Winthrop House.

Tom was the only person I knew, until much later, who was openly gay. He had a biting, flamboyant sense of humor, and brought what Susan Sontag had termed *camp sensibility* — which grew out of the gay subculture — to the Harvard class of 1969.

Tom's enthusiasm — whether it was about Jimi Hendrix's performance in *Monterrey Pop*, Bob Dylan's lyrics in *Highway 61 Revisited*, Lou Reed's discordant riffs in *The Velvet Underground and Nico*, or the opening scene in the movie *2001* — was infectious. It was from Tom that I learned about Andy Warhol's ironic glorification of popular culture. I recall that during the Warhol Film Festival he organized in the dorm, he showed a half-hour film in which a man sat in a chair the whole time eating a mushroom, and then Tom lectured to us about why it was brilliant cinematography.

Late in the spring, a WSA member named Reid Minot and I decided to write a pamphlet that would introduce the incoming Harvard-Radcliffe freshman class to SDS. When we were done with the draft, I showed it to Tom. Despite his political sympathies — he considered himself an SDS member — he often criticized the organization, especially for what he saw as our inability to get our message across in a lively, engaging manner. I told him that if he improved the writing for us, then even if his style was totally different from what WSA people felt comfortable with, I was sure that I could convince them to publish it. (And in fact, the booklet was later produced by the PLP's publishers in Chinatown in New York.)

Tom Hopkins rewrote the SDS booklet for incoming freshmen in a literate style that went way beyond what I could have done. For example:

> Daily Guernicas unleashed by American troops to prop up a faltering regime which totally lacked popular support, the heroic retaliation of the Vietnamese people against inhuman odds, the continual spectacle of American soldiers brought back in boxes from a war they had no interest in fighting — all these scenes combined in the early 1960's to make most of us now in SDS question the conception of the United States we had held since youth.

To explain what "imperialism" means, Tom included an English translation of "The United Fruit Co." by the Chilean poet Pablo Neruda. Tom liked to play around with the sounds of words; here is his description of the bust:

> Cops clubbed protester after protester and dragged them by their hair down the steps into paddy wagons and buses. The astonished crowd outside joined the chant: "Smash ROTC! No expansion! Smash ROTC! No expansion!" They had seen Chicago, they had seen Mayor Daley, they had seen his pigs. Now the pigs were Pusey's.

The pamphlet ended with a warning about anti-working-class attitudes:

> No matter how good our intentions or how progressive our views, most of us are still disturbingly like the young Mao: "I began life as a student and at school acquired the ways of a student. I then used to feel it undignified to do even a little manual labor.... At that time I felt that intellectuals were the only clean people in the world, while in comparison workers and peasants were dirty." (*Talks at Yenan Forum on Literature and Art*)
>
> Mao overcame his prejudices. So must we. If we become stronger and increase alliances with working people in Cambridge and Boston, the [Harvard] Corporation will have to grant all our demands, and many more. It all depends on us.

The inside back cover had my favorite quotation from Mao Tse-tung: "Where do correct ideas come from? Do they drop from the skies? No. Are they innate in the mind? No. They come from social practice, and from it alone." A friend of Tom who designed the pamphlet laid it out on the page to resemble a poem.

A couple of years later I learned of an unexpected way that Tom was inconvenienced because of his association with me. At some point in 1969 Tom's telephone had been disconnected because he had gotten far behind in paying his bills. He needed to get a new account with the phone company, and I was one of the most stable (or at least financially solvent) of his friends. So I let him get the phone in my name.

He later told me that during a period of several months in 1970 his phone was malfunctioning, and at first he thought there was something mechanically wrong with it. Then at one point he was talking to a leftist friend who had emigrated from South America, and she suddenly exclaimed, "Tom, your phone's being tapped!" (In those days it was apparently easy for someone with experience with police surveillance to recognize telltale signs.) Tom's phone problems stopped abruptly at the end of the year, which coincided with my release from the Army. It seems that Army Intelligence had found that there was a phone in Cambridge under my name and had been tapping it in order to learn what radicals were planning to do at Fort Dix and Fort Eustis. I can only guess what they must have thought as they tried to decipher Tom's long-distance calls to his friends at the Warhol Factory and his fellow activists in the emerging gay liberation movement!

I last saw Tom in the early 1970's and lost track of him soon after. I believe that he wound up teaching film studies at New York University. He died of AIDS in 1985.

The last major event that I participated in as a member of Harvard-Radcliffe SDS was the SDS national convention, held in Chicago from June 18-22, 1969. This was the meeting that split SDS into two antagonistic factions and soon after led to the demise of the organization.

At first I had little interest in going to the convention. SDS functioned in a decentralized way, campus by campus. Unlike some liberal anti-war groups, in the late 1960's it did not organize national marches or demonstrations (although SDS'ers sometimes participated in them). Despite my activism in SDS, I did not know the names of any of the national officers, and I had vaguely heard of but never read the magazine *New Left Notes*, which was put out by the SDS national organization.

The national office of SDS was firmly in the hands of the Youth Culture wing, which the WSA people insisted on calling the "right wing" of SDS. I disliked this terminology, which I thought was unnecessarily divisive. I knew several people around Harvard in the anti-WSA faction — such as Mike Kazin (son of a famous literary critic I had heard of) and Michael Ansara (who was responsible for several leftist publications in the local alternative press) — and I thought that they were intelligent, honest, and progressive, even though they might not be as militant as WSA people.

Of course, with WSA solidly in the majority in our chapter, the Youth Culture people had to be relatively mild and collegial in their critique of the caucus. I was even sympathetic to some of their criticisms — for example, that WSA used too much off-putting rhetoric, was too antagonistic toward leftists with whom they disagreed, and was too harsh on youth culture.

But after the events of April, I had tremendous respect for the WSA activists, and they gradually convinced me that the SDS national convention was important. They were campaigning to get students to become paid-up members of SDS, because the number of official members would determine the number of votes of Harvard's delegation. They succeeded in getting 300 paid members, so the Harvard-Radcliffe chapter was entitled to send a delegation of 60. That turned out to be one of the largest from any campus — over 4% of the total number of delegates nationwide.

At an SDS meeting late in the spring, the Youth Culture faction tried unsuccessfully to argue for proportional representation. Not surprisingly, they were outvoted, and the entire slate representing our chapter was chosen by WSA. Although I wasn't a WSA member, I was asked to be in the delegation because, first, I had promised definitely to go to the convention, and second, they knew that I would always support WSA positions.

About 1500 students, most of them delegates from the many campus chapters, converged on Chicago for the meeting. Harvard-Radcliffe SDS chartered buses for the long, tiring trip from Boston. When we arrived, I, being more bourgeois than most, paid for a bed in the YMCA hostel a short walk away rather than sleeping on a floor somewhere.

The convention was being held in the Coliseum, and the atmosphere was tense. Police cameras were photographing everyone as we entered. The national SDS office had established security details to check ID's and search everyone for weapons. An early vote resulted in reporters being barred, except for those from the leftist and alternative media.

It was fascinating to meet people who'd participated in campus disturbances that I'd only heard about through the newspapers. I also enjoyed seeing what was at the tables set up by various leftist organizations.

I remember an amusing conversation I had with another delegate from Harvard. He mentioned that his father, a conservative pro-war Harvard alumnus, had had no objection to his participation in the SDS national convention. He had explained to his father that he had been chosen to be part of the Harvard delegation, which would be one of the largest and most influential. His father's pride in his *alma mater* trumped his politics, and he said that by all means his son could go — even though in this case the cause of Fair Harvard coincided with that of the left wing of SDS.

Early in the conference it became clear that WSA had out-organized the anti-WSA "Revolutionary Youth Movement" (RYM). There must have been about 600 strongly pro-WSA delegates, and about half of the remaining ones were uncommitted, affiliated with other groups, or mildly sympathetic to one side or the other, depending on the issue. For WSA the most important vote of the first day concerned the amount of time in the agenda that would be devoted to workshops. They knew that in small groups the WSA delegates, most of whom came from very active chapters such as Harvard's, would be able to convince most of the middle group to support WSA positions and probably vote for WSA candidates. Compared to the RYM people, who were anything but united among themselves, WSA could offer a coherent, well-argued platform. When WSA easily won the vote on workshops, it became apparent to the RYM leaders that they were heading for defeat.

After my earlier skepticism about WSA's characterization of the "right wing" of SDS, the convention in Chicago was an eye-opener. I was utterly disgusted by the behavior of the RYM leaders — people like Mike Klonsky, Mark Rudd, and Bernadine Dohrn — who made up in bravado what they lacked in coherence. They would swagger up and down the aisles, adopt postures that were perhaps based on the popular posters of Che Guevara, and imitate Black Panther jargon ("you jive-ass motherfuckers..."). They would ramble on about the revolutionary potential of the *lumpenproletariat*, and an extreme group within RYM even circulated a leaflet praising the Manson Family murders.

I decided not only that WSA had been correct in their disparagement of the SDS national office faction, but that the RYM spokespeople were actually conforming to the worst media stereotypes of the youth movement — that we were a bunch of crazed, anti-social misfits acting out some form of adolescent rebellion.

As the convention wore on, the RYM leaders felt that they had to resort to drastic measures. They knew that the PLP had criticized the Black Panther Party as part of its increasingly negative take on all types of nationalism. In fact, the PLP almost never offered uncritical support to other groups on the left; it had, for example, criticized both the Vietnamese and Cuban communists for "revisionism" (i.e., their increasingly pro-Soviet orientation). The Panther leaders were furious at PLP and its allies within SDS. What they had been used to from the student left was sycophantic adulation, and this suited them much more than PLP's analysis.

So the RYM leaders asked two representatives from the Black Panther Party to speak at the convention and attack the PLP. That was their only hope for embarrassing WSA and driving a wedge between them and the majority of the delegates. But the two leaders of the Illinois Black Panther Party — Chaka Walls and Jul Cook — were disasters. On the very day when delegates had been discussing SDS's role in the women's movement both Panther speakers launched on a misogynist diatribe, saying that they were for "pussy power," which meant that "the position for a woman in the movement is prone." These insults were met with a chorus of chants "Fight male chauvinism! Fight male chauvinism!" which most of the delegates, not just WSA, joined in.

After their last tactic had failed, Bernadine Dohrn announced a walk-out of RYM delegates to caucus in the next room. Ultimately almost 500 (roughly one third) of the convention left with RYM. The following day they returned to the main meeting room, and Dohrn declared that her faction was the true SDS and that they were expelling from SDS all members of PLP and WSA and anyone else who did not uncritically support all national liberation struggles around the world. That included, according to Dohrn, not only the Black Panther Party, the Young Lords (a Puerto Rican group), and the National Liberation Front of South Vietnam, but also the governments of China, Cuba, North Vietnam, North Korea, and Albania (!). Then they left.

Although the delegates remaining in the room made an effort to finish the convention and work on establishing themselves as the true SDS, over the course of the next two years it became clear that SDS as such was no longer a viable organization. As for the RYM version of SDS, they quickly split into RYM I — which renamed itself the "Weathermen" and degenerated into some petty attempts at terrorism — and RYM II, a more moderate group led by Mike Klonsky. The Klonsky group also soon stopped using the name SDS, and shortly afterwards they went defunct as a national organization.

Postscripts

Several years ago at the University of Washington, where I teach, the dean's committee to recommend a new math department chair was talking individually with the professors about our thoughts. In addition to finding the next chair, they were charged with drawing up a list of people who might be good prospects in future years. They apparently thought that my comments about my senior colleagues were well-balanced and sensible, and so they broached the question of whether I might be interested some day in being chair. This I didn't want, so I replied, "Well, I don't know. My outlook on how to deal with university administrations comes from my formative years as a student radical in the 1960's, when we would seize their building, evict them from their offices, and issue non-negotiable demands." My name was not put on the list of possible future chairs of my department.

Finally, I want to comment on a historiographical issue related to researching the era of the 1960's. Since the University of Washington libraries, disappointingly, do not have old issues of *The Harvard Crimson* either in print form or on microfiche, I was at first relieved that the *Crimson's* website has free online access to all their issues since 1873. And indeed, I was able to check a few facts and dates in that way. But then I discovered that I could find none of the articles that I wanted from 1968-1969 that covered SDS actions.

The *Crimson* online archives, it turns out, are selective. For example, I searched for "Paine Hall" from December 10-17, 1968. (The Paine Hall sit-in against ROTC occurred on December 12.) The *Crimson* article reporting on the demonstration was missing. There were several letters and columns by conservatives criticizing it. And an article titled "Christmas Gifts for Each and Everyone," describing some of the nice gift items available that year at the Harvard Coop, had been digitized and included in the archives.

I remember clearly that SDS was on the whole pleased with the *Crimson* coverage of our protests. The general tone was sympathetic and they got the facts right. Those are the articles that are missing from the *Crimson's* website.

Not surprisingly, the *Crimson's* online archives were not set up primarily in order to serve scholars. Their main purpose, it seems, is to support fundraising — either for Harvard or for the *Crimson* itself — among alumni. So I suppose that one has to expect that the articles chosen would be the ones that nostalgic alumni with a lot of money would be most likely to enjoy reading. Apparently the December 13, 1968 article reporting on the Paine Hall sit-in is not among them.

My wife Ann, who is a historian, has commented that students today have a naive confidence that online sources are reliable, complete, and unbiased. That confidence is misplaced.

CHAPTER 4 **THE ARMY**

B etween roughly 1966 and 1968 a large amount of anti-war activity centered around the draft resistance movement. In the last chapter, for example, I mentioned the "We Won't Go" statement of spring 1967 that was signed by thousands of students nationwide, including me.

Throughout the country anti-war people set up draft counseling centers to help young men avoid military service. Advice was given on a wide range of possible strategies — from applying for religious Conscientious Objector status to making sexual advances on the sergeant at the induction physical. Many guys from affluent families temporarily went under the care of a psychiatrist, who would build up documentation of a psychological condition that made it impossible to accept authority figures.

And one could always go to Canada. Canadians have a long humanitarian tradition, best exemplified by having provided a safe haven for runaway slaves from the U.S. The Canadian border was the gateway to freedom for tens of thousands of slaves who fled north on the "Underground Railroad."

In a similar spirit, during the Vietnam War the Canadians welcomed over 50,000 American draft-dodgers and deserters. Near the end of the War, in August 1974, I went to Vancouver, Canada for the International Congress of Mathematicians. While I was there, an incident occurred at the U.S. border forty kilometers away. A draft-dodger intending to slip back quietly into the U.S. for a visit approached the border point. He noticed two men looking at him and immediately started running back into Canadian territory. The two Federal Bureau of Investigation (FBI) agents tackled him after a chase and dragged him across the border into the U.S. A photographer captured the incursion into Canada. As a result of the ensuing outcry and an official protest of the violation of Canadian sovereignty, American authorities decided to return the young man to Canada after a few days in custody. This incident got little coverage in the U.S. press, but a lot in the newspapers in Vancouver.

I wish I could say that the tradition of sanctuary in Canada is still strong in the 21st century. However, the current Canadian government has refused to adopt a policy of granting asylum to American soldiers who desert from the war in Iraq — despite the overwhelming opposition of the Canadian people to that war.

In 1968 I seriously considered the option of going to Canada after graduation. The University of Toronto was reputed to have the best mathematics department in Canada, so I sent away for material on applying to graduate school there. I also collected information from draft resistance organizations about immigration to Canada.

But when I started to get politically closer to SDS in late 1968, I abandoned that plan. About a year earlier SDS had split with the draft resistance movement, which the increasingly radical SDS activists saw as elitist and counterproductive. The Progressive Labor Party especially was guided by the example of the Bolsheviks, who had organized soldiers at the front during World War I and whose slogan had been "Turn the guns around." Several leftist organizations, including the PLP, decided that, rather than keep people out of the military, they should send their members into the Army to organize dissent from within.

In a dramatic reversal that perhaps sounds counterintuitive to someone today, SDS came out in strong opposition to student deferments. Then as now conservatives often accused the anti-war movement of being motivated by cowardice — by a fear of being called to military duty. If that were true, then I wonder how they would explain the fact that students were not put off by the new SDS position on draft resistance. On the contrary, the explosive growth in SDS's influence and membership occurred *after* the organization announced its opposition to student deferments.

As I became more active in SDS and closer to the PLP, I realized that I might have to go into the Army. From the beginning of my first year as a Princeton graduate student the draft loomed large in my mind, even though I didn't learn what my own fate would be until December.

I remember gathering on December 1, 1969 with other Princeton students in a big room in front of a TV. For the first time since World War II, a lottery had been organized to decide the order in which young men would be drafted. The lottery determined an ordering of all possible birthdays (including February 29 for leap years). Three hundred sixty-six blue plastic balls, each with a date, had been placed in a large glass container. The dates were drawn one by one, and my birthday, December 24, was the 95th to be chosen. The rate at which local draft boards went through the birthdays depended on the proportion of young men with various types of deferments. My home address was in relatively affluent Westchester County, New York, where deferments were plentiful. For this reason it was likely that the draft in my district would get through birthdays numbered well into the 200's, and someone with birthday 95 could expect to be drafted before the middle of 1970.

Incidentally, Mark Kac, the mathematician who had helped me since my high school days and who was an expert in probabilities, later was on a commission that investigated the lack of randomness in the 1969 drawing. It turned out that the container had not been shaken enough, and so a lot of December dates remained near the top and ended up with low draft numbers.

During the year before I was drafted, I studied some math, but my heart wasn't in it, and I wound up devoting most of my time to political organizing. That is not to say that I accomplished much.

Princeton SDS had been largely led by a group of undergraduates affiliated with the Revolutionary Youth Movement faction. When I arrived, the Worker-Student Alliance people were all graduate students, led by a PLP member named Robby Nurenberg. The group was pretty stodgy and not very active. Although we went through the motions of claiming to be the true SDS chapter, no one bought it.

The previous summer many WSA students had participated in a "work-in" where they got industrial jobs and interacted politically with the workers there. This was quite successful as an educational experience for the students — which was actually its central objective. The idea was to follow it up during the school year with a "campus worker-student alliance."

By then I had finally joined the WSA caucus. I took a part-time job as a waiter in Prospect, which was Princeton's faculty club. I didn't need the money, since I was fully supported by a graduate fellowship from the National Science Foundation, and I continued to translate Russian mathematics at $6/page for the American Mathematical Society. Rather, the purpose was for me to see what the university looked like from the viewpoint of a campus worker. I got to know some of the full-time employees, who were very friendly toward the student workers. Most of them (and virtually none of the students) were black.

I also did some canvassing of undergraduates. Two of the freshmen I met — Dan Lichty and Alex Farquhar — later became good friends and key organizers of radical activity on campus. Alex, who was the son of a minister in London, Ontario, had a humble, self-effacing demeanor and was a good listener. He was the most hard-working and effective student organizer I have ever known.

The most significant anti-war activity at Princeton that year occurred in May 1970. That was the time of the U.S. invasion of Cambodia and the killing of four student demonstrators at Kent State University by the Ohio National Guard. The focal point of protests at Princeton was the Institute for Defense Analysis (IDA), which was located right next to campus and had some administrative overlap with the university.

That protest, which was led by RYM, at times had several hundred students around the IDA facility. However, they did not attempt to blockade it or shut it down, and they avoided any clash with the police. Rather, they projected slide shows on the side of the building and engaged in various antics that were probably influenced by Abbie Hoffman and the "Yippies." It seemed that their objective was theater, not confrontation. After a few days the protest petered out.

I went to those demonstrations, but more as an observer than a participant. My mind was elsewhere. I had gotten my induction notice and was due to report in a few days.

I was very apprehensive about going into the Army. I was convinced that the PLP had the right strategy — in fact, it was becoming obvious to many people that the quickest way to end the Vietnam War would be to spread the anti-war movement to the soldiers themselves — but that didn't mean that I had any confidence in my own ability to organize within the military.

I had become close politically to the PLP and had had long talks with Alan Gilbert, Robby Nurenberg, and others about different aspects of the "party line." The only PLP position that I found completely unacceptable was their evaluation of Stalin. Following the Chinese line at the time, they regarded him as basically a positive figure who made some mistakes. In contrast, I thought of him as a tyrant who betrayed socialist principles and committed genocide against the Chechen-Ingush and the Crimean Tatars. According to the Leninist principle of "democratic centralism" under which the PLP operated, joining the Party meant agreeing not to publicly voice any disagreements with Party positions. I decided that this would not be a problem, since it was extremely unlikely that historical questions on the Stalin period in the Soviet Union would come up in discussions, especially in the Army.

Much more importantly, I felt that I needed the guidance and discipline that the PLP could provide if I were to have any chance of success as an organizer in the Army. In early 1970 I joined the Progressive Labor Party.

On May 12, 1970, Dan Lichty drove me to the pickup point in Trenton, and I got on a bus with about twenty other inductees. For the first time in seventeen years, I was no longer a student.

The bus took us to Fort Dix, which was less than an hour away. Before Basic Training got under way, we went through preliminary processing. Soon after getting our clothing issued and our heads shaved, we took a battery of multiple-choice tests to see what Military Occupation Specialties (MOS) we were suited for. I did well on all the tests, including ones on such subjects as auto mechanics, about which I knew nothing. I didn't even know how to drive a car, let alone what the various things under the hood were for. But I had gone to privileged schools and, unlike most of the other recruits, had taken dozens of multiple-choice exams. I had excellent test-taking skills and so did much better on the auto mechanics questions than most guys who'd been tinkering with cars since their early teens.

The secretary at the testing center said that I'd done so well that perhaps I should consider Officer Candidate School. Oops! I had done too well — this was definitely not where a PLP'er would want to go. Before I had time to come up with a plausible reason why I wasn't interested, someone at the next desk said, "No, he really can't. There aren't any openings right now."

Because of my high test scores I could choose from a broad range of MOS's. The reason I was given a choice at all was that I'd voluntarily "re-upped" for a third year, thereby changing my status from inductee to enlistee. The mandatory military service was two years. But if an inductee agreed to a third year, he would be allowed to choose a specialty within the Army. In that way he could, for example, avoid the infantry.

The PLP policy was for all members to re-up for a third year. The reason was that otherwise most soldiers coming in with college background were given desk jobs. That usually meant that they were in close proximity to the "brass" (officers) and isolated from the rank-and-file. As PLP organizers, we needed to choose a blue-collar MOS that would put us in contact with working-class draftees and low-ranking enlistees. I chose helicopter repair. That meant that after eight weeks of Basic Training I would be sent to Fort Eustis, Virginia, for another couple of months of Advanced Individual Training (AIT) in the maintenance and repair of helicopters.

I found Basic Training to be extremely unpleasant. Even though I had started jogging with some other graduate students at Princeton, I was in pretty bad shape and my performance as a soldier was embarrassing. I was consistently in the bottom group of recruits in the physical drills; I had particular difficulty with the overhead hand bars.

My record on the rifle range was equally unimpressive, except that I actually received a commendation ribbon for good marksmanship. The explanation for the apparent contradiction is that we were required to stay at the range practicing until we achieved a certain minimal level. The guys keeping score for us were privates from other units who were awaiting their orders to ship out. It was mid-summer, and as the hot New Jersey sun got higher in the sky the soldiers who were keeping score would start to cheat for us so that we could all get out and go to lunch. In my case, in their impatience they got a little carried away and scored me much higher than the minimum necessary. In any case, according to official Army records I'm pretty good with an M-16.

Much as I hated being in the Army, the military way of doing things was fascinating to observe. In Basic the idea was first to break us down as civilians and then to build us up as soldiers. We were forbidden any contact with the outside for the first two weeks, and after that it was severely limited. The drill instructors (DI's) were constantly screaming at us and berating us, and for the first few weeks we were sleep-deprived. They would rush us from place to place, herd us into "cattle cars," and tell us to press up so close to the guy in front that "you make your buddy smile." (This was one of many homosexual innuendos that the DI's loved.) We were told to forget about our girlfriend back home, since she was already with another guy.

As a PLP member, I almost immediately started talking politics with the other guys. Most were cynical about the War. When we were marching, often they would replace the songs we were supposed to use to mark cadence by the words of Country Joe and the Fish:

And it's one, two, three,
What are we fighting for?
Don't ask me, I don't give a damn,
Next stop is Vietnam.
And it's five, six, seven,
Open up the pearly gates,
Well there ain't no time to wonder why,
Whoopee! we're all gonna die.

In private, when guys were angry about harrassment by an NCO (non-commissioned officer), they'd make comments such as, "Just wait 'til he gets to 'Nam." The soldiers had heard that on the battlefield gung-ho officers would often get shot by their own men. So I guess the old Bolshevik slogan "Turn the guns around" wasn't far from what was going on in Vietnam in the later stages of the War. A few months after that, when I met returned soldiers from Vietnam in the stockade, I learned that the preferred weapon for killing one's own commanding officer was the fragmentation hand grenade; this was called "fragging."

In addition to talking to the other soldiers, in private I gave PLP publications to some of them. The Party had prepared a special flyer with the banner headline *Smash the Bosses' Armed Forces*. (The PLP could never be faulted for excessive subtlety.) It was illegal for me to be passing out this material.

Out of the 200 soldiers in our company I encountered only one who was enthusiastically pro-war. He was a Cuban exile from Miami, and he would complain that he was the only patriotic American among us all, and he wasn't even an American yet. He was the one who turned me in to the brass; this happened near the end of Basic Training.

The captain who ran our company had me brought in for questioning. He told me that, even if I hadn't been turned in, he had just received a report from Army Intelligence, and they were onto me. Most likely a background check had revealed my arrest in Cambridge the previous year.

As part of the induction process I had had to sign a couple of loyalty oaths. Those sworn statements were not, however, effective in catching real communists who entered the army in order to organize and who happily signed them, although they occasionally snared a civil libertarian who had philosophical objections to them. Because of the odd phraseology of the oaths, it is possible that I didn't even have to lie in order to sign them. One of them was an affirmation that I did not belong to a list of communist and communist-front organizations that was provided. The list turned out to be badly out of date — the Progressive Labor Party was not on it, but it did include a number of groups from the 1930's that had been formed in defense of the Spanish Republic against Franco.

The captain had to confer with his superiors, who would decide what to do with me. The decision was to do nothing, at least not for now. The problem was that Fort Dix was an open base that the public could visit freely, and it is located near New York, Philadelphia, and other population centers that had large anti-war movements. In the past there had been major demonstrations at the base, and the brass had come to realize that it was not a good place to hold political trials.

In view of the MOS I had selected, my next assignment would be to AIT in helicopter repair at Fort Eustis, which is located near Newport News, Virginia. Such a setting in the conservative heartland would be a much more propitious locale for a political trial. So the best course for them was to continue processing me in the usual way and then to alert their colleagues at Fort Eustis that they should prepare to take legal action against me.

At the end of July I was shipped to Fort Eustis, where I started taking classes related to my MOS. The atmosphere was more relaxed than in Basic Training, and the soldiers had more free time. I talked with many of them about politics and continued distributing *Smash the Bosses' Armed Forces*.

I was not the only PLP member at the base. There were four of us, and we soon got a small discussion group going with a handful of other soldiers. I increasingly felt, however, that it was very difficult to make progress. Most of the guys were happy to agree that the War was awful, that the Army sucked, that rich people in America exploited the working class, and that Country Joe and the Fish were right-on. Some even said that the communists were probably right about a lot of things. But they didn't see any percentage in confronting the Army. The universal slogan was CYA — "cover your ass"; just put in your time and get out.

No one had arcane arguments in support of conservative positions, as the pro-ROTC people had had at Harvard, and no one argued that communism was bad because the Soviets had mistreated Pasternak. The issues were totally different now. The big question was: Who could stand up to the brass? What could anyone do that wouldn't just end up getting oneself screwed?

I recall a conversation I had with one guy who simply said that, much as he agreed with us, he wouldn't get involved because he was afraid; he was totally honest about it. I was struck by the contrast with the academic world. In campus politics one never hears anyone say they won't take a stand against the university administration because "I'm scared shitless I won't get my degree / get promoted / get a salary raise if I do." No, it is always because they are too busy or have some principled disagreement with the wording of the statement or the political philosophy of the activists.

And compared to the university setting, the danger of repression was much more severe in the Army. On campus a student or professor could usually get away with quite a bit of troublemaking without suffering any serious reprisals. In the military, however, the stakes were higher.

I knew that I was being observed by the Fort Eustis authorities. On one occasion an officer saw me hand a leaflet to another soldier, and he took it. After about two weeks I was arrested, along with a comrade named Steve Wenger.

Just before that, I had been thrown out of AIT. The lieutenant in charge of the classes told me that as soon as he heard about me, he pulled me out immediately, and that there was no way he'd allow me back into the program. "I would never want to fly a helicopter," he said, "that had been repaired by an anti-American agitator." Strictly speaking, he was wrong in his supposition that I would deliberately do something to his helicopter. The PLP had a firm line against terrorism and sabotage. Acts of violence by individuals or small "vanguard" groups were called "adventurism" by the Party; Lenin had called that approach "left-wing infantilism." The PLP was certainly not opposed to violence, but it had to be violence by the masses — like the violence that almost took place against the police and strikebreakers at General Dynamics — and not by individuals.

On the other hand, despite my sterling performance on the multiple-choice tests three months before, I was ill-suited for helicopter repair. Like many mathematicians, I have always been rather incompetent in dealing with anything mechanical. During Basic Training I had had difficulty with the drill that consisted of cleaning and reassembling an M-16. And I found the material we were supposed to be learning in AIT at Fort Eustis hard to sink my teeth into.

Thus, the lieutenant had reached the right conclusion, but for the wrong reason. Flying a helicopter that had been repaired by a communist would not necessarily have been a mistake. However, flying one that had been repaired by a mathematician would definitely have been a bad idea.

On September 11, Steve Wenger and I were tried on several counts of "unauthorized distribution of publications." We were given a special court-martial, which is the middle level. A "summary" court-martial is a quick trial before an officer in which the maximum sentence allowed is thirty days, a "special" court-martial involves Army lawyers and has a maximum sentence of six months, and a "general" court-martial is for serious crimes and can result in long prison terms or the death penalty.

Even though we had a court-appointed military lawyer, we insisted on conducting our trial according to the principles of the PLP. That meant turning the trial into a political demonstration — no legalistic or civil libertarian arguments, but only bold political statements. The problem, however, was that, despite having circulated a leaflet calling on people to attend the trial, we had been unsuccessful in our organizing — the hoped-for crowds of supporters were conspicuous by their absence. We gave our political speeches, but, except for one or two friends in the audience (and my mother and sister, who had flown down from New York to be there), we were speaking only to ourselves. We were, of course, admitting our "guilt"; at one point our exasperated defense attorney just put his head down on the table.

We were both given the maximum sentence — six months of hard labor. The "hard labor" part was an anachronism; it was more like a sentence of six months of boredom.

In another sense, however, the sentence would turn out to be "hard time." Military prisons, like their civilian counterparts, are depressing places. The main problem is not necessarily the bad facilities, nor is it that all the prisoners are terrible people. On the contrary, during the Vietnam War most of them were in the stockades for offenses that in civilian life would not have landed them in jail, such as going AWOL (Absent Without Leave) or being caught smoking marijuana.

But there were a certain number of prisoners who were true criminals — there for drug dealing or for violent crimes. These were the *lumpenproletariat* who had been held up as revolutionary role models by the RYM faction at the SDS convention. Some — especially the ones who had been in Vietnam — boasted of their plans to make a lot of money as pimps or drug dealers when they got out. Several had grown their fingernails very long. I had never seen this on a man before. The purpose was apparently similar to the reason why the custom was introduced for upper-class women — to advertise the fact that they had no intention of doing manual work.

Just as in civilian prisons, the tendency was for the worst elements of the inmate population to acquire power and authority. They also knew when it was in their interest to work with the guards. The soldiers who were in the stockade for drug offenses had a strong incentive to get on the good side of the jailers. If they were awaiting trial, they knew that the sentence could range from very strict to very lenient. If they had already been sentenced, they knew that if they cooperated with the guards and officers they would be released and discharged from the Army well before their sentence was over.

At the Fort Belvoir stockade outside Washington, D.C., where Steve and I were initially incarcerated, as good PLP members we soon started talking with the other inmates, trying to prepare the groundwork for organizing something. After a little over a week of this, we were attacked. Earlier in the day I had seen a guard talking intently with a few of the inmates who were in on drug-related offenses. Soon after, they cornered us and beat us up. After a while the guards came in and took us to solitary confinement "for our own protection."

We spent nine days in solitary. There was little to do other than listen to the radio that was piped in. They would play Jimi Hendrix songs again and again in tribute to the great rock musician who had died on September 18. Shortly before we left, the radio announced that Janis Joplin had also died.

In early October, Steve and I were moved to another stockade. Located in rural Pennsylvania, the Indiantown Gap facility was designed for difficult cases and overflow from other prisons. For example, some of the inmates had been transferred from the "LBJ" — the Long Binh Jail in Saigon.

Unlike other stockades, where there was at least the pretense that many of the prisoners would be returned to active duty after their sentence was up (although in practice few were), at Indiantown Gap everyone knew that incarceration would almost always be followed by discharge from the Army. The only reason I could think of why the Army didn't save itself the trouble and expense and discharge everyone right away was that large numbers of soldiers would have deliberately gotten themselves court-martialed if it had been that easy to leave the Army. So it was important to make us pay a steep price first. As in civilian prisons, that price was months of degrading treatment by the jailers and their proxies among the prisoners.

By late October I realized that I was unable to function effectively as a PLP organizer in the stockade, and I did not know how to deal with the continuing difficulties I was having. My parents at one point even contacted the office of their congressman, and his staff made an inquiry at the Pentagon about my mistreatment in the stockade. My parents also hired a lawyer named Jeremiah Gutman, who belonged to the American Civil Liberties Union (ACLU), to try to help me. A little later Mark Kac wrote to the Commandant of the stockade urging him to process me for a discharge as soon as possible.

I was now thinking a lot about my return to civilian life. Even though I'd been in the Army only a few months, it had been almost two years since I had last studied mathematics intensively. I knew that my return to graduate studies at Princeton, which had the country's top graduate school in mathematics, would require some effort. In order to prepare, I asked my parents to bring me Serge Lang's *Algebra*, which was one of the best graduate-level textbooks. There is an old tradition in American prisons of allowing even prisoners in solitary to have one book — usually the Bible. For me, an atheist and aspiring mathematician, the "bible" was Lang's textbook. I carefully read every page and solved every exercise problem. I had no TV to watch, so this was a form of escapism for me — I could immerse myself in abstract algebra and temporarily forget about the real world around me.

That is not to say that there weren't some nice moments while I was in the stockades at Fort Belvoir and Indiantown Gap. I preserved a sheaf of letters that people wrote me. I got a long, friendly letter from a young woman with whom I'd once had a date. After an article about my court-martial appeared in a Westchester County newspaper, old friends of my family contacted my parents and offered support. Vera Rony (to whom I had written the detailed letter describing my experiences in 1966 in Grenada, Mississippi) visited me at Fort Belvoir. A letter from my old high school English teacher, Doris Breslow, was especially touching. She wrote, "You have probably heard the Emerson–Thoreau story so often that it is a chestnut, but I'll risk reminding you of it anyway and of Thoreau's marvelous retort when his friend scolded him for being in prison — 'And why are you *out there?*'"

I also made some friends among the other prisoners and was able to join them in various acts of defiance from time to time. I served as a "jailhouse lawyer" at a summary court-martial of a black soldier who had taken a swing at a guard. I had read a book about soldiers' rights under the UCMJ (Uniform Code of Military Justice), and since childhood I'd been interested in the law. (It has been said that the structure of legal arguments resembles that of a mathematical proof.) So I was reasonably good as his lawyer: he ended up with fifteen rather than thirty days added to his sentence.

There were racial tensions between white prisoners and black and Puerto Rican prisoners that the guards knew how to exploit; they understood the strategy of "divide and conquer." (The inmates who attacked Steve and me in the Fort Belvoir stockade were black.) One day in November the Indiantown Gap stockade was preparing for an inspection by a colonel. Part of the cleanup was to remove some graffiti. An inmate who had since been released had painted "Pigs Die Now" in boot polish on his cell wall. This had to be removed — I guess the guards were worried that the colonel might take those words personally. At first the sergeant in charge of the cleanup asked some black inmates to do it. They said that the guy who had painted the slogan was a brother, and they had no intention of removing it. So the sergeant asked me to do it. I refused, because I knew that the effect would have been to set me against the black guys. I said that if it was put there by their friend and they didn't want to remove it, then it wasn't my business to. I offered to do another cleaning chore instead, but the sergeant pressed charges against me for disobeying a direct order.

On November 20, I had a summary court-martial, which my father attended with a tape recorder. Either because of his presence or because of my argument about "mitigating" circumstances, I was sentenced to an additional fifteen days rather than the maximum thirty. By then I knew that I was going to be discharged long before the six months were up, and the fifteen days added to my sentence wouldn't make any difference. In fact, I later learned from my lawyer Gutman that the second court-martial probably caused the Commandant to process my discharge papers a little faster.

By the end of November my mood had improved a little. Steve Wenger had just been released. In addition, I was encouraged by news from the other two PLP organizers at Fort Eustis — Art Small and Paul Weinberg. Art wrote my parents about their trials for "unauthorized distribution of publications" that took place in November. They had both been acquitted. In the time since Steve's and my court-martial, their organizing efforts had finally gotten off the ground. Over thirty soldiers had come to each trial to offer support. In addition, the PLP, presumably in response to the fiasco at our earlier court-martial, had softened its stand on presenting legalisms. Art and Paul had allowed their lawyer to argue that the statute was "unconstitutionally vague," and that argument, along with the impressive show of support from other soldiers, persuaded the jury of officers to acquit them. Art sent the records of the trial to my parents so that they could use them in appealing my conviction.

Other people in the Party had clearly had much more success in the Army than I had. Pete Bilazarian, whom I had known since he came to my freshman dorm at Harvard to clean the toilets, turned out to be a forceful organizer. A big demonstration at Fort Dix that he helped lead ended with protesting soldiers fighting with the MP's (Military Police). As a result he was given a general court-martial and a full year's imprisonment at Leavenworth. I visited him there in September 1971.

On December 9, 1970, I was released from the stockade and discharged from the Army. For the previous month or so my mind had been on getting back to civilian life, and I had not been acting as any kind of organizer. I felt that I had failed as a radical in the Army, and I immediately resigned from the PLP. I had been a member less than a year.

I had learned the hard way that what I was cut out for was the life of an academic mathematician, not that of a political organizer, let alone that of a professional revolutionary. I reverted to my way of thinking before 1969, when I was an SDS fellow-traveler, but not a member. I would keep my radical opinions, but would remain largely on the sidelines, or so I thought.

At the time I viewed my seven months in the Army as a nightmare that was best forgotten. I did not keep contact with anyone I had met there, although I made abortive attempts to correspond with Steve Wenger and one other guy from the stockade. Most likely they felt the same way I did — none of us wanted reminders, even in the form of correspondence, of that period in our lives.

Much later, however, I came to realize that, despite everything, there had been some value in my experiences at Fort Dix, Fort Eustis, Fort Belvoir, and Indiantown Gap. I matured, and my political perspective broadened. In a way that's hard to explain, after the Army I was better able to focus on what was really important.

Many of my contemporaries had experiences that had a similar effect on their lives. It could be military service, imprisonment, or the Peace Corps. Especially for those of us who had grown up in a sheltered, affluent environment, it was necessary at some point to come to understand a different reality. In the words of a Russian proverb that I learned from a poem by Pasternak, *zhizn' prozhit' ne pole pereiti* — "To live a life is not to cross a field."

Five years later the story of my relationship with the U.S. Army had a strange epilogue. Jerry Gutman, the ACLU lawyer my parents had hired when I was in Indiantown Gap, was pursuing appeals of both my discharge status and my original court-martial conviction. I had received a "general discharge under honorable conditions," which was the mildest — and the easiest for the Commandant to process quickly — of four less-than-honorable discharge categories. (Yes, the Army used the term "less than honorable" discharge "under honorable conditions" to describe the type of discharge I had been given.)

By the time the case finally reached the Discharge Review Board, I already had a Ph.D. and was teaching at Harvard. A hearing at the Pentagon was scheduled for October 23, 1975. I went with Gutman and my sister Ellen, who was then a law student at Yale.

What made my case strong was that on July 12, 1973 there had been a big fire at the Army's National Personnel Records Center in St. Louis, Missouri. Almost all of the Army's records, which in those days were still being stored in paper form, were destroyed. Other than the bare facts of my time of service and incarceration, the Discharge Review Board had nothing to go on, so they were completely dependent on the version of events that Gutman and I presented to them. Needless to say, our narration of the circumstances of my arrest and trial was sanitized, and we neglected even to mention the second court-martial (of which they had no record). The colonels on the Board wanted to appear liberal and fairminded. They ruled in my favor, upgrading my discharge to fully honorable.

There are certain occupations for which it would have been a major victory to get one's Army discharge upgraded. However, the professoriate was not one of them. Even in the most retrograde, backward college in the country the administration would not ask a faculty job candidate about his discharge from the Army. At most the conferring of an honorable discharge by the Review Board was some kind of moral victory — although if I were asked to explain exactly what kind of moral victory it was, I wouldn't be able to. Perhaps for my parents — more than for me — it provided a kind of closure and made it easier to put into the past the memories of that horrible autumn of 1970.

A few months later my special court-martial conviction reached the Military Court of Appeals, which voted not to overturn it. In the time between the Review Board hearing and the Court of Appeals ruling there had been a U.S. Supreme Court decision sharply curtailing First Amendment rights of soldiers. The military courts take their cues from the civilian courts, and so the Military Court of Appeals saw no reason to overturn my conviction. Thus, I am in an extremely unusual category among veterans — two convictions, including a special court-martial conviction that was upheld on appeal, and at the same time a fully honorable discharge after seven months of service.

If the special court-martial conviction had been overturned, that would have had a minor practical value for me. The eighty-nine days that I had been imprisoned would have been converted to "good time," and I would have been eligible for veterans' benefits (which require a total of 180 days of good time). Too bad! I guess I can't really call myself a veteran now.

Postscript

In 2005 I had a student in one of my classes who at first seemed to me to be perhaps mildly autistic. Alan (not his real name) spoke out at unexpected times and occasionally in inappropriate ways. I didn't mind — indeed, I was glad to have student participation in any form, even if it was awkward and slightly off topic — but the other students clearly found him annoying.

Soon after classes began Alan came to my office with forms from Disabled Student Services requesting special accommodations because of mental disabilities. I wouldn't have pried into his story — that would have been a violation of medical privacy — but without any prompting he started to tell me his background.

In early 2002 Alan was in U.S. Army Intelligence in Afghanistan. His job, working with a translator, was to determine who among the captured insurgents was cooperating and who was not. Those in the latter category, he said, were often summarily executed by U.S. soldiers or Afghanis who were working for them. He said that on one occasion they cut off a man's genitals in front of the other prisoners, who watched him bleed to death. This "softened" them up, and then they eagerly told the U.S. soldiers whatever they thought they wanted to hear.

Alan said that soon after that he decided that he couldn't take it anymore and was able to get a medical discharge. Since that time he has had severe mental problems, has been taking medication, and has been attending therapy sessions at a Veterans Administration hospital.

He seemed to feel at ease talking with me. I knew the Army lingo, which hadn't changed much in thirty-five years, and I belonged to a generation that had lived through an earlier war. He mentioned that in the counseling sessions he was the only one under 30 years old; the others were Vietnam veterans, with whom he felt he had a lot in common despite the age difference. I had the impression that one of the reasons for his difficult adjustment to college life was that he felt alienated from the other students his age.

I found it unsettling to talk with him: it brought back long-past images and buried memories. At the end of our conversation I alluded to the news reports about prisoner abuse at Abu Ghraib in Iraq. Alan said that he hadn't had much of a reaction to that. I was surprised, and then he explained what he meant with a question that has haunted me ever since: "Do you really think that that's the worst that we're doing over there?"

After our conversation I tended to discount what he had said. I mentioned it to a couple of people, but always with the disclaimer that he suffered from mental illness and I had no way to corroborate what he had told me.

I was wrong to have doubted him. Within a few months we all read in the newspapers about increasing evidence of large-scale U.S. atrocities in Iraq — marines shooting groups of prisoners, and four Army men raping a 14-year-old girl and then killing her family to cover it up. My student, mentally ill or not, had had a clearer understanding than the rest of us of what was happening in Iraq.

CHAPTER 5 **SPRING OF 1972**

Soon after my discharge from the Army in December 1970, I returned to Princeton to make arrangements for resuming graduate studies. The math professors, some of whom knew that I'd been in the Army and in jail because of political activity there but didn't know the details, went out of their way to make the readjustment easier for me.

Professor Edwin Nelson, who was in charge of graduate students, told me that he was sure I'd be able to get back into my studies, since I had an unusually strong undergraduate background. And he agreed to count my earlier year of enrollment as if it had been just a semester. This meant that I wouldn't be under such pressure to take the Qualifying Exam (an oral exam that must be passed before starting on thesis work), and the department would find financial support for me for an extra year.

When I eventually took my Quals in January 1972, my examining committee bent over backwards to be easy on me. There were two senior professors — Phillip Griffiths and Kenkichi Iwasawa — and a junior faculty member whose name I don't remember. At one point the junior guy asked a question that Griffiths thought I might have a hard time with: he asked me to give an example of a continuous function that goes from the origin to the point (1,1) and is almost everywhere horizontal (that is, it has zero derivative except on a set of measure zero).

Griffiths immediately interjected, "Oh, that's a tough one! I'm not sure I could answer that." But I said, "That's okay, I can answer it." The question was an easy one for me because I had studied Lebesgue measure and integration with Mark Kac when I was in high school, and in fact had solved that very problem eight years earlier. It's actually pretty simple if you know about the Cantor set.

At the same time as I was arranging my return to graduate studies, I also met and was very attracted to a Princeton sophomore named Ann Hibner. She was a friend of Dan Lichty and Alex Farquhar, whom I had met when doing political canvassing among freshmen before I was drafted.

Ann was in the first class of women admitted to Princeton. That year there were about a hundred young women among the thousands of male undergraduates. Most felt awkward, on display, and over-protected by the

Princeton administration and campus police. However, a few who, like Ann, were from families of modest means had to work in Commons, the student cafeteria. They were the lucky ones, since they were able to make friends with the other student workers in a congenial environment. That's how Ann had met Alex, who also worked in Commons.

Although I was not quite as awkward as I had been at Harvard, I was almost totally inexperienced in dating. I was in the generation of "sex and drugs and rock 'n roll," but two out of the three had passed me by (and rock music had not become a passion of mine until late in my senior year of college). Ann was very tolerant of my social awkwardness. She was an avid reader, had a huge vocabulary, and viewed my choice of conversation topics — politics, mathematics, Lou Reed, and Andy Warhol — as charming rather than nerdy.

Ann was young — just eighteen — and when she told her mother that she was dating a graduate student, her mother was not pleased. She thought that Ann was being preyed upon by a "cradle-snatcher" — until she learned that I was only three-and-a-half years older than her daughter.

In late February 1971 Ann moved in with me, although for a while she was able to hide this from her parents. Ann's mother was quite angry when she eventually learned that the dorm room they had been paying for was unoccupied while Ann lived "in sin" with her boyfriend, and that this boyfriend was also luring her into political protests that were bound to lead to trouble. Without notifying Ann in advance, Mrs. Hibner came to our rooming house on University Place. I was furious that she had descended on us without warning. When she asked me "Where are you leading my daughter?" I responded, "Down the primrose path to hell." In retrospect, that was not a very politic thing to say to one's future mother-in-law. Fortunately, she pretended not to hear. Not long afterwards Ann and I became formally engaged, and Ann's parents from then on related extremely well to me and my whole family. Ann's family was Polish Catholic and mine was Jewish, but neither her parents nor mine had any objection to the mixed marriage.

Ann taught me how to drive, and in 1971 we took many leisurely excursions together in her grandmother's old Buick — to the Jersey shore, the Delaware River, and even Fort Dix. After my discharge from the Army I had started to lift weights at the gym in Princeton, so I stopped the car at some overhead horizontal bars of the sort that had caused me so much difficulty in Basic Training and showed off to Ann that I was now able to go across them hand over hand.

Neither of us was under much academic pressure — I was just studying and reviewing in preparation for my Quals — and very little was happening on campus politically. That year was a time of simple pleasures and getting very close to one another — a tranquil period that for me was in stark contrast to the traumas of the previous two years.

In the autumn of 1971 Ann and I, along with Dan and Alex and several other campus radicals, organized a group that we called University Action Group (UAG). We were still in touch with people in the Progressive Labor Party and the old Worker-Student Alliance caucus of SDS, and were trying to find some way to campaign on behalf of campus workers. But it was clear that a lot of research and groundwork would have to be completed first.

We chose not to call ourselves SDS because of unpleasant memories of some of the behavior of leaders of the Princeton SDS chapter who were in the Revolutionary Youth Movement. In fact, one of the central principles of the UAG was that there were no leaders. Meetings had rotating chairs, and the overblown rhetoric and "ego-tripping" characteristic of many of the RYM-type radicals were strongly discouraged.

When we got the application to become a recognized student group, we found that the university required us to designate at least three officers of the organization. Since we had no officers — but needed university recognition in order to reserve meeting rooms, get an office, and use university mimeograph machines — we decided to fill out the form in a satirical vein. We went around the table, and everyone got to choose a fanciful "office." The first person was Keeper of the Seal, the next was Cleaner of the Seal's Cage, and Ann and I opted to be Duke and Duchess of Lichtenstein. Those were our "officers."

In December Alex, who was a far better grassroots organizer than any of the rest of us, told us that we should forget about the campus worker issue (or non-issue). The central question was still the War, and a lot of students he was talking with were upset about reports that Princeton was planning to bring back ROTC (which a year-and-a-half earlier the Board of Trustees had decided to phase out).

On January 15 the Trustees had a scheduled meeting in Nassau Hall, the administrative and historic center of Princeton University, at which they were going to discuss ROTC in a closed session. We organized about 75 students to link arms and block the entrances to Nassau Hall. Even though student opposition to ROTC was strong, as an organization the UAG did not yet have much support or visibility on campus. Considering this, our choice of tactics was rather bold — we could have been brought up on serious charges. However, after the campus police broke up our blockade, the administration did not take any disciplinary action. This was the first major protest after a long quiet period at Princeton, and Dean of Students Neil Rudenstine (who later became president of Harvard) probably thought that we would be less likely to attract a lot of attention and build up momentum if they responded in a low-key way. And in fact it was another month-and-a-half before anti-war activity built up in earnest.

The next month the UAG got embroiled in a controversy not related to the War, and we ended up having a major impact. In February a progressive psychology professor named Charlie Gross told us that his colleagues had invited Harvard's Richard Herrnstein to give a talk at Princeton.

A few months before, Herrnstein had published an inflammatory article titled "I.Q." in the *Atlantic Monthly*. He suggested that modern America is a meritocracy in which the individuals — and races and ethnic groups — at the top are there because they are genetically endowed with superior intelligence. Herrnstein was elaborating on a thesis that had been put forward most infamously by Arthur Jensen in a 1969 article in the *Harvard Educational Review*. Jensen's piece, titled "How much can we boost I.Q. and school achievement?" had argued that compensatory education programs such as Head Start were doomed to fail because children's achievement was largely determined by their genes.

The Progressive Labor Party, Worker-Student Alliance, and other progressive groups at Harvard had been conducting a boisterous and confrontational campaign against Herrnstein, and our friends in the PLP urged us to take some action when he visited Princeton. We called a UAG meeting, which Ann chaired. There was a lot of anguished discussion about what we could do. Clearly something strong was called for. On the other hand, the issue was an unfamiliar one for most Princeton students, who would be much less willing to support militant action over Herrnstein than over the War.

The topic of Herrnstein's invited talk in the psychology department concerned some earlier work of his on pigeons. We decided to organize a large number of students to go to Herrnstein's lecture, and at the end we would insist that he respond to our questions about his "I.Q." article and listen to our angry objections to the role he was playing in American education. We would refuse to let him leave until he did this, and we would physically block the exits if necessary.

We knew, of course, that this plan was extremely risky. It's one thing to blockade a Dow Chemical recruiter or Defense Secretary McNamara's car, as had been done five years earlier at Harvard. But in the eyes of most Princeton students and faculty, it would be another thing entirely to use a similar tactic against a psych professor — especially since there was much less awareness of issues of race, class, intelligence, and educational policy than there was of the Vietnam War.

When Herrnstein heard of the planned protest, he canceled his visit, thereby unintentionally doing all of us in UAG a big favor. In fact, we breathed a sigh of relief when we learned that he wasn't coming. However, the chair of the psychology department, Leon Kamin, was furious with us. He said that by threatening disruption and creating a hostile environment we had deprived Herrnstein of academic freedom.

I knew that Kamin was a liberal who had been a communist in his youth and had been fired from Harvard during the McCarthy period. I didn't want him to be our enemy, and went to see him in the hope of moderating his position. He told me that what he objected to was that we had intimidated Herrnstein and interfered with his free speech. I answered that we had not threatened to do anything until after his invited talk was over, and that we were merely insisting on holding him accountable for his public writings in the *Atlantic Monthly*, which would have a horrible effect on the education of minority children. Kamin said that the way to fight against wrong ideas is to publish refutations, not to intimidate the professors who expounded them. I said that that was fine for purely academic disputes. But the problem with what Herrnstein had done was that his inflammatory statements and outrageous claims got extensive coverage in the news magazines and on TV, whereas the refutations were either ignored or relegated to the back pages of newspapers. So the liberal response was ineffective — we needed a much noisier, more confrontational approach. Our discussion was cordial, but neither of us changed our views.

A little later Kamin telephoned me and said that he was still very bothered by what had happened and was organizing a special forum on the subject of academic freedom. He wanted to know if we would participate. I answered that we believed the central issues in the Herrnstein matter to be racism and the societal responsibilities of a scholar. We would not come to a forum that was devoted exclusively to discussing academic freedom, but we would participate if its title were broadened to include those other issues. Kamin agreed to our condition, and so we came in large numbers to his event.

The forum on March 9 that was organized by the psychology department was well attended. Even the university president Robert Goheen came. A cross-section of viewpoints were represented, and most people put a lot of thought into what they chose to say. I spoke immediately after a notoriously right-wing representative of the Young Republicans named Ira Strauss. I said that we in the UAG didn't mind being on the opposite side of this issue from Ira Strauss, but I didn't think that we should be on the opposite side from liberal faculty members, who presumably shared with us a strong opposition to racism in America. I pointed out that Herrnstein's article in the *Atlantic Monthly* was read by hundreds of thousands of people, including numerous school teachers who would get the message that many of their pupils — especially the minorities — are uneducable. If people on the faculty didn't like our militant tactics in responding to this, then it was their obligation to come up with something else that would be effective.

Several of the radical students made similar appeals. Instead of our being on the defensive because of our choice of tactics, we wanted to make the professors feel guilty for not having done anything to oppose the likes of Jensen and Herrnstein. Aside from a couple of right-wingers like Ira Strauss, just about everyone on both sides felt that it was a worthwhile meeting. Disagreements were aired in an atmosphere of mutual respect.

In order to raise awareness of the Herrnstein issue, we organized a talk by the famous Harvard philosophy professor Hilary Putnam, who was still an active radical (later he did a political about-face, recanted his former views, and became anti-communist). Putnam gave an excellent overview of the issues, clearly explaining, for example, the logical fallacy in extrapolating from differences between individuals to differences between groups. Even if intelligence were largely hereditary, he argued, it would not follow that a difference between racial/ethnic/class/gender groups in average performance on intelligence tests would have any genetic component at all. He explained this by analogy. Suppose that there's a large genetic component in the size and height of a certain type of animal. If the animals are randomly divided into two groups, one of which is then fed well and the other poorly, there will be a big difference in their eventual average sizes, and it will be entirely for reasons of nurture rather than nature.

We felt that our anti-Herrnstein efforts at Princeton had been moderately successful. We'd bitten the bullet with our threat of militant tactics, which we never had to carry out. Kamin's forum in the psychology department had gone well, and our relations with the faculty were largely free of antagonism. On the other hand, most UAG members would have said that among all of our actions in the spring of 1972 the one for which we had the least support on campus and with which we accomplished the least was the Herrnstein episode.

However, history sometimes surprises us. A couple of months later Charlie Gross told us that we'd really made an impression on Kamin, who was devoting a lot of time to investigating not only Herrnstein's writings, but all the hereditarian work on intelligence.

When Herrnstein and others claimed that intelligence is largely determined by heredity, the main justification they cited was the identical-twin studies of Sir Cyril Burt. During a long and prolific research career Burt, who is often regarded as the father of educational psychology, influenced generations of psychologists and educators in Great Britain and the U.S. The idea of his twin studies was that identical twins who were separated at birth would have the same genes but would be raised in different environments, so that the correlation between their performance on I.Q. tests would give a measure of the contribution of heredity rather than environment. Burt calculated that I.Q. is about 80% genes and 20% environment.

Kamin went back to Burt's original papers and tried to locate his raw data. Almost immediately he found striking irregularities. For example, Burt published three reports in 1955, 1958, and 1966, during which time the numbers of identical twins reared apart and reared together increased, presumably because more data were coming in. But the reported correlation of I.Q.'s in both of the tested groups remained identical to three decimal places! The more Kamin examined Burt's work, the more evidence he found of fabrication.

After the publication of Kamin's book *The Science and Politics of I.Q.* in 1974, the results of his investigation became widely known among educators and psychologists. Then on November 28, 1976 a front-page article appeared in *The New York Times* with the heading "Briton's Classic I.Q. Data Now Viewed as Fraudulent." At about the same time Herrnstein lashed out at Kamin, calling the charges that Burt had deliberately faked his results "outrageous and incompetent." Interviewed by *The Harvard Crimson*, Herrnstein further commented about Burt's studies that even "if he did fake his data, then he faked it truly." At that point Herrnstein and his ilk did not have much credibility.

In the late 1970's my mother, who was active in the leadership of the National Education Association (the largest teachers' union in the U.S.) and whom I had told about the whole Herrnstein controversy at Princeton, arranged for Leon Kamin to receive the NEA's Human Rights Award for his work in debunking and refuting racist pseudoscience. (I don't think he ever knew that I had had anything to do with his getting that award.)

By that time Ann and I were no longer at Princeton, and the UAG no longer existed. Only then did we realize that the anti-Herrnstein campaign of February–March 1972 had succeeded beyond our wildest dreams. We had stirred up the complacent psychology professors at Princeton and had prodded Kamin to do something about this issue that had a tremendously positive impact. There's an obvious lesson here about political judgments: it's not possible to predict what the long-range effects of some action might be. A campus crisis that seemed important at the time might look quaint and trivial many years later, whereas a militant action (or in this case just a threat of militant action) that seemed only marginally successful might turn out much later to have had a significant impact.

That does not, of course, mean that Kamin's exposé was the death knell of racist pseudoscience. Herrnstein himself later acquired even more notoriety when, shortly before his death in 1994, he and Charles Murray published *The Bell Curve*. It is an unfortunate fact of academic life that there are a small number of professors, often at the most famous universities, who repeatedly use their prestige and some type of bogus methodology to launch attacks against ethnic and racial minorities.

In 1972 the Nixon administration was pursuing a policy of "Vietnamization" of the War. The idea was to shift some of the fighting from the American troops to the Saigon regime's military forces. That would decrease the rate of U.S. casualties, allow draft calls to be lowered and, they hoped, reduce the intensity of domestic opposition to the War. In addition, the U.S. strategy was increasingly to rely on high technology (the "electronic battlefield," as it was called) and air power rather than ground combat.

Although the U.S. government tried hard to convince its own people that the War was "winding down," it certainly was not winding down for the Vietnamese. On the contrary, the U.S. unleashed some particularly vicious attacks during that year, including the mining of Haiphong harbor and increased bombing of North Vietnam, culminating in the "Christmas bombings" of December 1972. The purpose of the accelerated attacks was to pressure the Vietnamese side at the Paris peace talks into agreeing to terms that were more favorable to the U.S.

In early March we learned that Admiral Thomas H. Moorer, Chairman of the Joint Chiefs of Staff, was going to visit Princeton on March 14 and speak to the Whig-Clio Society. The highest-ranking officer in the U.S. military establishment, he was personally responsible for authorizing each bombing raid on North Vietnam. When he came to Princeton to speak, he would be accompanied by a film crew from the Pentagon. The purpose was to produce propaganda showing that the student anti-war movement was dead.

We saw this as a provocation and felt that it was crucially important to organize a militant response. The evening of the 12th, which was a Sunday, we held a UAG meeting that was not well attended — there were only fifteen of us, mainly the "hard core." We voted to shout down Admiral Moorer and prevent him from speaking. At the same time we realized that if we were the only ones doing this, then we would fail, and all we would accomplish would be to get ourselves thrown out of school. So we decided to call another meeting the next day, which would finalize the decision on tactics for the Moorer visit. The plans for the second meeting were announced in Monday's *Daily Princetonian*, which also reported that our first meeting had reached a tentative decision to shout down the speaker.

Approximately 95 people came to the second meeting about the UAG response to Moorer; it was standing room only. Most were anti-war liberals who were upset that we were going to interfere with Moorer's "free speech." Ann chaired the meeting; somehow Ann's turn as chair coincided with the two most difficult UAG meetings that spring, the ones on Herrnstein and on Moorer.

Emotions were high on both sides of the tactical debate, and it seemed that everyone wanted to speak. Ann was scrupulously fair in recognizing everyone in turn and not favoring the radicals who had been at the earlier meeting. We were outnumbered by the moderates, few of whom had had any previous connection with UAG actions and none of whom had been planning to do anything about the Moorer visit until we announced our plan for action. It was clear that we were likely to be outvoted at our own meeting, but at least we could say to the liberals — much as we had done at the psychology department forum on Herrnstein — that if they disagreed with our tactics, then they were morally obliged to find something else to do that would be effective. At the end of an exhausting and sometimes heated discussion, the meeting voted not to prevent Moorer from speaking. Rather, we would organize a large number of people to come to the Moorer speech and protest in whatever peaceful way they wanted — turning their backs, holding up signs, or shouting objections to what he said.

When we lost the vote to shout down Moorer, a few of the radicals felt upset and betrayed. Dan Lichty thought that Ann had been a weak chair and should have tried to skew the discussion by recognizing the radicals more often. Another friend, Pat Koechlin, pointed out that the outcome of a vote depends on how the alternatives are perceived. When people think that the radical tactic is to shout Moorer down, the other extreme is to do nothing, and the middle alternative is to have a big non-militant protest, they'll vote for the latter. If we had posed the question differently — pointing out that Moorer, who had committed horrific war crimes, deserved to be shot, that a less radical alternative would be to beat the shit out of him, and the least radical tactic would be to shout him down — then the liberals would've happily voted to shout him down. Pat's comment was meant as a joke, of course, but Ann and I liked it so much that we have since referred to it as the "Pat Koechlin principle."

I told Ann that she'd done an excellent job, and I thought that the meeting had truly been a success. At first she assumed that I was just saying that because she was my girlfriend, and I knew that she was upset at the reaction of Dan and some of the others. But I really meant it — I felt that we had lost the battle but won the war. The liberals who attended our meeting had been impressed with the democratic way we ran it and with our willingness to be outvoted and to abide by the outcome. Moreover, they felt pressured into putting their energy where their mouths were and doing something effective to protest Moorer's visit. It was clear to me that as a result of the meeting something big was going to happen.

A broad coalition of anti-war people organized for the event. About twenty gathered in a room on campus for a marathon session of poster-making. Nancy and Fred Damon, a couple who lived in the same graduate student rooming house as we did and whom we'd never thought of as activists, stayed up making signs literally the whole night before Moorer's visit.

The Corwin Hall auditorium held only about a hundred, so many of the protesters arrived an hour early to be sure to get in. A half hour before Moorer's arrival there were already a few hundred people waiting, many chanting and carrying signs. University officials realized that there'd be pandemonium if all but a hundred were barred from Moorer's speech. So they moved the event to McCosh 50, one of the largest lecture halls on campus.

That caused a further delay, because Moorer's advance team had to make a security inspection of the new room with bomb-sniffing dogs. Finally they opened the doors, and over 500 people, most of whom were protesters, filled the lecture hall. After the tension and the long wait, Moorer entered with his aides and camera crew. As soon as he started speaking, people from all directions started shouting and heckling in response to almost every word. One person in the audience — a veteran dressed in combat fatigues — yelled insults nonstop. There was no camera angle that wouldn't show anti-war signs. After about fifteen minutes, Moorer stopped and left the room, followed by the cameramen. The event had been a fiasco for the Pentagon.

Despite the claim of a group of conservative students and alumni that we had prevented Moorer from speaking and should be punished, the university administration ruled that no regulations had been violated. In any case, there was no feasible way that they could have disciplined hundreds of people for their spontaneous outbursts and heckling.

The protest was a tremendous victory for UAG, and not only because we had thwarted the Pentagon's plan to use Princeton to demonstrate the demise of the anti-war movement. We had also earned the respect of campus liberals, and that proved to be crucial in our later actions. Finally, since we had avoided university discipline, we had a clean slate; if we had already been on probation for the Moorer protest, we would have been in a weaker position to have a militant anti-ROTC protest the following month when the Board of Trustees met.

There are two fundamental mistakes that radicals often make in their relations with liberals. The first is to kiss up to them, agree to wishy-washy ineffective tactics, and suppress any desire to do something more militant. The other extreme is a sectarian strategy of cutting oneself off and needlessly antagonizing liberals and moderates. Radicals who go to either extreme are generally ineffectual. The best approach is to provide a type of tactical and political leadership that the liberals are willing to follow, in part because what the radicals are goading them into doing is what their consciences are telling them they should be doing anyway.

It should also be stressed that the differences between the liberal and radical wings of the anti-war movement were not just a matter of tactics. Most liberal activists believed in a single-issue coalition to end the War, and they thought that they could best forge such a coalition by talking almost exclusively of U.S. deaths in Vietnam and the damage to the U.S. from the War. In contrast, our UAG chapter consistently spoke about the Vietnamese victims, lauded the Vietnamese for their struggle first against French colonialism and then against the U.S., and combined anti-war activities with protests against other forms of social injustice, especially domestic racism. For example, we had gone pretty far out on a limb in protesting Herrnstein because we felt strongly about the effects his writings would have on the education of minority children. UAG also passed out leaflets criticizing Princeton's gentrification policies in the town and explaining that Princeton's once-vibrant African American neighborhood was being forced out. And late in the spring we again confronted race issues after our blockade of the Institute for Defense Analysis.

The next major challenge for the anti-war movement at Princeton was to organize a protest at the Board of Trustees meeting on April 15. We asked to be allowed to argue against ROTC there, but the meeting was a closed one and our demand was rejected. We knew that a poll of students and faculty had shown that solid majorities of both groups were opposed to bringing back ROTC. So the time was right to take action.

As in January, the Trustees were to meet in Nassau Hall in the heart of campus. A national historic landmark, Nassau Hall was the site where the

Continental Congress met in 1783, making it the U.S. capitol for a short time.

After our demonstration in January, we knew that the campus police would be there in force, and we certainly had no desire to get into a fight with them. We had to think of a tactic that would be militant but nonviolent. UAG decided to call for a large rally in front of Nassau Hall at 2 p.m. on Friday the 14th, followed by an overnight sit-in so that we would be there for the Trustees the next day. We also voted to form a small group within UAG to work out tactical details. That committee's decisions would be kept secret from the administration and police.

A tense moment occurred when someone who was active in student government volunteered to be on the tactical committee. This woman was known to be on friendly terms with several university officials. We didn't trust her, but at first none of us wanted to confront her. Then Ann said what needed to be said. She looked at her directly and explained, "Marcia, you're too close to the administration. We can't be sure that you won't reveal our plans to them." So she was barred from the group that would decide tactics.

The tactical committee came up with a simple plan. We would phone another couple of dozen trusted activists, and all thirty of us would enter Nassau Hall at noon, two hours before we'd be expected. The campus police would not yet be in position, and we'd already be in the locale of the Trustees' meeting. Then others could join us as they arrived for the 2 p.m. rally.

The plan worked like a dream. No one betrayed us — this presumably meant that UAG was not infiltrated and our phones were not being tapped. (Some activists had been a little paranoid about such matters after it was learned that the chief of the campus police had been in the FBI.)

Shortly before noon thirty-five of us occupied the main meeting room in Nassau Hall without incident. During the next half-hour or so before the police arrived, several more joined us, including Ann, who had had a class until 12:20. When the campus police came, they were dismayed to see us already inside the building. At first they sealed it off so that no one else could come in, but they soon realized that that would be counterproductive, and they reopened Nassau Hall. In late afternoon the deans came in to tell us that the Trustees' meeting would be moved, and that anyone who didn't leave immediately would be punished. Those who didn't want to face university discipline left, but most stayed. Seventy of us spent a largely sleepless night on the benches in Nassau Hall, and another twenty-five joined us early the next morning. The administration sealed the doors at 8 a.m. on Saturday, at which point fifteen more climbed up from the lawn outside and crawled through the windows. At 11 a.m., 110 protesters marched out of the building and held a brief but noisy demonstration in front of President Goheen's house, where the Trustees were meeting.

Out of eighty-eight students brought up on disciplinary charges for the occupation of Nassau Hall, ten were graduate students. Interestingly, five of them (including me) were in mathematics. The other four math students had probably made the same calculation that I had made at Harvard — they knew that their careers would not be adversely affected by taking part in radical protests and as a result felt a special obligation to put themselves on the line. We were aware that leftist grad students in many fields — particularly in the non-sciences — faced constant discrimination and reprisals from conservative professors, especially if they participated in campus protests. But in my experience there have been few mathematicians, whatever their political views, who have let extraneous considerations cloud their professional judgment of graduate students and junior colleagues. That in part explains why five out of the ten grad students to face discipline for the Nassau Hall sit-in were in the math department.

The administration had brought photographers into Nassau Hall so that they could later identify the protesters. They got the names of only a little over half of the group in this way. Those they had missed felt left out, and thirty-five of them went over to the campus police office to try to pick themselves out of the photos. According to a staff person for the disciplinary committee who was quoted in *The Daily Princetonian*, "We have four written statements from people who say they were there, but who cannot identify themselves from the identification pictures. These people are just begging to be charged, but the committee doesn't have the authority to do so."

At our insistence the disciplinary committee held the three-day hearing in a big auditorium, and we organized supporters to attend. We formed a committee to present a comprehensive legal and political defense. Our legal argument was that it is permissible to commit a minor infraction if necessary in order to prevent a much greater crime from occurring. We gave evidence that officers trained in ROTC went on to commit war crimes in Vietnam. A former Army officer who had come through ROTC testified that, before he grew disillusioned with the War, he had participated in Project Phoenix, which arranged the assassination of village leaders who were suspected of being sympathetic to the guerrillas. We also showed an anti-war documentary and a slide show about the "electronic battlefield" and its horrible consequences for Vietnamese civilians.

The chair of the disciplinary committee was a professor of international relations and former labor arbitrator named Frederick Harbison. He seemed almost to enjoy the process of negotiating with our defense team about the conditions for the hearing. At the outset we had said that we wanted to (1) show a film and slide show, and (2) call at least two Trustees as witnesses. (We wanted to call Trustees who were on the boards of directors of companies that were profiting from war contracts and question them about their complicity in war crimes.) Harbison initially ruled that we could not bring in broader issues related to the War. We caucused and came back to announce that we would walk out of the hearing and attack its legitimacy unless we were allowed to do either (1) or (2). Harbison consulted with his committee and ruled that they would not agree to call any Trustees as witnesses, but they would let us show the film and slide show. So we didn't walk out, and the hearing proceeded.

If we had had the movie *Apocalypse Now*, which was not made until seven years later, as part of our defense we would have definitely shown the scene where the Robert Duvall character orders the destruction of a village, exclaims "I love the smell of napalm in the morning," and then plunks down his Princeton mug.

The disciplinary committee was not unsympathetic to our presentation, despite the fact that none of the truly progressive faculty members were on it. But they still decided to put everyone on probation.

Near the end of the semester we organized a blockade of the Institute for Defense Analysis (IDA), which at that time was still located adjacent to the Princeton campus. (A couple of years after our protests it was moved a few miles away to a far less accessible site.) IDA was a quasi-private research lab that was contracted by the Pentagon to develop communications technologies and electronic devices for battlefield use in Vietnam.

The immediate impetus to take action against IDA was President Nixon's May 8 announcement of increased bombing of North Vietnam and the mining of its waters. For four days between Wednesday, May 10 and Monday, May 15, we blocked the entrances to IDA and effectively shut it down. Almost 200 anti-war protesters were arrested, some two or three times. Never in Princeton's history had so many students been arrested; the massive protests dominated campus news for the next two weeks and got a lot of coverage in the outside press as well.

Ten Princeton faculty members — including Arno Mayer and Stanley Stein in history, Malcolm Diamond in religion, and Charlie Gross in psychology — joined our blockade and were arrested and booked. Ann recalls seeing Arno Mayer, who had been her professor in a European history course, after he had been fingerprinted. He was shaken — he had never been arrested before, and he found the whole experience frightening and unnerving.

Yet he and a broad cross-section of anti-war liberals and radicals had been willing to join in a strong UAG action against IDA. At the time I was struck by the contrast between the attitudes of the Princeton faculty — virtually none of whom condemned us and many of whom supported us — and those of the Harvard faculty in 1969, who had given a standing ovation to Gerschenkron after his hysterical tirade comparing SDS to the Nazis.

The story of one of the graduate student protesters particularly touched us. The first day of the blockade he came to support us from the sidelines. He explained that he could not get arrested, since later in the day he was going to his swearing-in ceremony to become a U.S. citizen. The next day he returned to IDA, told us he was now a citizen, and got arrested with us.

In the spring of 1972 I was still wearing the boots and raincoat that I'd been issued when I was in the Army. I didn't see any reason not to — they were of good quality, and I hated to spend money on clothes and shoes. After the IDA blockade, a conservative student wrote a satirical column in *The Daily Princetonian*, in which he changed my name to "Kibitz" and snidely referred to my army-style raincoat that I'd "purchased at Saks Fifth Avenue." The anti-semitism was clear, but, compared to other things that were on my mind, it didn't bother me much. Institutional anti-semitism, which had once been endemic at Princeton, had disappeared long before I got there. However, there were still conservatives on campus who seemed to feel that any essay ridiculing leftists would be incomplete without a dollop of anti-semitism.

On the second day of protests I had a bullhorn but wasn't actually in the blockade, since I had already been arrested twice. I was only a few feet away when the police started clubbing a student who had his arms linked with the others. Then I noticed that the student whom they were hitting again and again was black. I started screaming "racist, racist, racist!" At that point the cops grabbed my bullhorn and arrested me.

Jim Hinton, the student who was beaten, had several bruises and two broken teeth. He had not been resisting arrest, but had clearly been singled out because of his race: he was the only black student arrested and the only protester who was beaten. The whole Princeton community was outraged at the racist brutality of the police.

The people arrested at IDA decided not to attempt a legalistic defense against the charges. The punishment for our "disorderly persons" offense was only a 100-dollar fine. Under New Jersey law we had the option of refusing to pay the fine and serving a day in jail for every ten dollars. Thirty-two of us — about one-sixth of those arrested — decided to do this.

Ann read a statement to the court on behalf of those who were going to jail. She said that instead of paying thousands of dollars to the court system, we would give the amount of the fine as a donation to the Medical Committee for Indochina, which was sending medical supplies to North Vietnam.

It was clear to us that our action would greatly increase the publicity and impact of our protests. It was probably the first time that a group of Princeton students had ever gone to jail over political principles. We also liked the idea of thumbing our noses at the criminal "justice" system and sending the money to the Vietnamese instead.

However, the decision to choose incarceration over paying the fine was not an easy one. I thought for a long time, agonized over it, and had a long conversation with my parents (whom I normally did not consult before taking a political action) before deciding to push for a group of us to go to jail. I felt a special responsibility because I was the oldest of the "core" group (having reached the ripe old age of twenty-three!), and I also was the only one with the experience of having been behind bars. I had never before had to lead people into such a potentially dangerous situation, and I kept turning it over in my mind. But I would always come back to the thought that going to jail would be by far the most effective step that we could take. And I hoped that the special circumstances — a lot of newspaper publicity, the large number of us, the Princeton affiliation, and the very short sentence — would make it unlikely that the jailers would permit any violence to occur. So in public I energetically lobbied for as many of us as possible to refuse to pay the fine. But my decision in May 1972 to argue for going to jail was the most difficult political decision that I have ever made.

The county prosecutor tried to talk us out of the jail option. He focused his efforts on the women, of whom there were fewer than there were men (since in 1972 Princeton still had a disproportionately male student body). He said that the women would be in cells with murderers and child molesters.

The twenty-five men and seven women who ultimately chose to go to jail were a diverse lot. Some of the core group of organizers, including Dan Lichty and Alex Farquhar, didn't go. Ann and I were a little disappointed, but we weren't nearly as upset at them as they were at themselves. They had heard about the horrors of prison, they admitted that they were simply afraid, and it seemed to us that they were excessively beating up on themselves about it.

On the other hand, some of the people who went to jail with us had been only on the periphery of UAG and considered themselves liberals rather than radicals. They saw going to jail as "bearing moral witness" in the pacifist tradition. This was a vivid lesson to me that there is not always a correlation between degree of radicalism of people's ideology and their level of true political commitment.

The men were sent to the Mercer County Workhouse, located a few miles outside Trenton. We were put in a large barracks-style room that was separate from the other inmates' living quarters. At first we would see them at meals. However, three of the Princeton students refused to allow their beards to be shaved, as the admission procedure required. They were thrown into solitary confinement, and the rest of the group announced a hunger strike in solidarity with them. As a result we were totally isolated from the other inmates for the rest of our time in the Workhouse (and we weren't let out to do any "work").

I wasn't too thrilled about making an issue of body hair, but once the three students were in solitary there was no choice but to support them. I also was not a big fan of hunger strikes, but again I had to go along with the majority sentiment. After three days without food I was completely ennervated and listless. I got up from my cot and promptly fainted. Others had similar reactions. This was ridiculous, we decided. We'd made our point, and we called off the hunger strike. Food was then brought to our room, but we were still kept away from the other inmates.

Two students — Dan Friedan (son of the famous feminist author Betty Friedan) and our friend Marty Bachop — decided to continue the hunger strike for the full ten days. I never really understood why they did that — their way of looking at political protest was philosophically alien to me.

The seven women, who were taken to the Trenton County Jail, had a much different experience. They mingled with the other inmates, most of whom were black and many of whom were in pre-trial confinement. For example, the "murderer" the prosecutor had had in mind, a woman named Calethia, was awaiting trial for the shooting death of her husband. She had an alibi, but needed a private attorney to carry out the necessary investigation to establish her innocence. She hadn't had the funds to pay for both bail and a lawyer, so she opted to forego bail and use her money to prove her innocence. Calethia was eventually acquitted, after having spent two years in jail awaiting trial.

Ann has often said that the ten days in Trenton County Jail were an invaluable experience for her. She saw close up the role of race and class in the criminal "justice" system. Some of the black women were incarcerated for offenses such as shoplifting, which almost never would have resulted in jail time for a white person. Even the women who were guilty of a crime seemed in most cases to be more victims than villains. In her women's studies courses at Arizona State University (where she has taught since 1998), Ann has frequently drawn upon her experiences in the Trenton County Jail to illustrate some of the complexities of the treatment of women, especially women of color, in our society.

Soon after the Princeton students arrived at the jail, the inmates had a scheduled weekly meeting with a black psychologist. He was popular among the women because he was handsome and young, and liked to joke with them. He introduced the meeting by saying that its purpose was to straighten out their heads, because the fact that the women were there meant that there was something wrong with them that needed to be talked through. Immediately the Princeton women objected. They pointed out that they were in jail for having protested U.S. government atrocities in Vietnam, and there was nothing wrong with their heads for having done this. They also reminded him that some of the women had not yet been tried and should be presumed innocent until proven guilty. They further said that it was society that had failed the women, and their presence in jail was an indictment of an unfair system.

The other inmates seemed impressed by the words of the well-educated young women, and the psychologist couldn't come up with an effective response. A week later — a couple of days before the Princeton protesters' release — it was announced that because of "sickness" the psychologist had had to cancel the next session, but would be back the following week.

Over the weekend Ann's parents came to visit her in jail. At first they were quite upset about their daughter being behind bars. But while they were there a New Jersey state legislator of whom they had heard was also paying a visit. He talked with them, told them that he admired what the students had done, and said that they should be proud of Ann. His words affected them greatly.

In the summer of 1972 when I thought back on our activities over the previous six months, I felt that we had accomplished more than I would have thought possible. I did not have the feeling of frustration and inadequacy that I had had after the Army and after the debacle at the SDS convention in 1969.

And some of the effects of our actions would become apparent only later. Leon Kamin's work on Cyril Burt, which was a direct result of the UAG campaign against Herrnstein, had not yet made it to the pages of *The New York Times*. In addition, it wasn't until Ann and I started visiting Vietnam six years later that we fully appreciated how much the student anti-war protests had meant to the Vietnamese. During the darkest moments of the War, they found encouragement in the intensifying dissent within the United States, which signaled to them that the War would not go on indefinitely. The main reason why we have encountered no anti-Americanism in Vietnam during our many visits there between 1978 and the present is that the Vietnamese news media always stressed the distinction between the U.S. government and the American people, and gave ample coverage to the protests.

Many visitors to Vietnam, including us, have found that the generation of Vietnamese who lived through the War tend to express an unwarranted degree of gratitude to veterans of the anti-war protests and have an exaggerated impression of the sacrifices we made. In 1989, on the occasion of the twentieth anniversary of the Harvard Strike, one of the speakers, Aldyn McKean, who had been a soldier in Vietnam, quoted someone he had known in Pleiku. He knew that the man was with the National Liberation Front (NLF), although of course this was hidden. Here (slightly edited) is what McKean quoted the man as saying:

> We in Vietnam have no choice. If you are the child of a farmer in South Vietnam, the only possibility of having a future where you can keep some measure of respect is to join the NLF. I have read about things that happened in the United States, where students at the most prestigious universities — Columbia, Berkeley, Harvard — gave up futures that were sure to bring them what many people in the world would consider beyond their wildest dreams, in order to show that what was happening — that what their country was doing — was wrong. They gave that up. That took courage, and we were heartened by it.

Postscript

In Chapter 3, I mentioned that when a colleague gave me a 1970 vintage University of Washington SDS leaflet, I was amazed at how well-written it was in comparison with student writing nowadays. I had a similar reaction when, scrolling through microfilm of 1972 issues of The Daily Princetonian,

I came upon the following column written by Ann, who was just nineteen years old at the time. It was a response to an angry letter by Professor Leon Kamin, and it appeared on March 9, 1972, the day of Kamin's public forum on the Herrnstein affair.

Smokescreens for Racism

by Ann Hibner

I was sorry to see that Professor Kamin based his letter of March 7 on one or two false notions which he and any who wish to discuss the Herrnstein issue intelligently must speedily rid themselves of.

First of all, Herrnstein was *not* "prevented from speaking at the University of Iowa by a demonstration." Rather, he himself refused to enter the room in which the speech was to be held. He was in no way subjected to "physical violence" as Professor Kamin erroneously claimed. Iowa students had no intention whatsoever of preventing him from speaking.

The tactics agreed upon in Iowa were essentially the same as those planned by Princeton's University Action Group (UAG) and students working with us (such as the Women's Center). To accuse us of intending to prevent Herrnstein from speaking is grossly inaccurate. The action proposed and unanimously agreed upon by UAG members was that Herrnstein be allowed to deliver his talk uninterrupted, after which we would use the question period to challenge him on his racist, sexist, and elitist theories. If he refused to debate or answer questions, we would have blocked the doors of the room until he consented to speak on the I.Q. controversy.

All questions of tactics aside, however, it seems that Herrnstein has a very strange notion of academic freedom. In the interview quoted in the *Prince* (March 6), he said that Princeton should "assure that I will be able to speak" and that "it would be enough for me not to come if they had placards on the wall." What Herrnstein is talking about here is not academic freedom, but rather the freedom to deliver his speech in an atmosphere of respect, cordiality, and at most mild disagreement. This new kind of "academic freedom" could only be attained by suppressing the placards and "verbal violence" of groups like UAG.

Finally, Herrnstein has said that since the I.Q. issue is "so easy to misinterpret" and must be "brought to the public with the greatest caution," he has avoided public discussion. Very strange statements from a man who cared so little for caution that he published his original article in the *Atlantic,* a magazine which makes no pretentions to scholarly impartiality and has a very wide circulation among primary and secondary school teachers.

It would be unfortunate if Herrnstein succeeded in diverting our attention from his own academic irresponsibility. "Academic freedom" is a smokescreen from behind which Herrnstein can throw his grenades of racism and sexism without fear of reprisals. He can rally many liberals who, while they disagree with his ideas, will support him because they are deceived by his smokescreen and by his appeal to them on the "issue" of academic freedom.

Herrnstein, and those whom he has rallied to his side, choose to ignore the other side of the coin — namely, academic responsibility. Having written his article, he now refuses to discuss its content or debate its consequences with anyone in a public situation. *We* are not the ones who are stifling debate. The sooner the Herrnstein debate returns to the substantive issue of racism, the better.

CHAPTER 6 **ACADEMICS**

The time I spent on political organizing during the spring of 1972 did not cause me any academic difficulties. Professor Phillip Griffiths, to whom I was initially assigned as an advisee, was about to move to Harvard; and in any case my main interest was number theory, which was not his field. He suggested that the best person for me to work with was Nicholas Katz, a newly-tenured expert on arithmetic algebraic geometry — a field that uses algebraic techniques and geometric intuition to solve problems in number theory. Katz was in France in 1971-1972, and I wrote to him asking for suggestions for what I should read so as to be better prepared when he returned. He sent me a list of articles, a few of which I read quickly and several of which I found to be too difficult. Other than reading what I could from Katz's list, I had no academic obligations until the autumn.

Ann, however, was not so fortunate. She neglected most of her course work and flunked out of Princeton. Politics was the primary reason, but not the only one. We were planning our wedding for September, and this also served as a major distraction.

During Ann's first years at Princeton, she did not have ambitious career goals. She used to tell me that she'd be happy working some day as what was then called a "gal Friday" (the term in our politically correct and bureaucratized age would be "administrative assistant") at some business. Although her mother was a bookkeeper, she had given her daughters an ambiguous message about professional ambitions. When Ann was a girl, her mother had warned her never to let boys know that she was smarter than they were, or else she'd never be able to catch a man. Notice that she didn't tell Ann that she shouldn't be more intelligent than the boys — only that she should not deprive the boys of the satisfaction of thinking that they're the bright ones.

Ann followed her mother's instructions closely. When we were dating, I thought that I was the smarter of the two of us. It was only after we were married that I found out that the reverse is true.

After Princeton suspended Ann for academic reasons, she spent the second half of 1972 working as a lingerie salesperson in Bamberger's department store in the Princeton Shopping Center. The pressures of that job —

staying on her feet all day long and dealing with the customers, including a creepy underwear fetishist who came all the way from Pennsylvania to shop anonymously for women's undergarments — were enough to provide Ann with a powerful incentive to finish her Princeton education.

Ann was readmitted in February 1973, immediately immersed herself in her studies, and soon became a very serious student. Much later, when Ann herself was teaching and had many non-traditional students, some of whom had had a rocky start to their studies, she was able to empathize with their situation. Based on her own experiences, she would encourage them to have confidence that they could catch up with and surpass their peers who had had an easier time of it.

I was nervous about meeting Nicholas Katz in September. Much of the reading that he'd recommended had been too difficult, and I had heard rumors about how intimidating he could be. According to legend, once while teaching an undergraduate course he learned that one of the students was from France, at which point he gave the rest of his lecture in French. Another story was that when introducing freshman calculus students to conditional convergence of series, Katz had told them, "This isn't the p-adics, you know." (A series of p-adic numbers always converges whenever the individual terms approach zero, which is not the case for ordinary real numbers; but students usually do not learn about p-adic numbers until graduate school.) As it turned out, my experience with Katz was totally different from what these rumors had led me to expect.

While Ann was selling lingerie at Bamberger's, I was trying to solve the thesis problem that Nick Katz had given me. The question related to equations in several variables that are considered over finite fields. A finite field is a number system with properties similar to those of the usual real numbers, except that there are only finitely many elements in the system. In any such field a certain integer p, where p is a prime number called the field's "characteristic," is equal to 0. I had to show that for most such equations an object called the Hasse-Witt matrix is invertible, a fact that would have some implications for the modulo-p properties of the number of solutions of the equation. (An integer's modulo-p value is the remainder you get after dividing the integer by p.)

At first I was stumped, but after several weeks of getting nowhere I had the idea of using a mathematical induction argument in which the simple linear equation of a hyperplane could be used to make the crucial step. The trick worked, and I'd solved the problem.

Katz was pleased and encouraged me to carry the work further. After the modulo-p study of these objects, one can ask more difficult questions related to their *p*-adic properties. The idea of a *p*-adic approach to a question is to simultaneously look modulo prime powers p^n for $n = 1, 2, 3, \ldots$ A basic concept is that of the *p*-adic "metric," which is a notion of distance between two numbers whereby two fractions are considered to be close to one another if their difference has a numerator that is divisible by a large power of *p*. Just as the real numbers "fill in the holes" between the fractions in the usual sense of distance on the number line, similarly the *p*-adic numbers fill in the holes between the fractions in the sense of the *p*-adic metric.

Even though I had solved the mod-*p* problem and could have conceivably received a Ph.D. for that work, it would have been a mistake to rush to get my degree in 1973. Katz encouraged me to stay for another year, improve and expand upon my results, and at the same time broaden my knowledge. Not only did I end up with a much better thesis, titled "*p*-adic variation of the zeta-function over families of varieties defined over finite fields"; I also spent a lot of time listening in on the mathematical conversations that Katz and a group of junior faculty would have. I would tag along with them to lunch at Prospect, the faculty club where I had worked as a waiter four years earlier, and try to understand as much as I could. During the two years under Katz's guidance I learned how to think like a research mathematician.

Katz's generosity continued after I was no longer his student. In 1977 he paid a visit to Harvard, where I was an instructor. I told him that I was looking for a *p*-adic formula that I thought should exist for Jacobi sums. Jacobi sums are number-theoretic analogies of the classical beta-function that arises when evaluating certain definite integrals. I'd noticed that a modulo-*p* formula of the right form existed and thought that that formula should extend to a full *p*-adic one. Katz said that he had done some calculations based on work of T. Honda that might be useful, and he sent them to me.

As soon as I received Katz's calculations, I saw that a small step would transform them into the formulas I wanted, so in a short time I had the result I was looking for. A little later, a brilliant graduate student at Harvard named Benedict Gross found a nice way to go from these formulas to a *p*-adic formula for Gauss sums, which in a sense are even more basic than Jacobi sums. Gross and I published a paper in the *Annals of Mathematics*, and the result became known as the "Gross-Koblitz formula," although in reality much of the hard work had been done by Katz.

I had a similar experience with the intellectual generosity of senior mathematicians toward young colleagues when I was studying in Moscow the year after I received my Ph.D. The eminent Soviet number theorist

Yuri I. Manin suggested to me that there should be an indirect way using mod-p formulas to prove that certain functions could not have algebraic integer values outside their "critical strip." I did the calculations and then wrote up the result and put both of our names on it. Manin insisted that I take his name off, even though the main idea had been his.

The tradition of established mathematicians sometimes letting junior colleagues take credit for what was really joint work is the opposite of what happens in many other areas of science, especially in experimental fields. Often directors of institutes and laboratories expect their names to be included among the authors of papers that come out of their institution, even if they had no intellectual role in the work. Getting their names on literally hundreds of papers is a sort of perk of their administrative position. It was arguably poetic justice when in a few notorious cases a paper turned out to be fraudulent, and then the lab director, who was often a well-known scientist, regretted having been a coauthor.

Soon after Ann was readmitted to Princeton in February 1973, she started thinking about her senior thesis. The topic she chose was a sophisticated one that combined her interests in history of science and in political and social issues. She traced the history of the American eugenics movement of the early 20th century and of the social Darwinism that immediately preceded it.

Ann has always been a voracious reader, and for her thesis she read and thought critically about roughly a hundred books and many articles. Earlier she had studied the scientific developments — genetics and the theory of evolution — that had been carried over to the social realm in order to justify extreme inequalities between races, ethnicities, classes, and the sexes. Many of the books she read for her thesis treated the social and political history of the two movements that claimed to be based on these scientific theories.

As I typed her thesis for her on my old Smith-Corona electric, I became tremendously impressed with the work she had done. She presented the viewpoints of leading historians such as Richard Hofstadter, along with her own critiques of some of their conclusions. She delved into the complex relation between the two movements — the earlier one based on an "optimistic" laissez-faire philosophy and the later one playing on fear and advocating a draconian social agenda — and she argued that the link between them had been overstated by earlier writers.

Ann also investigated the extent to which the scientific pioneers, such as Charles Darwin and Charles Davenport, bore responsibility for the political movements that used their work to buttress a reactionary ideology. She was particularly harsh in her judgment of Darwin, and found in his writings a contempt for non-European people that contrasted with the respectful tone and egalitarian philosophy of the other great evolutionary biologist of the time, Alfred Russell Wallace.

In 1974, when Ann had to defend her thesis, it was not considered proper to make accusations of racism against the great icon of biology. Her committee included the famous historian of science Charles C. Gillispie, who was an expert on Darwin and regarded him in an almost hagiographic light. Gillispie also had well-known prejudices against women and had been opposed to Princeton's decision to admit females in 1969. Gillispie initially doubted that Ann would pass her defense. He showed her a shelf with Darwin's collected works and demanded that she give some textual evidence to back up her accusation against Darwin. When Ann fumbled nervously and couldn't find what she wanted, he said, "You don't have any, do you?" She replied that she was looking for a letter to the American social Darwinist William Graham Sumner and also passages where Darwin commented on the cultural level of natives on the west coast of South America. Gillispie knew which letter and passages she meant and found the references, which indeed had disparaging remarks about the culture and intellect of non-European peoples. Gillispie grudgingly admitted that Ann deserved to pass.

Ann graduated in 1974 at the same time as I received my Ph.D. During my post-doctoral year in Moscow, she applied to various graduate schools in history. She was not accepted at Harvard or other prestigious schools, but was at Boston University — although only into the Master's program, with the option of transferring to the Ph.D. program if she did well. (This was tantamount to provisional admission.) Since I had accepted a job as an instructor at Harvard, BU was the obvious choice.

To me it was clear that Ann should go to graduate school. Her senior thesis not only had been a well-researched and sophisticated treatment of a complex subject, but had included her own independent evaluation of Darwin's role in fomenting racism. Several years later it started to be generally acknowledged in the academic community that certain iconic figures of the 19th century such as Darwin and Freud had social attitudes that deserved to be sharply criticized. However, in 1974 it was almost heretical to use a word such as "racist" in reference to Darwin.

In addition, Ann was developing a strong interest in Russian intellectual history of the same period and thought that it would be fascinating to compare and contrast the Russian Darwinists and social Darwinists, such as the famous paleontologist V. O. Kovalevskii and the sociologist M. M. Kovalevskii, with their American counterparts. During our year in Moscow she laboriously read through a long turgid essay by M. M. Kovalevskii; her knowledge of Russian was still so rudimentary that it took her a couple of hours with a dictionary by her side to get through each page.

For many years when she thought back on her early academic career, Ann attributed her decision to go to graduate school entirely to a type of feminism on my part. She pointed out that I seemed to be the only one who was convinced that she could have a career as a scholar. Because I came from a family where the women were accomplished professionals — in the mid-1970's my mother was a teacher in one of the country's best schools and was active in the National Education Association, and my sister was a top student at Yale Law School — I assumed that my wife would also get an advanced degree and become a leader in her profession. So I stubbornly refused to go along with other people's low estimation of her professional capabilities.

I never completely agreed with Ann's interpretation. After all, during the first two years I was with her, I never objected when she said that she'd probably become a "gal Friday" in some business — I did not feel disdainful about that type of work. Whatever snobbism I had had as a Harvard student — and my letter to Vera Rony about my time in Grenada reveals that I did have a fair amount — had become greatly diminished as a result of my experiences with the Worker-Student Alliance caucus of SDS and in the Army. In addition, I already knew enough about the academic world to appreciate the fact that there are plenty of people with Ph.D.'s who, by any reasonable definition of intellect, are less intelligent than some of the secretaries and janitors.

In reality, it was not until I read and typed Ann's thesis that the thought entered my head that she was cut out for academic work. I could see that it not only was well written, but also showed a level of scholarship and independent thought that is unusual at the undergraduate level. That is why I wanted her to apply to graduate schools.

One evening almost thirty years after Ann graduated from Princeton, we were reminiscing about that period and trying to decide why I had been the only one urging her to go to graduate school. Out of curiosity, I went into our basement and rooted around in the dust and spiderwebs until I found her undergraduate thesis. We both looked through it, and then for the first time it occurred to us that what was really hard to comprehend was not my belief in Ann's potential as a scholar, but rather the complete failure of any of her Princeton professors to see from that thesis that she was a student who should be encouraged. In retrospect my enthusiastic response to Ann's thesis now seemed much easier to understand than her professors' total lack of interest in her scholarly potential.

None of them spoke to her about continuing in the field, and when she went to one of them — Gerald Geison — for a letter of recommendation, he said that he simply couldn't write one if she were applying to a history of science program. When she explained that she was applying "only" in history, not history of science, he reluctantly agreed to write something. (Years later he must have been surprised when Ann was awarded a prize by the History of Science Society at its 1990 annual meeting for an article she wrote in its premier journal *Isis*.) Clearly none of her professors — Michael Mahoney, Dorothy Ross, Gillispie, or Geison — had anything strong to say for her; she was barely admitted to Boston University's Master's program.

Nowadays the perspective of Ann and her colleagues in our generation of university professors is very different from that of her Princeton professors in the 1970's. If someone has a student who read and thought deeply about a hundred books for her thesis (well, these days even twenty would be enough!) and came up with some provocative and independent-minded observations, they would consider it irresponsible in the extreme not to encourage such a student to go on to graduate school. So why were the attitudes of Ann's professors so different in 1974? Of course, we cannot be sure; we suspect that there were several explanations. For Gillispie gender was undoubtedly an important consideration; at that time he tended to believe that women by nature could not be true scholars. For the others it might have been Ann's radical politics and her use of words like "racism" — in reference to Darwin, no less — that they would have regarded as "radical rhetoric." In addition, her lower-middle class origins and ethnic New Jersey accent seemed a little out of place in Princeton's aristocratic setting. (Ann noticed that several Princeton professors, including one who had also been born in New Jersey, had adopted British or continental European accents so that they would fit in better.) Finally, she had pointedly dedicated her thesis to the memory of a junior faculty member named Charles Culotta, "the most understanding and best teacher I have ever known." Culotta, who had been Ann's instructor in a seminar on Darwin and the Darwinian Revolution, had been a visiting assistant professor who, like Ann, had come from lower-middle class background and did not affect an upper-class accent. He committed suicide shortly after his stint at Princeton, and Ann and her fellow students thought that the condescending and unsupportive attitude toward him by other faculty had contributed to his depression.

After Ann and I discussed the possible reasons for her professors' low opinion of her scholarly potential, what I found odd was that it wasn't until three decades after Ann's graduation that it occurred to us to analyze their reluctance to encourage her. I believe that there were two reasons why we had never thought about this before.

In the first place, Ann had liked her history of science courses at Princeton and had felt that the faculty on the whole were good teachers. Michael Mahoney, for example, had excellent rapport with the students, and they enjoyed his lectures. Thus, the deep-seated insecurities that some of these professors might have had would not have been apparent to the undergraduates.

Mahoney, a former student of the famous historian of science Thomas Kuhn, had been a beneficiary of the "old-boy network." He was given tenure (that is, lifetime job security) at Princeton on the basis of his biography of the mathematician Pierre de Fermat; a rather thin list of scholarly accomplishments did not prevent him from being promoted to Associate Professor and later full Professor. For someone outside the old-boy network, Princeton's standards for tenure and promotion would have been much higher.

Mahoney's book on Fermat was in its final stages when Ann and I were dating in 1971. When she told Mahoney that her boyfriend was a graduate student in number theory, he initially expressed interest in having me read his manuscript with special attention to the mathematical details. I would have been happy (and flattered) to be asked to do this. However, he changed his mind because he was in a hurry to get the book to the proof stage, presumably because his tenure at Princeton depended on it.

I later was glad that I had had nothing to do with Mahoney's biography of Fermat. Shortly after it appeared, the famous number theorist Andre Weil, who was a permanent member of the Institute for Advanced Study in Princeton, published a scathing review in the *Bulletin of the American Mathematical Society*. Weil, who in old age was devoting much of his energy to the history of mathematics, had found a number of serious historical errors and mathematical misunderstandings in Mahoney's book. Mention of Weil's review was taboo among the history of science faculty in Ann's program, but undoubtedly everyone knew about it.

The second reason why Ann over the years was not inclined to think about the Princeton professors' failure to support her in her desire to go to graduate school was that she felt no bitterness about it, because in fact everything turned out for the best. If they had written strong letters for her and she had been admitted to graduate school at Harvard, then she of course would have gone there. Given her interests, most likely she would have been assigned to Richard Pipes as an advisee, and that would have been a disaster. Pipes was notoriously right-wing (he once accused the CIA of being too soft in its evaluation of the Soviet threat), and he had a bad reputation with graduate students. Someone who was politically leftist would have had no future with him at Harvard. In contrast, at Boston University Ann liked her course work and her professors. Life is full of ironies, and for Ann one of them was that she ended up with a much better experience at a less prestigious graduate school than she would have had if her Princeton professors had written enthusiastic letters in support of her application to Harvard.

Despite the happy ending for Ann, it is important to recall that in the non-sciences it was common for faculty to reflexively give low evaluations to politically radical students. For example, a leftist graduate student in Romance Languages whom we knew at Princeton learned that a professor had described her as a "juvenile Marxist" in a letter he had put in her file.

When Ann entered BU, the history department had about ten first-year graduate students in the European history track. Only two of them — Ann (now at Arizona State University) and Eric Weitz (now at the University of Minnesota) — ended up getting doctorates and having successful academic careers. Interestingly, they were also the only European history students that year who had been initially admitted just to the Master's program rather than the Ph.D. program. Eric, like Ann, had been a radical activist at his undergraduate college. Most likely the letters of recommendation in his graduate school application had been no more enthusiastic than Ann's had been. It is hard to escape the conclusion that both Ann and Eric had been discriminated against by their professors because of their leftist politics. In any case there can be no doubt that in the 1960's and 1970's some faculty viewed student protesters as a threat to the sanctity and security of academic life and saw nothing wrong with letting those feelings color the evaluations they gave in letters of recommendation.

Postscripts

In January 1983, Ann defended her Ph.D. dissertation — a biography of the Russian mathematician Sofia Kovalevskaia — at Boston University. One of the members of her committee was Professor Thomas Glick, who happened to be an expert on Darwin. In the course of an informal conversation afterwards, she was asked about her undergraduate thesis. After she recounted the controversy at her defense nine years before, Glick asked her if it would surprise her to know that in recent years many scholars were less willing than before to acquit Darwin of complicity in racism. Although Darwin's defenders pointed out that he had opposed one of the extreme racist views of his time — namely, that the different races should be considered to be different species of humans — historians had become increasingly sensitive to issues of race and cognizant of the fact that there was more than one way to be racist. And there could be no doubt that Darwin believed in a racial hierarchy and the cultural superiority of Europeans. Ann said that she was not surprised, since in 1974 that had been clear enough to her, and she had assumed that others would soon come around to a harsher view of Darwin's stance on non-European peoples.

In 1998 I was invited to give a short course on cryptography at the University of Chile, and in December of that year Ann and I visited Chile for a month. Although we stayed in Santiago most of the time, on one occasion we flew to Puerto Montt and traveled south by rented car to the island of Chiloe.

A colleague at the University of Talca let us stay in a vacation cottage that she and her husband owned. It was a beautiful, bucolic setting. The location is essentially symmetric with respect to the equator to our own Puget Sound and San Juan Islands near Seattle. The mists coming in from the sea, the dense vegetation, the fractal shoreline, and the sea lions reminded us of what we love about the Pacific Northwest in the U.S. Chiloe was much more remote and undeveloped, however, than the islands near our home in Seattle.

In Ancud, the largest town on the island, we went to a museum. The historic visit of Darwin was one of the main themes, because that was what first brought the island to the attention of Europeans, and the Chilotes (as residents of Chiloe are called) were proud to have been part of Darwin's voyages. The museum tried to put everything in as positive a light as possible, but the Chilotes were clearly embarrassed by the actual text of Darwin's comments. Reading those derogatory remarks from *The Voyage of the Beagle* once again, Ann was reminded of her thesis defense a quarter century earlier and of her good fortune in having remembered those words of Darwin in the face of Gillispie's hostility.

CHAPTER 7 **THE SOVIET UNION**

During the eleven years from 1974 to 1985 I lived in the Soviet Union for a total of about two years and Ann for about three. In 1974-1975 my post-doctoral year in Moscow was arranged through the International Research and Exchanges Board (IREX). Ann and I lived in a dormitory in Zone V of the gigantic wedding-cake style building of Moscow State University. We returned for the spring semester of 1978, this time on the National Academy of Sciences – Soviet Academy of Sciences exchange program; and in 1985 we again spent the spring in Moscow, I on the NAS exchange and Ann on IREX. In addition, Ann lived in Leningrad during the academic year 1981-1982 (with one month in Moscow) on IREX and Fulbright grants; and on about a half dozen occasions between 1973 and 1989 one or both of us made visits of one to several weeks. Finally, in January 1992 Ann took a group of students on a study tour of the then-collapsing Soviet Union.

Our first trip together for nearly four weeks in the winter of 1973-1974 was sort of a trial run to see whether Ann, who had never been overseas before, liked it and could handle living there for a year. I needn't have worried. Ann's first impression of the country was formed when the passport inspector looked at her passport and at her and commented to me, *Khoroshá!* (Pretty!) From that moment on, Ann always had good rapport with passport and customs inspectors. Even though in 1973 she knew hardly a word of Russian, her mannerisms and appearance reflected her Polish parentage, and Russians often thought of her as one of their own.

In Moscow there were many visitors from the provinces who thought that Ann, with her fashionable clothes (by mid-1970's Soviet standards anyway), must be a Moscovite. So she was frequently asked directions. I had taught her how to say "I don't speak Russian," but her accent in Russian was so good that they thought she was just being rude. So I jokingly told her that she could instead say "I am a stupid American tourist," and I taught her those words. That then became one of the three or four sentences she had memorized.

A few days later we were at dinner at the Kirillovs. (A. A. Kirillov, a prominent Soviet specialist in representation theory, would be my formal sponsor the following year, even though he was not a number theorist.) I asked Ann to show off the Russian that she knew; I expected her to say the first sentence that she had learned, "I'm going to the post office to send a letter." Instead she said, "I am a stupid American tourist." Yuri Manin, who was there, was scandalized and scolded me for having taught her such a thing. But Kirillov just burst out in a loud belly-laugh.

Manin, who was the main person I wanted to come to work with, introduced us to two of his graduate students, Volodya Berkovich and Anas Nasybullin. That year they were living in a large dorm room in graduate student housing (by the following year they had moved to Zone B of the main building of Moscow State University), and we went over for an evening with them and their roommates. At one point I referred to them as "Russians," and Volodya pointed out that, although they were all Soviet citizens, none of them were Russian. He was Jewish (which was classified as a non-Russian "nationality," i.e., ethnic group), Anas was Tatar, Saidakhmat was Uzbek, and Galim was Kazakh.

Early in the evening they asked me what my impressions were of their country. I wasn't sure what I should say, so in the tradition of husbands everywhere I hid behind Ann. I said that this was already my third visit, so I couldn't speak of first impressions. But it was Ann's first time in the Soviet Union, so I'd let her answer, and I'd translate. Ann said that she really liked the metro and was impressed with how clean and safe the city was. On the negative side, she noted that it seemed that the snowplow drivers were all men, while the people doing back-breaking work shoveling the sidewalks (probably for less pay) were all women. This led to a lively discussion, with the all-male group of math students on the defensive. In the U.S. that year was a high point of "second wave" feminism, but there was little evidence that the women's movement had reached the consciousness of Soviet men.

By the time we returned to Moscow in September, Ann had attended Middlebury College's intensive summer school in Russian. At first her comprehension was far better than her spoken ability. She impressed the Soviet grad students as an ideal wife — she listened carefully, prepared delicious food, and said almost nothing. But after a few months of this, Ann forced herself to overcome her inhibitions about speaking. The final impetus to do this came during a discussion of women's roles. After hearing some typical Soviet male viewpoints, she blurted out an angry and understandable (though ungrammatical) response. The men decided that perhaps she wasn't such an ideal wife after all.

Most of our friends that year lived in Zone B and would come to our room in Zone V to socialize. We had more space, because we had been given a whole "block" with two small bedrooms. We moved both beds into the same room and used the other one as a living room. In addition, the graduate students were all bachelors, and Ann even then was an excellent cook. She quickly picked up on the Russian style of preparing several salads with meats and potatoes and some vegetables from the farmer's market. Like a true Russian hostess, she was ready to spread the table with salads, drinks, and hors d'oeuvres whenever anyone dropped by unannounced. Cooking became almost a hobby for Ann, who enjoyed the challenge of making something delicious under a new set of constraints.

On one occasion I had to get something from Zone B, and I dropped in on our friends in their room for a change. This was the first time that they'd seen me socially without Ann. I ended up staying much longer than I had intended, as they told me all the dirty jokes that they couldn't tell in front of a woman. When I returned and Ann found out what had happened, she was furious at having been left out.

When she was at Middlebury, as a device to motivate the students her teacher, Frank Miller, had given them a list (with approximate translations) of the filthiest and most insulting words and expressions that he knew in Russian. Ann had brought the list, and the next time our friends came over she showed it to them as proof that she was fully capable of listening to their off-color stories.

They took one look at the list and were incredulous that a foreign woman would have been taught such expressions. I remember Misha Vishik (another student of Manin) literally rolling on the floor with laughter after seeing an expression on the list which means "ass on wheels." From that point on the guys were willing to tell most of their stories in front of Ann. The majority of the risqué "anecdotes" they told were pretty juvenile, but a few were funny.

During the year 1974-1975 I regularly attended the Manin, Shafarevich, and Gel'fand seminars. These weekly meetings lasted at least two hours, often with no break. This took some getting used to — American mathematicians are conditioned to get restless if a talk goes over fifty minutes. Each seminar was known by the name of the senior mathematician who ran it rather than by the subject matter. At this time Moscow had the highest concentration of research mathematicians anywhere in the world, and even a narrow specialized talk could draw an audience of forty or fifty. Typically the most prominent senior mathematicians would be the center of a "school" that consisted of their students, former students, and junior colleagues. Since the best researchers almost always had studied in Moscow and remained in Moscow, many of these mathematicians had been close colleagues and friends since student days.

Each week I had to choose between the Shafarevich and Gel'fand seminars, because they took place at the same time on Mondays. The time conflict had a deeper significance — the two mathematical schools, although they met in the same building, were at a great psychological distance from one another. First of all, their styles were in sharp contrast. Shafarevich was punctual, well-organized, and efficient. He was a superb lecturer and expositor (it was from a textbook by Borevich and Shafarevich that I first learned number theory). The Gel'fand seminar was often disorganized and full of digressions and witty exchanges with the speaker. It was more entertaining than the Shafarevich seminar, but I probably learned more from Shafarevich's speakers.

The second reason for the distance between the two mathematical communities was never mentioned openly in my presence. Gel'fand and his school were mostly (though not exclusively) Jewish, while Sharafevich was a Russian nationalist. Although American newspapers tended to lump together Jewish dissidents and Russian anti-communist dissidents, in fact there was a big gulf between them.

Much later this became clear to mathematicians in the West when Shafarevich published some political, historical, and philosophical essays that were overtly anti-Jewish. At the urging of some Jewish emigrés there was even some discussion of expelling Shafarevich from the National Academy of Sciences (to which he had been admitted as a foreign associate). In 1992 the NAS went so far as to write him a letter asking him to resign. I was totally opposed to this for two reasons. In the first place, none of the Jewish mathematicians whom I had known in the Soviet Union had ever had anything bad to say about Shafarevich's behavior; to the best of my knowledge he had never been accused or suspected of having participated in discriminatory treatment of a student or colleague. In the second place, I thought that it was hypocritical of the NAS to try to get rid of Shafarevich, when the notorious American racist William Shockley had remained in the NAS until his death in 1989 while devoting his energies to vitriolic attacks on blacks and other minorities.

There was only one occasion when the Gel'fand and Shafarevich seminars combined forces for a joint colloquium. This was when the brilliant young Belgian mathematician Pierre Deligne — who three years later would receive the Fields Medal for his proof of the Weil Conjectures — visited Moscow. Deligne was known to be a difficult speaker to follow, because he went fast and also spoke English in a very soft voice. Before introducing Deligne, Shafarevich commented that it might not be necessary to have a translation, and he asked for a show of hands of those who needed to have Deligne's words translated into Russian. In those circumstances hardly anyone had the nerve to vote for a translation. Gel'fand interrupted with the comment, "Well, at least it would be nice to have a translation from English into English," but he was ignored. Shafarevich obviously wanted Deligne to say as much as possible without losing time waiting for a translation, and he got his way. Needless to say, hardly anyone followed Deligne's lecture, and after that there was enough negative feedback so that Deligne's subsequent talks were all translated by Manin. As we left the seminar, I remarked to a friend that Shafarevich had shown up the sham nature of bourgeois democracy when he'd asked for a "vote" on whether a translation was necessary.

A big project during 1974-1975 was to translate Manin's textbook *A Course in Mathematical Logic*. Manin came to visit regularly to go over the translation with us. An unusual feature of the book was the large number of digressions relating notions of formal logic to other fields such as physics, medicine, and poetry. It was challenging to translate it all, and Ann helped a lot, especially with the literary passages.

Ann often cites that experience as an example of her (and my) low level of feminist consciousness at the time. Manin joked that because of her help he would bring a bouquet of flowers for her for every fifty pages. She accepted that and thought it was sweet. He didn't offer to include her name on the title page as a translator, and neither she nor I thought of asking him to do that.

A tricky issue arose when I came to a digression that Manin wanted me to translate that was called "Women's Logic." He had made up a set of absurd rules for the way that women supposedly argue. It was offensively sexist, and it wasn't even funny. I tried to convince him that we should delete it, but he thought that I was just being humorless, and he wanted to keep it. I didn't feel that I could refuse to include it, since that would have seemed like Soviet-style censorship. And I didn't see how to convince him to leave out that section.

When Deligne came over to our place for dinner, I had this very much on my mind, and I asked his advice. When he heard what the digression said, he had no hesitation in saying that for Manin's own good I should bring it up with him again and try harder to convince him that it would only make him look bad to have such a passage in the English translation of his book on logic.

By the next time I saw Manin I'd figured out a new tack. I asked him to consider an analogous situation. When I was a high school student I had overheard the following joke: "How was the Grand Canyon formed? A Jew dropped a dime in a gopher hole." What would he think if this "joke" appeared in a Soviet geology textbook? Manin immediately switched to a serious demeanor and said that if discrimination against women was as sensitive an issue in the West as anti-semitism was in the Soviet Union, then let's just drop the section on "Women's Logic." I breathed a sigh of relief.

When Ann and I returned to the U.S. at the end of the summer of 1975, I had a big scare. I had mailed all four thick notebooks containing the translation, along with other papers and preprints, in a large box through the U.S. Embassy mail. What arrived at Harvard was a broken box with only part of the contents — and a book in German about soccer. There had apparently been a mishap in the transshipment center in Frankfurt, several boxes had broken apart, and the contents had been lost or mixed up. Only one of the four notebooks was there. I was distraught. There was no way to recover all that work, and I had foolishly not made a carbon copy (photocopying machines were not available in Moscow at that time).

Several days later I received a letter from a graduate student of Franz Oort in the Netherlands saying that a package of material with no address had been delivered to their institute, and after examining its contents, they thought that it might belong to me. It turned out that the German postal authorities had noticed several preprints of Oort (some of whose work was closely related to my Ph.D. thesis) with the address of his institute on the covers. Since that was the first address they could find, they sent everything there. I telephoned the Netherlands and had them ship the material to me immediately. To my immense relief all three of the missing notebooks were there. I was tremendously grateful to Oort and his student for figuring out that the package was mine and sending it to me.

The graduate student was Hendrik Lenstra, Jr., who would later become one of the leading number theorists of his generation and whose work would have a major influence on my own direction of research, as I'll explain in Chapter 14.

In May 1975 I ran in a marathon — for the first and last time. The occasion was Victory Day, the thirtieth anniversary of the defeat of Nazi Germany. I wasn't really in shape for a marathon, having been running seven or eight miles (twelve kilometers) on alternate days. But I didn't realize that from the standpoint of endurance twenty-six miles is of a different order of magnitude than seven or eight.

It was no simple matter to enter. The Soviet approach to sports was that everyone was assumed to belong to a club — no one who was the least bit serious would simply jog independently. An official or doctor associated with the club would sign a document vouching for the physical fitness of the runner. I, however, did not belong to any club, so I had to have a special physical exam in order to qualify. The doctor, not wanting the responsibility in case I collapsed in the twenty-fifth mile, said that he'd agree to my participation only in the ten-kilometer version — which I knew was open just to women and old men.

I went to the organizer of the race and told him what had happened. He was a Komsomol (Young Communist League) official for whom the marathon was a civic duty. I was the only American who wanted to race, and that type of foreign participation would enhance the prestige of the event. It was especially fitting to have me in the marathon because America had been a key ally of the Soviet Union during World War II, and Victory Day had been a time of joint celebration. So the Komsomol guy went to talk with the doctor and strong-armed him into agreeing to give me an okay for the full marathon.

Well, my first and only experience in international sports competition was a bust. I had to quit half-way — after 20 kilometers. The results were posted around the university and stayed up for weeks. If I had been a patriotic American (which, thankfully, I wasn't), I would have felt humiliated. My name and country were listed — I was the only one representing America — and instead of the finishing time that accompanied most entries, there was one stark word: *soshol* (gave up).

Although the Komsomol people were roughly the same age as the American graduate students, they sometimes had a hard time understanding us. Once at Moscow State University a grad student from Columbia who'd been chosen as the representative of the IREX group was attending some sort of formal event with long speeches. During the university president's talk he decided that he couldn't take it anymore, and walked out. A Komsomol person later criticized him for doing that and asked, "Would you have shown such disrespect to the president of your own university?" That was an unfortunate question to ask the American, who replied that he had participated in the student uprisings of 1968 and had the view that the appropriate way to relate to the president of his own university was to seize several administration buildings and issue non-negotiable demands.

Once during Ann's year in Leningrad, the graduate students in her dormitory had a potluck banquet. The residents in the dorm came from many parts of the world, and each national group was asked to put their country's flag beside the dishes they brought. The Americans didn't have a U.S. flag and — as was typical of our generation — did not have much interest in displaying patriotic symbols anyway. So they steamed off a label from a Pepsi bottle and put that on their table instead; that would make it clear enough what country's dishes were there. The Komsomol people who had organized the potluck were scandalized, and didn't think that was appropriate. They had no tolerance for that type of irony or satire.

Each month graduate students from a different country would prepare a display for the central bulletin board in Ann's dorm. When the Americans' turn came, they made up a collage with newspaper headlines and magazine pictures. Prominently displayed in the center was the famous *Mother Jones* satire of the "Uncle Sam Wants You" recruiting poster: "Join the Army! Travel to exotic, distant lands. Meet exciting, interesting people, and kill them." The Soviet student leaders in the dorm were not amused, and took down the caricature.

When we started going to the Soviet Union, we were enthralled by Russian literature and history, and we were opposed to the Cold War anti-communism that dominated American journalism and political discourse. However, we did not idealize the Soviet system and did not consider it any sort of model of socialism. On the contrary, our wing of the student movement had taken China's side in the Sino-Soviet split and had regarded the Soviet Union as a conservative superpower that had sold out revolutionary principles. Many leftists in the West were much more inclined to look for models of revolutionary idealism in the Third World (China, Cuba, and later Nicaragua) than in the Soviet Union. Because we started out with a rather dim view of the Soviet brand of socialism, we did not get disillusioned later when we personally encountered its flaws.

Some other anti-war activists were closer than we were to leftist groups that looked approvingly at Soviet socialism. When they had occasion to actually live in the U.S.S.R. for a while, they were shocked at what they found, and often felt betrayed. For example, one of the Princeton graduate students who got disciplined for the Nassau Hall sit-in was Ben Eklof, who later became a history professor at Indiana University. He was active in the movement at Princeton and played a key role in organizing our defense strategy at the disciplinary hearings for the sit-in. He went to Moscow on IREX, married a Russian woman, lived there for three years, and was friendly with many dissidents. When we saw him near the end of that period, he had become quite anti-Soviet. It seemed to us that if he had started out with lower expectations, he wouldn't have been so bitterly disillusioned.

In addition to my political reasons for not expecting to find any sort of workers' utopia, I also knew about Soviet mistreatment of mathematicians. In 1968 the mathematical logician and political dissident Esenin-Volpin (son of the famous romantic poet Sergei Esenin) was arrested and sent to a mental hospital. In response, a group of about a hundred prominent mathematicians signed a petition asking for his release. They thought that the country was gradually liberalizing, and such a petition could possibly be effective. They were wrong. There was an immediate clamp-down on dissident mathematicians, and most who had signed the petition were subjected to reprisals. The main punishment was that for many years they were barred from most foreign travel. Yuri Manin, who was one of my mathematical heroes, was one of them. For this reason before my post-doc in 1974-1975 he suggested that his name not be given as my supervisor in the IREX document sent to the Soviet authorities. Instead, he arranged with A. A. Kirillov for him to be my formal sponsor, although in reality I came to work in Manin's research group.

Although we had long known about the history of repression and bureaucratic ineptitude in the Soviet government's dealings with many of the country's leading intellectuals, that didn't stop us from approaching our year in Moscow in a positive spirit. We had numerous Soviet friends, and we spent almost all our time in a Russian-speaking milieu.

This set us apart from many of the Americans on the exchange program. In fact, several of our neighbors in Zone V thought that we were Canadians because we socialized with Soviet colleagues. Unfortunately, the Americans in Moscow tended to spend much of their time with one another. Ann and I would go regularly to the U.S. Embassy for mail and a few purchases in the commissary, and then we'd leave; but we noticed that other exchange scholars would routinely have long lunches in the Embassy snack bar. It most likely wasn't that they couldn't have survived the year without hamburgers (McDonald's had not yet come to Moscow), but rather that they simply enjoyed spending time with other Americans in a familiar setting.

We found it odd that so many of the young scholars whose field of work related to the history and culture of the country would want to be cut off from the life around them. In many cases their spoken Russian did not even improve over the course of the year. They worked in the libraries and archives, got access to enough sources so that their theses would be amply footnoted, and then went home.

Other Western visitors — the Canadians, British, and French, for example — did not form an insular enclave to the extent that the Americans did. There was a tendency for the Americans, more than anyone else, to internalize the Cold War viewpoint of their government and adopt almost a siege mentality while in the Soviet Union.

Why so many of the American academic visitors did this was always a mystery to us. Most of them in private were highly critical of U.S. government policies and made disparaging comments about various aspects of American society and culture. But on the rare occasions when they had serious discussions with their Soviet counterparts they felt obliged to defend the official American viewpoint. While in Moscow they tended to identify with the diplomatic community and share their negative attitudes toward the local and national authorities.

The blinders that many Americans wore when evaluating their host country sometimes reached comic proportions. One of the older IREX exchange participants was a professor of Russian history from one of the University of California campuses near Los Angeles. (I've forgotten his name and the college where he taught.) He was a pleasant fellow, and Ann and I had him over for dinner. He commented to us that when he returned to the U.S. he wanted to be able to convey to his students how poorly the Soviet system worked. In order to do this he had taken photographs of a car having to turn right in order to turn left off a major artery.

We asked him what he meant, and he drew us a diagram. At the intersection in question to make a left or U turn a car would take a right and circle around to a traffic light. When the light turned green, the car could go either straight (the result then was a left turn) or left (to effect a U turn). We heard his explanation, looked at his diagram, and said to him that that was the way left turns were made from Route 1 near Princeton, and we didn't think anybody had ever complained. Out of politeness we refrained from telling him that Los Angeles could learn a lot from Moscow and other European cities about how to handle public transportation. Between ourselves we remarked that it was particularly absurd for someone from southern California to ridicule the way traffic was managed in the outskirts of Moscow.

In reality, academic exchange visitors had little reason to feel antagonistic toward the Soviet authorities. The government adopted a very different stance toward scholars than toward Americans whom they viewed as essentially hostile to their interests. We lived in close proximity to our Soviet peers, whereas Western diplomats and journalists were required to live in special apartment buildings with guardhouses to discourage informal contact with ordinary Soviet citizens. The KGB would monitor their activities, and there was constant tension.

Formally speaking, exchange scholars, like other American residents, were forbidden from traveling more than 25 miles (40 kilometers) outside Moscow or Leningrad without permission. This was probably enforced somewhat for diplomats and journalists, but it certainly wasn't for us. We routinely ignored the 25-mile limit, and we knew of only one occasion when anyone got caught. This occurred during Ann's year in Leningrad, and the only reason it happened was that one of their party asked a policeman in badly-accented Russian to recommend a nearby restaurant. He did so, patiently waited outside until they finished eating, and then politely informed them that they were not supposed to be so far from the city. He escorted them to the station and waited for them to catch the next train back to Leningrad.

At the suggestion of my parents we contacted a *Washington Post* reporter named Kevin Klose, whose mother had been a friend of theirs twenty years earlier when we lived near Red Hook in rural New York state. His experiences and perceptions had nothing in common with ours. A few years later he wrote a tendentiously anti-Soviet book about his years in Moscow. He later became director of Radio Liberty — the same rabidly anti-communist propaganda network of which my high school Russian teacher's husband had been director many years before — and he is now President and CEO of National Public Radio.

On one occasion Klose's wife came over to visit without Kevin, who couldn't make it. She said that she wished she could write about some of the positive experiences that they had had, particularly with their children's schooling. She had been very impressed with her dealings with Soviet teachers and school administrators. However, she said that it would not be good for her husband's career for her to publish a positive account of life in the Soviet Union.

At the end of our stay in Moscow in 1974-1975 we wanted to visit some cities in the Caucasus and Central Asia and continue eastward, returning home via India, Southeast Asia, and Japan. It was my thesis adviser Nick Katz who had suggested getting a round-the-world ticket. Always a source of good practical advice — particularly on financial matters — he noted that the cost would not be much greater than the roundtrip fare to Moscow, and of course IREX would pay for that portion.

As mentioned before, several of our friends in Moscow — Saidakhmat, Galim, Anas, and others — came from non-European parts of the Soviet Union. They had friends or relatives who could show us around some of the cities in those regions. We decided to spend a couple of weeks visiting Tbilisi, Samarkand, and Tashkent. In addition, we had some hope of visiting Vietnam — which, it turned out, we weren't able to do until three years later — and so we had routed ourselves through Thailand and Laos, from which a trip to Hanoi would have been a short hop.

A problem with this plan was that we had to get special permission in order to leave the Soviet Union from a city other than Moscow or Leningrad. I went to the university official who was in charge of the foreign exchange people, and he said no. Refusing to take *nyet* for an answer, I went over his head and made an appointment with an official at the higher education ministry. He and I had a friendly chat, both of us showing off our diplomatic skills. I told him how much we had enjoyed our year in Moscow and explained why we wanted to see some other parts of the country and then leave for India. Soon after, I heard from the university official that he had changed his mind because of my *nastoichivost* (persistence).

We arrived in all three cities without hotel reservations. In Tbilisi and Samarkand we easily got a hotel room for which we were permitted to pay in rubles (tourists were required to use hard currency); in Samarkand the college student working at reception gave us the Soviet rate, presumably because we spoke to him in Russian, and that was only four rubles a night ($5 at the official rate of exchange, $0.80 at the black market rate).

In Tashkent a mathematician named Khodzhiev, who was a former Ph.D. student of Kirillov, arranged for me to give a talk and put us up in a dorm room. The hospitality of people in the southern republics was legendary, and we were very well treated.

The scientists' attitudes toward socialism were different from the cynical disapproval that we typically encountered in Moscow. Khodzhiev's wife, who was an agronomist, told us that in the pre-socialist days her grandmother had been barely literate and had been deprived of all opportunities. In the 1920's, under the influence of the Bolshevik revolution, some women of her generation threw off their veils and demanded equality. A few of them were stoned to death by the men. The young Soviet government promptly tried and executed the killers, and the stonings stopped. Now she and many other women were leaders in their professions, and they were deeply appreciative of what socialism had brought to their lives.

After Tashkent we visited Samarkand, where we saw the restored Bibi Khanum Mosque and the ancient astronomical observatory of the great Ulugh Bek. Our last evening in Samarkand we had to take the night train back to Tashkent to catch our flight out of the country the next day. We allowed plenty of time to get to the station, but we just couldn't find a taxi. The few that came were quickly grabbed by someone, and we were getting desperate. It would be a logistical nightmare if we missed the train and then our flight from Tashkent. At that point an Uzbek man came up to us and asked what the trouble was. When we explained our predicament, he went and found a taxi, talked with the driver and the other people who were waiting, put us in the taxi, and got in with us. He came with us to the train station, insisted on paying the driver, and personally escorted us to the right place on the tracks where our train car would stop. The reason he did all that was, in his words, so we wouldn't leave with a negative impression of his country.

Since there was no direct flight from Tashkent to India, we had to spend a day in Kabul, the capital of Afghanistan. The poverty and disease were a shock to us after Tashkent and Samarkand. In the Soviet cities we had just visited there had been no beggars, no children with distended bellies, no rampant tropical diseases.

Since we had only an afternoon to look around, we headed for the old city, thinking that that would be the most interesting part of Kabul. But we cut our excursion short when we saw the men's attitude toward Ann. Even though she was dressed modestly, she was unveiled and obviously a Western visitor. They would jostle her, spit in her direction, and make menacing gestures. We soon went back to our hotel.

Leaving the Soviet Union to the east gave us a different perspective. We saw a vivid contrast between two parts of Central Asia that until about a century before had had similar cultural and political histories. Then the southern part came into the sphere of British imperialism, and the northern part fell under the control of the czars. And for the previous half-century the northern part had been under Soviet socialism. It was almost a controlled experiment in what socialism as opposed to capitalism had to offer to the Third World.

Typically when people in the West compared the two systems, they looked at Europe and spoke proudly of the success of the capitalists in rebuilding Western Europe so that it became more affluent than the Eastern European countries that were under Soviet control. But to us that was only part of the picture. We also would think about the contrast we saw in our 1975 visits to Uzbekistan and then Afghanistan.

Our trip home through Asia was largely just a tourist excursion. We seemed to be followed by some odd coincidences, though. While we were in India — in fact, the day we visited the Taj Mahal — Prime Minister Indira Gandhi declared a state of emergency and assumed near-dictatorial powers. While we were in Burma there was a big earthquake, and when we arrived in the ancient city of Pagan most of the famous temples had been damaged or destroyed.

But Ann had a nice experience in Thailand on the night train back to Bangkok from Nong Khai (the border town from which we crossed the Mekong River to visit Vientiane, Laos). I was asleep, but she was not. The conductor went down the aisle telling all the passengers who were awake that they should look out at the sky, where two lights were coming together near the horizon. Ann looked and saw the link-up of the Apollo – Soyuz spacecrafts — the first joint mission of American and Soviet cosmonauts.

During the Cold War there was a tendency to sensationalize the life of Westerners in the Soviet Union. Even minor incidents received extensive coverage, and Americans back from a one-week package tour would tell people at home that they were sure the KGB had been following them and bugging their hotel rooms. Why the KGB would be so desperate to know about the shopping bargains they had found or their opinion of the Bolshoi ballet performance they had seen was never explained.

Ann and I had very few experiences that remotely qualified as Cold War melodrama. The closest was an incident that occurred because of a thoughtless misstep by our friend Dima Kazhdan.

Among the brilliant students and protegés of the legendary Soviet mathematician I. M. Gel'fand, Kazhdan was considered the best. He immigrated to the U.S. in the mid-1970's and immediately joined the Harvard math department. He was unusual among the Soviet Jewish mathematicians in that he was very religious, having converted to Orthodoxy not long before. When Ann and I were about to leave for the Soviet Union in 1978, he asked me whether he could send religious books to us through the U.S. Embassy mail for me to pass on to Orthodox friends of his in Moscow. I agreed.

A few weeks after our arrival in Moscow, I was summoned to the U.S. Embassy to meet urgently with the cultural attaché. He brought me to a special room and said that we should assume the room was bugged and that some of the discussion we had to have should best be done using a type of etch-a-sketch pad that he gave me. The previous week the Embassy had impounded a box that was addressed to me. It contained pro-Israel and anti-Soviet books in the Russian language, and the Embassy could not let me receive them because it would be in violation of the conditions governing use of the diplomatic mail. In addition, they were concerned that if I were to distribute such material, it would lead to an incident.

He then told me that the previous Friday the Embassy had received a telephone call from a Zionist organization in Great Britain asking to speak to me about the books they had sent. The secretary who answered the phone, who had been trained to handle provocations, said that she knew of no one with that name and the caller must have the wrong number, and then she quickly hung up. Such a call on an unsecured line to the American Embassy in Moscow might have been a deliberate attempt to get me in trouble and provoke an incident that would have served the purposes of the Zionist group. The attaché said that the U.S. Embassy in

London had spoken forcefully with the organization, and they would not do this again.

Apparently Kazhdan, instead of sending me a few books for his friends, had passed on my name and address to the British group. I told the Embassy official about having agreed to Kazhdan's original request, but emphasized that I never expected that it would lead to this. I did not want to receive those books and had no objection to the Embassy's confiscating them. I then asked whether this would affect the rest of my stay in Moscow, whether I should change my routine in any way, and whether there would be any danger to our friends.

He said that I should expect to be followed for a while until the KGB convinced themselves that I was not doing anything out of the ordinary; then it would most likely blow over. He didn't think that there would be any serious consequences, provided that there were no further incidents.

Ann and I decided to make only one change in our activities. We had been visiting the family of Boris and Natasha Kac (yes, their names reminded us of the cartoon spies we'd watched as children on the show "Rocky and Bullwinkle"), who had been refused permission to emigrate and join their relatives in the U.S. The case was starting to acquire a high profile in the West, so after the phone call to the Embassy we decided that only Ann, who was less likely than me to be followed, would continue to see them. Except for that, Ann and I socialized with our friends just as before.

As far as we could tell, neither of us was ever followed, and it is unlikely that our room was bugged. Ultimately we decided that what had saved us was the fact that the British Zionists had placed their call late on a Friday. No one in the Soviet Union worked on Friday afternoons. Most likely the KGB person who was supposed to be listening to the phone lines had left early to do her shopping for the weekend.

We had known the Kac family since our stay in Moscow in 1974-1975. We first met the oldest of three brothers, Victor, who was a prominent young mathematician who soon after immigrated to the U.S. and took up a position at MIT. In 1978 his mother Haika Bimuvna and his youngest brother Misha were also living in the Boston area. After I got to Harvard in 1975, I had worked with the admissions office to arrange for Misha to become a Harvard student as soon as he arrived — they treated this as an extremely unusual case. Haika Bimuvna lived near Central Square, and Ann used to stop by to see her once a week on her way back to Cambridge from Boston University.

The middle son Boris and his family had not been allowed to leave for the U.S. As in the case of many "refuseniks," Boris had worked in an institute that did classified research for the Soviet government, and they wouldn't let him go until the institute officials signed a document stating that any secrets he might have known were no longer of value if divulged in the West.

In the Soviet Union, as in the U.S., secret classifications were often put on things for bureaucratic reasons unrelated to national security. The designation "secret" or even "top secret" would be stamped on many projects and documents that in reality no "enemy" would have profited from knowing about. There was a popular joke at the time that went as follows. A Jewish computer scientist was denied permission to emigrate on the grounds that he knew secrets. "How could I know secrets," he asked, "when all our work with computers is fifteen years behind the Americans?" "That's the secret," came the response.

What complicated the case of Boris and Natasha and led to a flurry of publicity about it in the U.S. was that they had a baby named Jessica who had health problems. She was lactose-intolerant and initially was losing a lot of weight. Organizations concerned with the issue of Soviet Jews — such as the Hebrew Immigrant Aid Society (HIAS), which helped Jewish immigrants settle in the U.S. — saw the Kac baby as a poster-child for the plight of the refuseniks. American journalists eagerly wrote about poor Jessica, who in some accounts was near death because the Soviet government refused to let her family come to the West, where she would get proper medical care.

The delay in allowing the family to leave seemed to be unusually lengthy compared with similar cases. Finally the family was released and flew immediately to Boston. There was a media frenzy at Logan Airport when the plane landed. Senator Edward Kennedy was on hand, along with a throng of journalists and a group of wellwishers from HIAS. An ambulance with sirens wailing was let onto the tarmac to meet the plane and rush the child to the hospital.

Imagine the reaction when Boris and Natasha emerged from the plane with their chubby, healthy, laughing Jessica! Of course, no one could say that they weren't happy to see the baby in such wonderful condition. But it was certainly an embarrassing moment for all concerned.

What had clearly happened — and what explained the long delay — was that the Soviet government had kept the family in Moscow until Soviet doctors were able to completely restore Jessica's health. Once the case had become such a *cause célèbre*, they did not want the media to be able to say that Soviet medicine was inferior and that the child could only be cured in the West.

Reporters soon learned that some people close to the case (not us!) had known for months that Jessica's condition was improving, but had not wanted to tell the media. A HIAS person told a *New York Times* reporter that they thought there would be more pressure on the Soviet authorities to release the family if people in the West continued to believe that the baby's life was in danger.

The New York Times was furious that they had been lied to and manipulated. In a move that was probably unprecedented, a *Times* editorial apologized to the Soviet government for having contributed to circulating false reports on the case. The heading of the editorial was "Jessica Kac Riding Hood"; the editors had confused the two children's stories *Little Red Riding Hood* and *The Boy Who Cried Wolf*, perhaps because they were under deadline pressure and because both stories involved a wolf.

During the decade from the mid-1970's to the mid-1980's Ann and I were heavily involved with a large number of Jewish professionals — mostly, but not entirely, mathematicians and their families — who wanted to leave the Soviet Union. We were in a position to help in various ways that required little effort on our part but were important to them.

During our stays in Moscow in 1974-1975 and 1978 we were allowed to send and receive mail through the diplomatic pouch, and we often sent letters of Soviet colleagues through the Embassy and received replies for them. (Use of the diplomatic mail by non-official personnel was greatly restricted in our later visits.) People who were about to immigrate to the U.S. could write to colleagues about jobs and apply for academic positions through us. Refuseniks could use us to send messages to their supporters in the West who were pressuring for their release. For example, during the period when the prominent mathematician I. Piatetski-Shapiro was being denied an exit visa, he wrote many letters through us to his former student Novodvorski, who had emigrated earlier and was coordinating efforts on behalf of Piatetski-Shapiro.

In some cases colleagues hoping to leave wanted to meet with us to practice English and ask questions about academic life in the U.S. On occasion they asked us to correct and edit something they had written in English in connection with their future professional life in the West.

We always had mixed feelings about this part of our life in the Soviet Union. On the one hand, we were well aware of the anti-semitism in the Soviet academic world, which mainly took the form of quotas limiting the number of Jews admitted to prestigious universities and discrimination against Jews in academic hiring and promotion. So it was perfectly natural for them to want to leave.

We also knew that many non-Jewish academics would have left for the West if they had been permitted to. Tenured professors in the U.S. enjoyed much higher material standards of living than their Soviet colleagues and also unlimited access to travel, books, and movies, all of which were restricted and controlled in the U.S.S.R. The Soviet government, because of external pressure and its own desire to reduce the number of Jewish professionals, allowed large numbers of Jews but very few non-Jews to emigrate. Ostensibly it was a special arrangement for them to follow their religion and move to the Jewish state.

However, relatively few of the Soviet Jews were religious, and most had no interest in moving to Israel. So the procedure was that they'd get visas for Israel and then go as far as Italy, where they would apply for U.S. visas. Typically a couple of months after leaving Moscow they'd arrive as refugees in the U.S. They would be supported by HIAS until they found jobs, and those who couldn't find employment — mainly the elderly and people (such as poets) without any transferable skills — would go on welfare.

Most of the prospective emigrés we met were pleasant people, and they tried to find ways to show their appreciation for the help we were giving them. For example, after Piatetski-Shapiro was finally permitted to leave the Soviet Union, he visited Harvard, where the math department chair, Shlomo Sternberg, gave a big party in his honor. Piatetski-Shapiro took the occasion to thank us publicly and profusely for having helped him, even exaggerating (I thought) the role that Ann and I had played assisting him in his efforts to leave. I was particularly touched by this, since in Moscow Piatetski-Shapiro and I had had heated political arguments.

Ann and I were constantly exasperated with the social and political attitudes of most of the Jewish professionals we met who were planning to emigrate. The disparaging comments they frequently made about people of non-European races and ethnicities — especially those from Central Asia, the Caucasus, and Africa — were even worse than what we heard from the Russians. Their political viewpoints often seemed to be based on "reversing the signs" in *Pravda*. Whatever the Soviet government's position was, the reverse must be true. For example, in reaction to the official Soviet criticism of the racial situation in the U.S., some of those who were planning on emigrating would say that if American Negroes were anything like African students in Moscow, then they could sympathize with the segregationists in the American South. Perhaps the most unpleasant part of our social interactions in Moscow was having to deal with racist and offensive remarks of this sort. And most of the anti-communist Soviet Jews wanted the U.S. to prevail in Vietnam. (This had been the topic of my acrimonious discussion with Piatetski-Shapiro.)

When we tried to answer their questions about life in the U.S., much of the time they wouldn't believe us, because what we said didn't always jibe with what they had heard from friends and relatives who had already emigrated. We soon realized that the letters home from the Jewish refugees in the U.S. were often misleading in the extreme. The emigrés clearly wanted to give the impression that they were doing well in their adopted country. They would write, for example, that they had just bought a car. (A car was a luxury in the Soviet Union, where almost everyone used public transportation, but of course it was a necessity in the U.S. They also neglected to mention that they had bought the car on credit and would have to make monthly payments.) They would write that they'd gotten a job at Harvard or Princeton or Michigan or the Institute for Advanced Study, without mentioning that it was a one-year post-doc, not a permanent position. They would give their salary, which would sound astronomical by Soviet standards, but wouldn't mention what the cost of living was or whether or not they had job security. Those who were on welfare virtually never let word of that get back to their friends and relatives in the Soviet Union.

Beyond the material hardships that some emigrés experienced, there were also psychological barriers that they had not anticipated. For example, we often heard the complaint that Americans have a superficial notion of friendship. Haika Bimuvna (the mother of Victor, Boris, and Misha Kac) once said to Ann that "Americans are always saying 'Have a nice day!' But none of it is sincere." Many years after she immigrated, when Ann and I no longer lived in Cambridge, she telephoned Ann in Seattle just to talk with her; she told Ann that she was the only American who was a true friend of hers.

When I first visited the Soviet Union in the 1960's, I learned that I was overusing the Russian word *drug*. The literal translation of *drug* is "friend," but most American uses of "friend" should be rendered into Russian as *znakomyi* ("acquaintance") or else *priyatel* (more than an acquaintance but less than a friend). The English equivalent of *drug* is perhaps "close friend" or "confidant" (terms that Americans use in conversation much more rarely than "friend"). My linguistic misunderstanding when I started speaking Russian reflected a cultural difference between modern America and more traditional social systems. By the same token, among Russian (and other) immigrants who settle in the U.S. it is a common perception that American society values casual friendships, but not the type of deep friendships that were so important to them in the old country.

Many people we talked with in Moscow and Leningrad — especially those planning to emigrate — viewed the West through rose-colored glasses and didn't want to hear anything that contradicted their idealized view. They also tended to assume that, in addition to the well-advertised advantages of life under capitalism, the U.S. would also offer all of the amenities that they took for granted in the Soviet Union.

In 1978 Ann was exchanging language lessons with a woman named Lydia who, along with her husband Misha, was hoping to emigrate some day. Misha was an engineer, but fashioned himself as a creative writer whose ambition could not be fulfilled in the oppressive conditions of the Soviet Union. He said that once he got to the West he'd be able to work as a writer without any constraints. When Ann asked him how he expected to be able to support himself, he said, "I'd receive a salary from the Writers' Union, of course." We had to patiently explain to him that in the U.S. there was no Writers' Union that paid everyone's salaries, and that in fact there were very few creative writers who could support themselves exclusively from the proceeds of their books.

Misha also said that there were plastic cards that could be used to obtain most of one's needs and desires. He had heard some vague rumors about credit cards, but did not realize that eventually one must pay the balance — they don't enable one to acquire things free of charge! Again he was confusing the situation with that in the Soviet Union, where salaries were small but many material needs were either highly subsidized (public transportation, apartments, food staples) or free (medical care, higher education).

Another source of frustration was the exaggeration and distortion in people's accounts of their own treatment. There was enough true anti-semitism in the Soviet Union — we didn't see why so many Jews would claim to have encountered it in situations where it was not really there.

For example, on one occasion our friend Volodya Berkovich, a student of Manin (and now a professor at the Weizmann Institute in Israel), told us that he'd been victimized by anti-semitism at his place of work. Now Volodya was someone who normally was careful in what he said, and he also was unusual in his tolerant, non-racist attitudes toward his peers when he was a graduate student. His roommate and good friend for a while was Tatar, and when an Egyptian graduate student named Sami was having great difficulty, Volodya tutored him free of charge.

So we wanted to hear the details of Volodya's experience. At the engineering company where he was employed he had gotten into an argument with two coworkers who had been making derogatory comments about a performing artist who had defected to the West and who happened to be Jewish. When they said that this person had betrayed his country for the money, Volodya told them that what they had said was anti-semitic and sounded just like the way the Nazis used to speak about Jews. They took great offense — after all, being compared to the Nazis was no minor matter in a country that had lost seven million soldiers and twice as many civilians in World War II — and they sued him for slander. The case came before a magistrate (who, like most judges in the Soviet Union, was a woman). She ruled that the evidence concerning blame for the incident was murky, so all she would do was direct the administration to transfer Volodya to a different section of the company. This was done, and Volodya said that he was actually much happier in his new surroundings. But he felt that the fact that he had been hauled before the magistrate showed the poor state of free speech in the Soviet Union.

We said that this incident was not clear evidence of either anti-semitism or suppression of free speech. In the United States free speech does not mean that you can insult a coworker on the job without any consequences. In fact, a series of incidents of that type in an American workplace could have easily ended with Volodya's demotion or dismissal. I don't know where he got the impression that an employee in the U.S. has absolute free speech with no reprisals no matter what he or she says while at work.

When we arrived in Moscow in January 1978, we were put in the Academy of Sciences Hotel on October Square. We couldn't believe that we were really supposed to live there for six months. There were no kitchen facilities, and guests would have to register and leave their ID at reception.

I immediately asked to be moved to an apartment and was told no, couples with children had priority for the apartments. I said that if the Academy expected good scientists from the West to want to come over on the exchange program, then they'd have to provide reasonable accommodations, and living in a hotel for six months was not reasonable. Still the answer was no.

One of our Soviet friends suggested that we threaten to leave and go back to the U.S. if they didn't let us move. But I don't believe in threatening unless one intends to carry out the threat. Bluffing just results in a loss of credibility.

I asked for another meeting with the Academy official who had authority over our housing arrangements. I told him that I had had some serious stomach ailments in the past, and that if I didn't get the type of food that my wife knew how to cook, those ailments might recur. That's why it was so important for us to have cooking facilities. This argument (which was not really a lie, only an exaggeration) worked. I knew that Russians had a tradition of great concern and sympathy for the sick. The Soviet official could save face by making an exception to his usual policy for humanitarian reasons. And of course he didn't want the worry of having a sick American on his hands. So within a week they found us a small apartment on Gubkin Street, right around the corner from the Steklov Mathematical Institute. It was a nice apartment in an ideal location.

Over the years when we traveled to and lived in the U.S.S.R., we found that Soviet officials were usually reasonable if approached in the right way. For example, in 1974 Ann's parents sent us a big "care package" containing all sorts of necessities of life that they mistakenly thought we needed. Ann had told them about the restrictions on sending things to us and had left instructions with forms that had to be included; under the IREX agreement we could receive a duty-free shipment three times during the year if the correct procedure was followed. But the Hibners had forgotten about this.

We were summoned to the customs office, where an official went through the box with us. When he came to the food items, he was visibly insulted. "Don't they think we have food to eat in Russia?" he asked. I explained that Ann was their oldest child, and this was the first time she'd ever been so far from home. They were worried about her, and they knew nothing about the Soviet Union. We hadn't asked them to send anything, and we knew that we had no right to receive this package. We would understand if he decided to confiscate it. After hearing the explanation, he softened his tone. "Don't worry about it," he said, "you can just take it." So we got everything without paying any duty.

In 1975 we met the mathematician A. T. (Tolya) Fomenko, who besides being a well-known author of books on topology and differential geometry had a growing reputation as an amateur artist. His paintings and pen-and-ink drawings were surrealistic in the spirit of Dali and Escher. Inspired by real experiences, historical events, mythology, and literature — a cycle of about a dozen works were based on the Soviet satirical classic *The Master and Margarita* by Mikhail Bulgakov — his paintings were dramatic, provocative, and sometimes macabre (especially in light of the stories he would tell about them). They were not, however, in the Soviet "socialist realist" tradition, and it was tricky to get showings of his work. He used some of his drawings as fanciful illustrations in his math books — thereby making those texts into cult classics among Soviet students. While we were in Moscow, to our surprise he was able to arrange displays of his work at the main graduate student dormitory and at a bookstore.

We enjoyed spending time with Tolya and his wife Tanya, who had a great sense of humor, a fascinating circle of friends, and diverse interests. We tried to see them whenever we were in Moscow; and in the early 1990's, when Fomenko was regularly visiting the University of British Columbia in Vancouver, we drove up to see them on two occasions, and one time they came to Seattle.

In the 1970's and 1980's Tolya kindly drew a frontispiece for each of my first four books. The first two books dealt with p-adic numbers, the third was about elliptic curves, and the fourth was on cryptography. The drawings depicted, respectively, the 3-adic unit disc, the 2-adic numbers as an inverse limit, a family of interlinked tori, and a massive building with the digits of π along its sides.

In 1985 Ann and I interviewed Tolya Fomenko for *The Mathematical Intelligencer*. The magazine published the interview the next year, accompanied by photographs of eight of his drawings. In 1990 the American Mathematical Society published a book of Fomenko's artwork, titled *Mathematical Impressions*.

During our six-month stay in Moscow in 1978 we had a lot of visitors. Inexpensive one-week tour packages were plentiful, and we encouraged our friends and relatives to come while we were there, since we enjoyed introducing them to our Soviet acquaintances and showing them around town. The people who came to see us in Moscow included my sister Ellen and her then-boyfriend Alfred, Luther and Debby Ragin (Luther will figure prominently in Chapter 8), Eric and Carol Weitz (Eric, who went to graduate school with Ann, was mentioned at the end of the last chapter), and Bob Oliver (who was in my Ph.D. class at Princeton and was active in anti-war protests there).

Sometimes Ann and I would get a new perspective from seeing how our friends reacted to the sights and sounds and experiences in Moscow. Once when Bob Oliver and I were in the metro a boy about ten years old who heard us speaking English started talking to us, practicing the words he knew. Two middle-aged women in a nearby seat spoke to him, and I translated for Bob. They complimented the boy for having taken the initiative to try out his English with foreign tourists, asked him where he went to school, told him that he must have had good teachers, and advised him to continue to study hard. Bob asked me whether the women were his relatives or family friends. I said no, they're strangers. Then it occurred to me that this would seem odd to an American, because such a scene would be unlikely in the New York subway, for example. But in the Soviet Union complete strangers — especially middle-aged and elderly women — took a direct interest in other people's behavior and welfare. In a sense they were the enforcers of social norms. They would think nothing of scolding a young mother they felt had not been properly watching over her children or had not dressed them warmly enough. And they would similarly praise a youngster, even a perfect stranger, for having done the right thing.

On another occasion Ann, Ellen, and Alfred were on a bus near the center of town when they saw a startling scene at a crosswalk. A large crowd of people waiting to cross had formed because the young policeman who was supposed to be changing the lights seemed not to be paying attention. Suddenly a middle-aged woman came up to him, hit him hard with her shopping bag, and yelled at him. The policeman snapped out of his daydream, apologized hurriedly, and changed the light. Ellen and Alfred, both of whom were lawyers in the New York area, were stunned — they could not have imagined someone doing that to an American cop. The Western media always used to refer to the U.S.S.R. as a "police state," but in fact the police were anything but an intimidating presence in Soviet cities.

In 1978 the mathematics editor at Springer-Verlag, Walter Kaufmann-Buhler, wrote me about someone named Beatrice Stillman who was coming to Moscow and whom we might want to meet. I had just published my first book with Springer, and Kaufmann-Buhler had been largely responsible for establishing their reputation as the best publisher for mathematicians to work with. (Five of the six books I've written were published by Springer.) Stillman had just translated *Memories of Childhood* by the Russian mathematician Sofia Kovalevskaia, whose husband had been the prominent Darwinist V. O. Kovalevskii mentioned in Chapter 6. I had bought Ann a copy of the Russian version of those memoirs when they were republished three years before. At the time what I thought would interest Ann was the reference to Herbert Spencer, the famous British social Darwinist, with whom Kovalevskaia had had an argument about women's capabilities. But Kovalevskaia herself was not yet the focus of Ann's interests.

Stillman came over to visit and gave Ann a copy of the introduction she had written for her translation. Ann was disappointed in the piece and thought that because of her lack of historical training or knowledge of the period some of what she had written was just plain silly. Kaufmann-Buhler had mentioned that Stillman was planning to write a full-length biography of Kovalevskaia. Eric and Carol Weitz, who were also visiting Moscow at the time, later told Ann that they remembered her saying right after the evening with Stillman, "I could write a much better biography than she could." (Ann didn't remember saying that.) That was the first time that the idea of switching her Ph.D. topic to Kovalevskaia occurred to Ann.

Shortly thereafter Kaufmann-Buhler admitted to Ann that he didn't have much confidence that Stillman could write the biography (and, in fact, Stillman never got far into the project). He said that if Ann wrote her dissertation on the life of that fascinating mathematician, socialist, and feminist, he was confident that Springer or another press would be glad to publish it. So Ann decided to write her Ph.D. thesis on Sofia Kovalevskaia. Her book was ultimately published not by Springer, but by Birkhäuser (first edition) and Rutgers University Press (second edition).

Ann's year in Leningrad in 1981-1982 was a transformative experience for her, and it gave her some of her most cherished memories. Initially she had hoped to be placed in Moscow, not Leningrad. Most of the archival material she wanted was there, and our recollection from the 1970's was that living accommodations were much pleasanter in the capital than in the Leningrad dorms. When she learned that she would be in Moscow only for a month, she was disappointed. But as it turned out, she found plenty of useful material in the Academy of Sciences library in Leningrad (much of it relating more to what would become her second book than to her biography of Kovalevskaia).

Moreover, her living setup was ideal. She was on the fifth floor of a dormitory that had graduate students from all over the world. Hers was a women's floor, and she formed wonderful friendships with her neighbors.

When she arrived, the first person she met was her Soviet roommate, Tanya Pavlenko, who was from Siberia. They sat down and, in the Russian tradition, each pulled out a bottle of wine. After several hours the table had four "dead soldiers," as empty bottles of wine were called, and the two had become fast friends. Ann learned a lot about life in the Soviet Union from Tanya, who also helped her improve her fluency in Russian. Once in the course of a political discussion with the other students Ann referred to Ronald Reagan as *zhertva pyannoi okusherki* (victim of a drunken midwife), which was a colorful colloquialism for someone who's mentally defective. One of the women asked indignantly, "Who taught her that expression?" and Tanya sheepishly confessed.

Each floor had a person — usually a middle-aged or elderly woman — who served as caretaker and watchwoman. On Ann's floor it was Tyotya Zhenya ("Aunt Zhenya"). One day her friend Tyotya Valya from the second floor was talking with her in the kitchen, where Ann and Tanya were preparing a meal. Aunt Valya asked for some sugar, and Aunt Zhenya pointed to a package that someone had bought. Valya asked, "Doesn't it belong to someone?" and Zhenya replied, "No, it's for anyone to use. Nothing here belongs to anyone. For us on the fifth floor communism has already arrived!" Ann and Tanya burst out laughing. Aunt Zhenya's remark was perfect, and its formulation (which is hard to translate) captured just the right tone of officialese. It echoed the slogan during the Stalin period "socialism in one country," but carried it a step further — "communism on one floor."

And in fact the slogan was in a way correct. Many of the young women in essence lived communally — everyone would pitch in and do what they could. If someone encountered a hard-to-find item in a store or at a streetstand, she'd buy several so that there'd be enough for everyone. Whoever got back to the dorm first would put a big hearty soup on the stove that would be ready when the others got home. It was a time of great fun and comradery — so much so that when spring came Ann wrote me that she was "cursing the lengthening days" because the approaching summer would mean the end of her year in Dormitory #2.

In 1975 we became friendly with a Bulgarian student of Shafarevich named Andrei Todorov. He was not only a gifted mathematician, but also a great storyteller and jokester. He especially liked to make fun of the Soviet and Bulgarian systems, even though both of his parents were prominent Communists. They had been active in the Party since its underground days during World War II, when Bulgaria was allied with Hitler. Andrei used to say that his mother was a "terrorist," because she had participated in a successful plot to assassinate the Nazi Deputy Governor of Plovdiv.

We visited Andrei and his wife Betty in Bulgaria three times in 1975, 1978, and 1982. On two occasions we went to the resort area near Varna on the Black Sea. It occurred to me that it was often easier and cheaper for Westerners to travel to Bulgaria than to the Soviet Union — in fact, the Black Sea coast had become a favorite destination for West German and British working-class tourists. Moreover, Soviet mathematicians who had been effectively barred from most foreign travel still could usually go to Bulgaria if they wanted. So I got the idea of setting up a retreat and guesthouse for small joint conferences of Soviet and Western mathematicians.

Andrei and Betty had close friends in high government positions, and in principle I probably could have gotten permission to buy a small beachfront cottage (which at that time would have been inexpensive). For a while in the early 1980's I was so obsessed with this idea that I insisted on visiting some seaside towns to see what might be available. I remember dragging Ann and Betty (Andrei wasn't with us this time) around the village of Shkorpilovtsy. We had bought a watermelon to eat later, and I was carrying it in a string bag of the sort that was commonly used among the peasantry. Betty commented that to local people we must have cut an absurd figure — househunting Americans with their watermelon in a string bag.

The Shkorpilovtsy project never went anywhere. But the idea of building a tiny conference center and then convincing American colleagues to visit the Black Sea coast of Bulgaria to meet with their Soviet counterparts wasn't really all that unrealistic. For me, in any case, it was a natural outgrowth of the dream I had had as a high school Russian student twenty years before of someday playing the role of an "ambassador" to the Soviet mathematical world.

My short-lived obsession with purchasing property in Bulgaria also presaged (although I never would have guessed it at the time) an interest that Ann and I developed twenty years later in buying historical mining claims in the American Southwest. That will be described in Chapter 16.

From a mathematical standpoint, my spring 1985 visit to Moscow was less useful than my earlier two stays had been. Manin and most of his students had switched their interests from number theory to mathematical physics. In addition, I was starting to work on applications of number theory to cryptography, which was only an area of secret research in the Soviet Union. I gave an invited talk on number theory and cryptography at the Moscow Mathematical Society (which I'll discuss in Chapter 14), but I had no real research collaboration with anyone.

I traveled through Moscow twice in 1987 and 1989, in both cases en route home from India. I spent only a few days there, but especially in 1989 I was starting to get a strong feeling that I would not want to return for another visit. I had no idea that the Soviet Union itself would collapse or that the conditions for research in math and science would go into a tailspin. But I had a sense that I tried without much success to explain to Ann that things were going in a bad direction. Perhaps my reaction was akin to animals sensing the approach of an earthquake.

Ann had a more productive visit than I did in 1985, and she returned for a month in 1986 to work in the archives. In May 1985 she was invited to the Institute of the History and Philosophy of Science to speak on her research on Sofia Kovalevskaia. She was very well received. A. P. Yushkevich, who was probably the leading Soviet historian of mathematics, hosted her talk. At one point Ann referred to the field of Kovalevskaia studies by appending the standard Russian suffix *vedeniye* ("study of") to Kovalevskaia's name. Yushkevich later complimented Ann's talk and added that she had "enriched the Russian language" with her new word *Kovalevskavedeniye*, so he would henceforth call her by her name and patronymic —

Anna Mikhailovna — in the Russian style. An article Ann wrote in Russian based on her talk was later published in the Institute's journal.

Ann visited for the last time in January 1992, when she accompanied a group of 18 college students on a study tour of Moscow, Leningrad, and Kiev. (Their trip was cut short when problems arose with their transportation to Central Asia.) In many ways it was a disheartening experience. She saw elderly people desperately trying to sell household items on the street — hyperinflation had rendered their savings and pensions almost worthless. And she heard stories of street violence. Nothing disastrous happened to her charges, but she talked to someone from an Australian group who said that one of their students had been raped.

A couple of years after that trip Ann wrote a paper (published in *Canadian Woman Studies*) comparing the effects on women of the free-market reforms in Russia and in Vietnam. In the former case she had observed a conscious and systematic effort to favor men. Many of the "reformers" explicitly identified socialism with a "feminization" of society and boasted that capitalism would be the man's way of doing things. For instance, the traditional word for "savings bank," *sberkassa*, which is feminine, was being replaced by the cognate *bank*, which is masculine; and no expense was spared in repainting all the signs to conform to the new usage. Moreover, as in many parts of the world, women were disproportionately represented among the have-nots in the new society.

Ann's scholarly interest in Russia continued long after mine had subsided. During the years 1989-1998 she taught Russian history (among other subjects) at Hartwick College in upstate New York. Soon after moving to the Women's Studies Program at Arizona State University in 1998, she was finally able to finish her second book, *Science, Women and Revolution in Russia*, which was published in 2000. It was only at that point that she felt free to leave the field of Russian studies and work entirely in other areas.

CHAPTER 8 RACISM AND APARTHEID

From 1975 to 1979 I was an instructor at Harvard, which for many years had been a center of research in number theory and arithmetic algebraic geometry. The late 1970's were a good time to be there. In addition to the stars of the field John Tate and Barry Mazur, there were many young researchers. For example, I got to know two of Tate's graduate students, Dick Gross (with whom I coauthored my most important work of that period, the p-adic formula for Gauss sums) and Joe Buhler (who subsequently visited and gave talks in Seattle when I was at the University of Washington and he was at Reed College in Portland, Oregon). The junior faculty included Andrew Wiles (who later became famous for proving Fermat's Last Theorem) and David Rohrlich (with whom I wrote a paper on quotients of Fermat curves).

Shortly after Andrew Wiles arrived in early summer of 1976, Ann and I walked with him down to the Charles River to see the Fourth of July fireworks, which promised to be especially impressive because it was the 200th anniversary of independence. Our conversation took a historical turn because of the significance of the occasion, and somehow the subject of British royalty came up. I asked, wasn't it true that Britain alternates between males and females, kings and queens? Ann was horrified at my ignorance and my willingness to reveal it to a British colleague. She later told me that she was sure that Andrew would use that when he told stories to the people back home about the "stupid Yanks" (although in reality I doubt that it would have been in his nature to tell such stories), and she wondered where I could have gotten such a foolish idea. We decided that most likely when I was a little kid I had asked my parents what determines whether England gets a king or queen. Not wanting to give a correct but complicated answer to the question, to shut me up they must've said, "They take turns." That notion lodged deep in the recesses of my brain, not to see the light of day until the subject of the British royal succession came up unexpectedly decades later.

Ann and I arrived in Cambridge, Massachusetts in August 1975 after our trip back from Moscow via Central, South, and East Asia. Soon after our arrival I took part in a political action that got me arrested.

The Progressive Labor Party no longer had influence on campus, but it had started a group called the Committee Against Racism (CAR) that was organizing a demonstration for the first day of public school in Boston.

The big political issue in the Boston area was busing — transporting children to schools in different parts of the city to achieve racial integration. Many whites were protesting — in some cases stoning buses that were carrying black children into their neighborhoods — and some local demagogues, such as the notorious Louise Day Hicks, were fanning the racism to advance their political ambitions.

On Monday September 8, CAR organized a convoy of vans to take people to demonstrate in favor of busing at a location by the bus routes in South Boston. Before we got there, the police stopped our vehicles, arrested all 74 of us for "disturbing the peace," and took us to jail to be booked. Their logic was that if we had demonstrated, we would have been attacked by anti-busing fanatics, and that would have disturbed the peace. They wanted to prevent a clash; most police sympathized with the anti-busing people, not with us; and there were fewer of us than of the racists. So the police simply prevented us from assembling. Eventually the charges were dropped, since the prosecutor realized that the rationale for the charge against us would not have sounded convincing in court. And in any case they had achieved their objective of blocking our protest.

Even though our demonstration never took place, I was glad that I'd gone and gotten arrested. *The Harvard Crimson* gave the issue more coverage than they otherwise would have because a Harvard faculty member had been involved. I suppose that if I had been in a field other than math, I might have wondered about the wisdom of starting my first year on the faculty with my picture on the front page of the *Crimson* for having been arrested.

Strangely enough, that was the last time I was ever detained by the police. Despite different types of political activity over the last three decades, neither Ann nor I have been arrested. In the late 1980's we needed to send in a report from the Seattle police in order to obtain visas for El Salvador, which at that time was putting obstacles in the way of visits by non-official Americans. Fortunately, we had spotless records in Seattle, since all of our arrests had been many years before in Princeton, Cambridge, and Boston.

On February 17, 1976, an African American cook in the Radcliffe dining halls named Sherman Holcombe was "suspended indefinitely" by Food Services, ostensibly because of an argument with his supervisor having to do with cauliflower. Just two hours before in his capacity as shop steward (union representative) he had submitted a list of five grievances related to safety and working conditions. The suspension also followed a week after Holcombe put out a newsletter addressed to students and workers objecting to a proposal by Food Services that would have cut back on breakfasts and resulted in layoffs of regular and student employees. The student-faculty Committee on Houses and Undergraduate Life had rejected the Food Services proposal on February 11, largely because of Holcombe's efforts.

Holcombe had been writing newsletters for three months. The first leaflets he mimeographed at his union office. I helped with the later ones: the writing was entirely his, but I retyped them, paid to have them photocopied, and in one case — the leaflet opposing the breakfast cutbacks — I found people to translate it into Portuguese, Spanish, and Greek, the most common languages among the foreign-born workers. (Fortunately, there was a graduate student in the math department from Greece who kindly agreed to translate Holcombe's words.)

There had been some small-scale student actions on his behalf. For example, at the beginning of December a group of about 25 students had marched in front of the main administration building to protest the warning slips that Holcombe had been given in reprisal for his union activities.

When Holcombe was fired, a major protest movement formed on campus. Students organized a 95% effective boycott of Sunday brunch at North House Dining Halls, where Holcombe worked. The Harvard-Radcliffe Committee to Defend Sherman Holcombe held rallies and meetings at which workers, students, and faculty spoke. It is probably safe to say that never in Harvard's history had there been such a massive response to the firing of a non-academic employee.

The Harvard Crimson carried a full-page opinion piece by Holcombe, titled "Blows Against the Empire," which carefully laid out the issues. He started by describing some of the complaints that he had made about unsafe procedures in the kitchen — for example, a method of cutting in half large cylindrical frozen blocks of ice cream using a long sharp knife while they rotate — and his dismay at finding that, after agreeing to change the procedures, the managers did nothing. He wrote about management's demeaning treatment of many of the workers, in particular, their habit of reprimanding those who were absent because of illness. He then explained what he viewed as the most sensitive issue that he had raised as union representative.

> Many people ask, "Why has Sherman Holcombe become so hated and abused by the management? What about the other stewards,

aren't they doing their job, too?" To this I answer, yes, they are, in fact many of them also have been hit with foolish warnings. But for the past year-and-a-half that I've been a shop steward, I have been raising an issue that the university is especially anxious to keep swept under the carpet. I am speaking of the grievances I have raised about the non-posting of job opportunities. I believe it is mainly because of raising this subject that I have become so unpopular with management.

Why is the non-posting of jobs so important? Article 13.2 of our contract states, "Any vacancies occurring in the Food Services Department of the university will be posted for five (5) days on the bulletin boards so that the present employees may make application for such positions if they so desire." The purpose of the article in the contract is to ensure that *all* employees will have an equal opportunity to be considered for promotion on the basis of *merit* and *seniority*. But I've seen this provision of our contract ignored time and time again at Radcliffe by assistant manager Montville. He has promoted kitchen men to cooks at least five times in my experience without posting the job, purely on the basis of favoritism. Even when the worker promoted really deserved the promotion, he is made to feel that it is being done as a personal favor from his supervisor and that he owes a debt of personal gratitude. This is insulting to the dignity of the worker promoted, as well as unfair to the workers who don't even have a chance to be considered for promotion.

The practice of favoritism also violates Harvard's affirmative action program. Usually Montville favors young white males. Women, older people, Puerto Ricans and blacks are at a disadvantage. In the three years I've been at Harvard, four or five women have left cooking positions at Radcliffe and all have been replaced by men, without the job being posted.

I think Harvard would like to be rid of me because I have hit a sensitive nerve which goes to the core of Harvard's unfair labor practices and violations of equal opportunity and affirmative action.

Holcombe went on to discuss the broad implications of the way in which Harvard handled his case.

At the first hearing over my recent suspension, the judge in the case was none other than Sweeney, *the manager who was under investigation at the time for having insulted me on the basis of my race.* [Italics in original; she had called him "damn nigger."] Why aren't employees allowed due process?...

Why are students and faculty allowed to speak out...while non-academic employees cannot do so without suffering reprisals?... Why do most workers only complain about their gripes in private, afraid to exercise their constitutional right to free speech in public?

Soon after Holcombe was fired, I organized a group of six junior faculty and two visiting scholars in the math department to write a letter to Harvard President Derek Bok requesting a meeting to discuss "certain complaints regarding treatment of non-academic employees that have been raised in connection with the Sherman Holcombe case." Bok had the university's General Counsel (chief legal adviser), Daniel Steiner, reply to our letter and arrange a meeting. Five of us met with Steiner on Friday, February 27. Before the meeting, I had prepared a detailed fact sheet so that everyone would know the full circumstances of Holcombe's firing and would understand what issues would be likely to come up in our discussion.

At a rally of Holcombe's supporters that was held the following Monday, I related some of what I had said to Steiner:

> I'd like to repeat to you what I told Mr. Steiner on Friday. Affirmative action means more than statistics, more than percentages. Affirmative action also means that the working environment and treatment on the job are no less congenial to minority and women employees than to white male employees.
>
> I told Mr. Steiner that I could imagine a situation a little bit different. I could imagine how I would feel if I worked for some university, and my supervisor called me a "damn kike," then the university took over four months to investigate the incident, and in the meantime appointed this same anti-semitic supervisor to sit in judgment on a charge that was being brought against me.

After less than two weeks, Harvard moved to defuse the issue by allowing Holcombe to return from his "indefinite suspension." They did not, however, overturn the suspension without pay. Moreover, the letter informing him that he could return to work — written by the same supervisor who had called him a racial epithet — contained a series of false and insulting allegations about his behavior. And a top personnel official told the *Crimson* that "if his activities over the past few months continue, more serious action will be in order."

Although Holcombe was never able to get the warnings, the suspension, or the insulting statements by management removed from his record, he was able to continue working in relative peace and functioning as a union representative. His supervisors knew that his treatment was being watched by the whole university community, so they would have to restrain themselves from harassing him further.

During the months that followed Ann and I socialized with Sherman and his girlfriend Cindy. The four of us found a lot to talk about other than the university and union issues. However, after our trip to the Soviet Union in 1978 we lost touch. On March 12, 1979, while crossing a street in Cambridge, Sherman Holcombe was hit by a truck. He died from his injuries on April 6.

Thinking about the Holcombe campaign many years later, I'm struck by how unusual it was for a large cross-section of a university community to mobilize in support of a non-academic employee. The success of the campaign must be attributed in large part to the leadership abilities of Sherman Holcombe himself. He had a way of reaching out across racial and class lines and explaining the key issues in words that resonated with students and faculty, as well as with union workers. Often someone who becomes a *cause célèbre* has difficulty keeping perspective and a sense of humility, both of which are necessary in order to be politically effective. Holcombe had neither of those problems — he was forceful and articulate, yet modest and supportive of others. Over a period of a few months in 1975-1976 he went from being a representative of his coworkers to being a leader of the broader community of all those at Harvard who believed in social justice.

A student leader of the Holcombe campaign named Luther Ragin later became a good friend of ours. After graduating from Harvard, Luther entered a joint program of the Harvard Law School and the John F. Kennedy School of Government and ultimately received degrees in both law and public policy. Since that time he has combined a career in the private sector — for several years he was a vice-president of Chase Manhattan Bank — with work for non-profit organizations and civil-rights activity. In the late 1980's he, his wife Debby, and two other African American professionals sued *The New York Times* for racially discriminatory real estate advertising in violation of the Fair Housing Act. After four years of litigation, in 1993 they won a $450,000 settlement from the newspaper; they donated their share to a fair housing group and the NAACP Legal Defense and Education Fund.

From casual conversations outside a professional setting one usually can't tell when someone has legal training from one of the best schools.

For example, on family occasions one would get no hint of the legal profession from listening to my sister Ellen or my brother Donald, who were top law students respectively at Yale and Stanford. However, in Luther's case even in informal discussions he reveals a careful, incisive, analytical approach to any issue that comes up that would make it easy for anyone to guess his profession.

Already when he was young Luther had to walk slightly hunched over because of back problems. But the hesitation in his gait did not seem like a physical impediment so much as the posture of a good lawyer who holds back for a moment before speaking to be sure that he has all his ducks in a row. Occasionally people might mistake Luther's contemplative demeanor for uncertainty or lack of confidence — much to their later regret. Luther could pounce on inconsistencies of argument or disingenuous statements with a swiftness that could be a little startling.

In 1979 I worked closely with Luther in one of the most important political campaigns ever to be organized at Harvard — the movement to pressure Harvard to sell its stocks in companies that profited from South African apartheid.

The late 1970's were a time of intense interest in South Africa on many American campuses. In 1976 the Soweto uprising had again put the injustice of apartheid on the front pages of the world's newspapers. And on September 12, 1977, the great anti-apartheid leader Steven Biko — founder of the South African Student Organization and originator of the "Black Consciousness" philosophy — was killed by the police while in detention in Port Elizabeth.

The general objective of the campus movement was to get the universities to break with the U.S. and British government policy of "constructive engagement" — of supporting the South African economy on the theory that "constructive" approaches would eventually promote greater democracy. Some people wanted the universities to use whatever leverage they had as shareholders — which was not much, because most votes were controlled by management through proxies — to try to pressure companies into either withdrawing from South Africa or at least adopting the so-called "Sullivan Principles" of fair labor practices that had been developed by Leon Sullivan, a minister and former civil rights leader in Philadelphia. Other anti-apartheid activists thought that the universities should go further and divest themselves of all stock in companies that did business in South Africa, thereby supporting the economic boycott of the apartheid regime that had been called for by the largest black South African liberation organization, the African National Congress (ANC).

On October 21, 1978 the library of the Kennedy School of Government, named after the South African industrialist Charles W. Engelhard, was dedicated. Engelhard, whose family had given a million dollars to the Kennedy School, had gotten his wealth from gold mining — directly from the exploitation of black South African workers. Four hundred protesters came to the dedication, chanted through Harvard President Derek Bok's speech, and after the official ceremony heard an impassioned and carefully reasoned speech by black student leader Mark Smith. The naming of the library was a hotly debated issue for several months; ultimately, the university essentially gave in, agreeing not to name the library after Engelhard, but only to put up a plaque in his honor to thank the family (to whom they did not return the million dollars).

Soon after the Engelhard protest I started attending the meetings of the main anti-apartheid organization on campus, the Southern Africa Solidarity Committee (SASC). At that time SASC had not yet decided what action it would demand that Harvard take with its investments. The Harvard-Radcliffe Black Student Association (BSA) was calling for Harvard to divest itself of all stocks in companies that did business in South Africa. However, South Africa was not a priority issue for the BSA, and they were not prepared to take any action to back up their demand. At the SASC meeting where the question was debated, the most articulate spokesperson for the total divestiture position was Mark Smith.

Several of the most active SASC members believed that it would be unrealistic to aim for the university to divest. Some of the largest U.S. corporations did business in South Africa, and Harvard would resist mightily such a drastic restriction on its investment portfolio. It seemed that a more achievable objective would be to ask Harvard to use its influence as a shareholder to pressure companies to leave South Africa, while at the same time selectively divesting from a few corporations, such as banks that made loans directly to the South African government.

I thought that the shareholder-lobbying and limited divestiture option was anemic and wishy-washy and voted for the stronger alternative. I didn't speak at the meeting, since as a faculty member in a largely student organization I thought it better to stay in the background. The vote turned out to be a tie — something like twenty-three votes for each choice. The group decided to talk over the issues some more and revote at the next meeting.

I had noticed that Luther Ragin, whom I knew from the Holcombe campaign, had voted for the more moderate shareholder-lobbying alternative. As we left the meeting, I commented to him that I was sorry that we were on opposite sides of this question, and that led to a long discussion.

Luther's basic position was that we should be pragmatic. He did not like the tenor of the arguments people were making for the divestiture demand. They seemed to feel that it was the "morally pure" thing to do, and that Harvard should not have "tainted" money. He pointed out that human rights and other non-profit organizations were always getting donations that, if looked at carefully, could be considered "tainted." He joked that the only thing wrong with tainted money was that there wasn't enough of it going to worthwhile causes.

I made several arguments for the divestiture demand, one of which Luther had not heard before. I said that since SASC was the main anti-apartheid group on campus, it should take a strong position. If others wanted to adopt a weaker position, then fine; but that's not what SASC should do. A lobbying group, like a union, should start out with a demand that is more than they might realistically expect to get. That way, if they have to compromise later, they'll at least get something reasonable. If, on the other hand, the initial position of the group is already a compromise, then what they ultimately get will be even worse. My argument was a variant on the "Pat Koechlin principle" that Ann and I had been so impressed with after the meeting at Princeton about Admiral Moorer's visit.

This argument — which was pragmatic, not moralistic — made sense to Luther. He said that he'd talk with some of the others and see if SASC could reach a consensus. Luther was extremely influential among SASC people, who thought of him as a moderate who had extensive knowledge of the legal and political issues. At the next meeting the chair started out by saying that he thought there'd been a lot of movement on the question of which demand should be adopted, and perhaps the group could proceed quickly to a vote. The vote was almost unanimous in favor of the total divestiture demand.

Nationally and internationally, the movement for divestiture and corporate withdrawal was paying close attention to events at Harvard for a number of reasons. In the first place, the dedication of the Kennedy School library — at which Senator Edward Kennedy had spoken and had asked the audience to remain to listen to Mark Smith's speech protesting the Engelhard name — had received extensive outside media attention. In the second place, President Bok, who had been a professor of law and seemed to enjoy debating, emerged as one of the most prominent and articulate opponents of divestiture. In the third place, Harvard's policies on the investment issue were among the worst. At the other end were institutions such as the University of Wisconsin, whose Board of Regents in February 1978 had voted to sell all stocks in companies with subsidiary operations in South Africa. Many universities took a middle position, keeping their stocks but sponsoring shareholder resolutions asking for companies to withdraw. Harvard was at the conservative extreme — its policy was essentially to do nothing.

In the fourth place, in order to improve its public image and defend itself against mounting complaints that it was shilling for apartheid, in 1978-1979 Harvard arranged for several South African dissidents to spend time in Cambridge. Although these visitors were grateful to Harvard for the opportunity, every one of them made it clear either publicly or privately that they agreed with the stance of SASC and BSA and not with that of Bok.

Their presence on campus was partly responsible for the intensity of anti-apartheid activity at Harvard that year. The newspaper editor Donald Woods, who had been one of the best known white opponents of apartheid in South Africa before he had to flee the country and who had written a widely-read biography of Steven Biko, spoke on many occasions during his year at Harvard as a Niemann Fellow. Dennis Brutus, an exiled poet, visited; his son Tony was a Harvard student and member of SASC. Dr. Deborah Mabiletsa, director of women's affairs for the South African Council of Churches and a top aide to Bishop Desmond Tutu, visited Cambridge in late May. (Because students were in the middle of exams, Ann and I ended up spending a full afternoon with her and speaking with her at length.) A South African exile named Chris Nteta, who represented the radical anti-apartheid group AZAPO, spent much of his time in Cambridge and spoke at rallies and meetings.

Finally, at the end of the year Harvard gave an honorary degree to Bishop Tutu, who was widely regarded as the most important South African leader of opposition to apartheid who was not in either prison or exile. Ann and I were among a group of SASC members who got to talk with him for about an hour. Harvard officials at first tried to prevent anyone from the campus anti-apartheid movement from meeting with him, and when we persisted, they insisted that a university official be present during the meeting. Our main purpose was to explain to Bishop Tutu Harvard's role in undermining the efforts of American anti-apartheid activists to support economic sanctions. The Bishop himself was very circumspect in what he said, in part because he was being hosted by the Harvard administration and in part because public statements of support for sanctions were considered treasonous within South Africa. His words to us were encouraging and supportive, however, and at one point in reference to his own anti-apartheid activities he commented that it was his wife who stiffened his resolve and would never let him back off from a strong stand.

On April 6, Derek Bok released a six-page single-spaced manifesto titled *Reflections on Divestment of Stock: An Open Letter to the Harvard Community*. This document read like a lawyer's brief, replete with footnotes and nicely constructed syllogisms. For anyone who had no independent knowledge of the issues it would seem to be a logical and convincing line of argument by someone who was sincerely opposed to apartheid but just thought that the campus movement was mistaken in its approach. Interestingly, despite Bok's anti-apartheid posture, the South African government understood his value for them. The magazine *South African Digest* that they put out for foreign distribution contained an article in its May 18 issue that approvingly described in detail Bok's opposition to divestment and corporate withdrawal. Donald Woods wrote at the time, "I feel that President Bok has underestimated...the extent to which the apartheid regime, with its state-controlled radio and television, gloats when people like him oppose divestment."

Luther was adamant that it was important to prepare a rebuttal of Bok's *Reflections*, and I agreed. Working with several friends and fellow activists, we wrote *The ABCs of Divestiture* in the form of questions and answers. Supplemented with photos, sidebars, highlighted quotations, a collage of newspaper headlines, and a cartoon, the document ran to eight newspaper pages. (Ann and I paid to have it published by the *Crimson's* printers.) It would take me too far afield to summarize all of the arguments we made. Perhaps it suffices to give some excerpts from the list of questions and then from statements by prominent South African black leaders:

> Why South Africa? Aren't there other equally repressive regimes?... Why would withdrawal of U.S. corporations from South Africa contribute more to the defeat of apartheid than an effort on the part of American companies to improve the wages, employment opportunities and social conditions of nonwhite workers?... Wouldn't economic sanctions, if effective in weakening the South African economy, hurt blacks the most by causing large-scale unemployment and loss of valuable training opportunities?... Wouldn't sanctions against South Africa merely result in a hardening of the Afrikaaner position and increase the likelihood of bloodshed?... Isn't the cost of divestiture prohibitive?... Can't Harvard be sued for breaching its fiduciary responsibilities [if it divests]?...

Among the viewpoints we quoted was the statement by Steven Biko that

> Heavy investments in the South African economy, bilateral trade with South Africa...are amongst the sins of which America is accused.

All these activities relate to whites and their interests and serve to entrench the position of the minority regime.

And John Gaetsewe, General Secretary of the South African Congress of Trade Unions, put it this way:

> The ending of foreign investment in South Africa is, of course, a tactical question; it is a means of undermining the power of the apartheid regime. But it is of such importance that there can be no compromise whatsoever about it from our point of view. Foreign investment is a pillar of the whole system which maintains the virtual slavery of the black workers in South Africa.

One of the people who helped us write and edit was an old friend from graduate school named Steve Weintraub. He had taken part in the militant anti-war actions at Princeton in 1972, and in 1979 he was in the math department at Louisiana State University. The opportunities for political activism at LSU were minimal, and he seemed to enjoy helping us with *The ABCs of Divestiture* while he was visiting.

We had our rebuttal to Bok printed one evening at the *Crimson* presses. Ann came over directly from Boston University, where that very day she had taken her Qualifying Exams. She, Steve Weintraub, Luther, two typesetters (who clearly liked what we were doing), and I stood by the presses meticulously proofreading the long document after it was typeset.

The ABCs of Divestiture turned out to be a key organizing tool in the Harvard-Radcliffe anti-apartheid movement, which reached a crescendo in the late spring of 1979. One of our demonstrations drew some 3,000 people in what was probably the largest campus protest since the Vietnam War. That was followed by a "sit-out" at University Hall that had hundreds of students barricading the stairs and entrances to that symbolic building that SDS had siezed and occupied ten years earlier.

Just as in the Sherman Holcombe campaign, the junior faculty in the math department were strongly supportive of the anti-apartheid movement on campus. All ten of us plus four of the senior professors (Raoul Bott, Heisuke Hironaka, Barry Mazur, and John Tate) signed a petition calling on Harvard to sell its South Africa-related stocks. The progressive politics of the younger faculty wasn't surprising, as we were all in the protest generation of the 1960's. Of the other instructors, for example, Haynes Miller had been one of the graduate students at Princeton who was disciplined for the Nassau Hall sit-in and was arrested at IDA; David Rohrlich had been a Ph.D. student of Serge Lang, and, like his thesis adviser, had participated in anti-war activities; and Andrew Wiles, whose father was an Anglican theologian, came from an anti-apartheid family in Great Britain.

On May 1 there was a special meeting of the Harvard faculty devoted to the issue of South African investments. It was only the second faculty meeting I attended. Although most of the junior faculty in the math department were officially denoted "Benjamin Peirce Assistant Professors," in reality we were non-tenure-track instructors barely out of graduate school, so it seemed a bit of a stretch to call us Harvard faculty members. Most of us would have felt out of place at a faculty meeting.

The only other time I went to such a meeting was when my old friend Carl Offner was being readmitted. Recall that he had been convicted of assaulting one of the deans during the SDS building seizure in 1969, had spent three months in jail, and had been dismissed from the university. He became a middle school math teacher, and at the same time continued working on his Ph.D. Finally, about eight years later he had done enough research to finish his dissertation. The Harvard math department wanted him to get his degree and prevailed upon the administration to let him do so. However, Carl had been expelled in a strong sense that required a two-thirds vote of the entire faculty to be readmitted. I attended the meeting at which this was done, which turned out to be largely uneventful, although a conservative historian named Oscar Handlin insisted that President Bok read the detailed list of charges against Offner from 1969 before the faculty voted on readmission. For me what was ironical — and a bit surreal — was being able to vote as part of the faculty in support of Carl in the very same room in University Hall where he and I and a couple hundred other radical students had met to discuss tactics during the takeover of that building on April 9, 1969.

The meeting on South Africa also took place in the Faculty Room of University Hall. The Harvard radio station was permitted to broadcast the entire debate, and the *Crimson* carried the full text. I spoke in support of divestiture, but my presentation was boring and not at all eloquent. I mainly talked about the University of Wisconsin's decision to sell its stock, which was based on the state attorney general's interpretation of Wisconsin's laws against racial discrimination as grounds for severing the university's economic ties to South Africa.

One of the nicest speeches was given by an elderly professor of political science named Karl Deutsch:

> I've spent a lifetime in trying to estimate or gauge political and moral changes in world politics.... What is happening now is we are

present at a certain shift in moral sensibilities.... There is in part a change in age groups. The British writer C. P. Snow writes about his children and their friends that they're very permissive about the sex life of their contemporaries but — I quote him — would not be caught dead serving South African sherry.

I confess I have kept some of the older views of legitimacy. I believe that, let us say, fidelity in love is a major value and that change is not always, let us say, a good thing in these matters. But I do feel that there is a shift that we cannot reverse. Racism has become one of the major glaring offences against decency in our time.

Not all of the faculty members played a good role in the debate. Hilary Putnam asked for the meeting to be cut short because he had a cocktail party to attend. After his break with the left, he had become a crude opponent of campus activism — much in the tradition of certain ex-communists and ex-socialists in the 1950's who jumped on the Cold War anti-communist bandwagon. Thinking about Putnam's objection to spending faculty time on South Africa, I recalled an occasion in 1972 when Ann and I had come to Harvard for a meeting of SDS (which by then was little more than a group of people close to the Progressive Labor Party). Putnam, who was still working with the PLP, tried to convince us to donate all our money to the SDS campaign against racism. We had accumulated roughly $2000 in our bank account, and he thought that we were being very bourgeois to want that level of financial security. Fortunately, we weren't persuaded by his arguments. Then I ran into him again soon after I arrived at Harvard in 1975. He asked me if I was still in the PLP, and I said that I wasn't, at which point he suggested we meet for lunch. Over lunch I told him that I had left the organization for personal reasons, had not had any political conversion to anti-communism, and felt no animosity toward my former comrades. Putnam then lost interest in talking with me, and we parted ways.

At the end of the faculty meeting I went up to talk briefly with President Bok about a minor point related to the legalities of divestiture. As we walked across Harvard Yard, a *Crimson* photographer snapped our picture. The front-page photo showed Bok scowling as I talked to him. Soon afterwards I received a phone call from someone in his office who explained that the expression on his face was a reaction to the reporters and photographer, not to me, and that they hoped I hadn't taken offense at the picture. I said that I knew that, my discussion with him had been cordial, and I had not been bothered by the photograph in the *Crimson*. In actuality, I was pleased that the photo made me look like a thorn in Bok's side, since, if anything, that image enhanced my credibility in SASC. I would not have wanted other activists to think that I was cozying up to Bok.

Several members of my Harvard class of 1969 were in the Cambridge area and involved in the anti-apartheid movement. We decided to bring up the South Africa issue at the class's Tenth Reunion in June. We contacted Joe Mullin, who was in charge of organizing activities, and he was extremely supportive and worked hard to implement our proposal. There was some irony here, since in 1969 he had been a fervent opponent of SDS. But in 1979 the radicals in our class were pleased with the job he did.

We originally proposed a debate between a SASC representative and Derek Bok. Mullin contacted Bok, who was unwilling to debate. However, a member of the Harvard Corporation named Hugh Calkins agreed to represent the university position. We remembered Calkins from 1969, when he had been the only member of the Corporation to speak publicly on campus at the time of the Strike. He was a corporate lawyer in Cleveland, and, like Bok, he fancied himself a good debater.

We asked Mark Smith to present the pro-divestiture argument. To prepare for the debate, Ann and I invited Luther and Mark to our apartment, and after dinner Luther and I peppered Mark with anti-divestiture arguments. Any time he wasn't sure how to give a forceful response to some claim that Calkins might conceivably make, we talked over with him the possible ways to rebut it. As it turned out, Luther and I knew the anti-divestiture arguments better than Calkins, who later made a weaker case for his position than we had made in our session with Mark Smith.

Mark was ready for the debate. For several months he had been a leader of the local movement; he had been the speaker at the protest at the Kennedy School the previous October; he and other SASC members had accumulated a file of newspaper articles and public statements by important figures in the international debate on the role of multinational corporations in apartheid; and Luther and I had been able to simulate a Hugh Calkins who was a more formidable debater than the real one turned out to be.

At the Tenth Reunion we had displays and literature tables with information on South Africa, as well as a "pledge" that we were asking people to sign committing themselves not to donate money to Harvard until it divested.

About 200 people, mainly alumni from our class, attended the debate. It soon became clear that Calkins was not well prepared and thought that he could speak in generalities and platitudes. When he had learned that he would be debating a black student leader, he apparently had assumed that Mark would just mouth slogans and give vent to emotions. The stereotype that he must have had was that Mark would be incapable of thoughtful, analytical arguments. But Mark had done his homework; he presented the pro-divestiture position thoroughly and convincingly, and replied effectively to everything Calkins said. For example, when Calkins praised the role of U.S. corporations in South Africa, Mark pulled out a newspaper article about the labor troubles of the Ford Motor Company there.

As the debate went on, Calkins became nervous and agitated. We could see his knees shaking and knocking together under the table. The audience was solidly on Mark's side, as was evident from the applause when he spoke and the hisses that were directed against Calkins. The first comment in the question period was from my classmate Phil Aranow, who said, "Mr. Calkins, I remember you from 1969, and you haven't changed one bit. You're feeding us the same stupid public-relations crap as you did then." I don't think I've ever seen such a lopsided debate — or such a one-sided audience response.

Part of the reason for the politically skewed audience was that the reunion organizers had scheduled an alternative activity at the same time — a panel discussion on "lifestyle choices" — for the less political people. Interestingly, *The New York Times* carried an article about our Tenth Reunion with the theme that the 1960's radicals had mellowed and become more concerned with personal than with political issues. They covered the "lifestyle choices" panel and ignored the South Africa debate. I suppose that the *Times* felt that a discussion among affluent 30-somethings about self-fulfillment was more newsworthy than a debate about corporate policy toward South Africa. Referring to the motto on the *Times* masthead, Ann and I derisively commented at the time, "All the news that fits, we print."

We also invited Senator Paul Tsongas, who was one of the few political figures in Washington who favored corporate withdrawal from South Africa, to meet with our class during the reunion. His background had included a couple of years in the Peace Corps in Africa, and that explained his strong anti-apartheid stand.

Shortly afterwards, I received a nice note from Joe Mullin. He appreciated the work we had put into bringing some important and unusual events to Harvard during the alumni festivities that June — it had been unlike any other Tenth Reunion that Harvard had ever had.

During the summer before we left Cambridge for Seattle, Ann and I participated in one more activity of the anti-apartheid movement. Luther Ragin had been following the public statements of Rev. Leon Sullivan with interest. On the one hand, activists widely regarded him as a sellout: the token black member of the Board of Directors of General Motors, he was instrumental in providing a plausible cover for corporations that wanted to continue operating in South Africa. His "Sullivan Principles" at first sounded like a reasonable set of guidelines for companies to follow in their treatment of workers. However, in practice they were full of loopholes and had no enforcement mechanism anyway. In addition, most U.S. companies (with a few exceptions, such as Ford) were capital-intensive rather than labor-intensive. So even if, hypothetically, they decided to treat their South African black workers far better than the norm under apartheid, relatively few workers would have been affected. Rather, the main effect of the continued involvement in South Africa of the signatories to the Sullivan Principles was to provide economic support and stability to the apartheid regime.

What Luther started to notice in May was that Sullivan appeared to be increasingly uncomfortable with the way his Principles were being used, and seemed to be contemplating a toughening of his stand. Luther proposed that a group of anti-apartheid activists from the Boston area meet with Sullivan with two purposes in mind. One was to learn what he might be planning. The other was to lobby him — to be sure that he heard an articulate and thorough presentation of the anti-apartheid movement's position on U.S. corporate involvement in South Africa.

Luther contacted Sullivan and arranged a meeting in July. Some of our group was dispersed during the summer, so we had a rendezvous in New Jersey at Ann's parents' house, where we had dinner and spent the night before driving to Philadelphia. I had flown down from Kingston, Ontario, where I had been attending a number theory conference. There were six or seven of us in all, including Luther, Tony Brutus, Ann and me.

The meeting was extremely interesting, but also frustrating. Sullivan was nervous, defensive, and evasive, and he was vague about any changes in his position that he might be considering. The only specific new information we got from him was that he was contemplating calling for universities and other institutions to divest their stock in companies that remained in South Africa without adopting the Sullivan Principles. (But I believe that he never did this.) He had an assistant named Roman present at the meeting, and that guy was a bit creepy. He remarked to Luther, "You know, in South Africa you wouldn't be considered black." Roman was dark-skinned, and he was alluding to the fact that under South Africa's racist classification system most American blacks (including both Luther and Sullivan, but not Roman) would have been labeled "colored," meaning of mixed-race ancestry, rather than black. What Roman thought that this had to do with anyone's credibility on the corporate withdrawal issue was obscure; his remark struck us as a particularly clumsy attempt to play some kind of race card.

It is hard to say whether anything useful was accomplished by the meeting, except maybe to pressure Sullivan a little. It was certainly one of the more interesting encounters that Ann and I participated in during the late 1970's.

Unfortunately, back in Cambridge a group of SASC members decided that the meeting with Sullivan had been a bad idea, and they chastised Luther for it. They accused him of not going through the proper process within SASC to have it approved. Luther was incensed. The meeting hadn't even been billed as a SASC activity, and everyone who went spoke as individuals. Moreover, whether or not anything worthwhile resulted, there was no way that any harm could have been done. Luther rightly saw that the criticism masked an attitude of resentment toward his initiative and leadership. I was reminded of what I had observed many years before in Grenada, Mississippi, where the white civil rights workers often tried to undercut the political authority of the local black leaders. Luther resigned from SASC; I also had no more active involvement with the organization — Ann and I were about to leave Cambridge anyway, since I had to be in Seattle in early September.

During the summer of 1979 I noticed an unhealthy trend among the SASC people. They formed a study group, lost interest in making tactical plans for the coming year, and seemed to be depressed about how things had gone in 1978-1979. They had put in a lot of effort and were disappointed that they had not met with unqualified success. I wondered if they could have seriously expected Harvard to have simply divested in response to their demands. Even in 1969, when the scale of the protests had been much larger — when several hundred Harvard students had had a direct confrontation with the deans and the police — we hadn't completely accomplished our goal to "smash ROTC, no expansion." ROTC left campus, but probably the gentrification of Cambridge was not significantly slowed by our protests.

I thought that SASC had been surprisingly successful, considering that, compared to the Harvard-Radcliffe SDS chapter of a decade earlier, it was a relatively small group (having a few dozen members, compared to 300 dues-paying SDS members in my senior year at Harvard). SASC hadn't even done anything truly militant during the year — certainly no one had been arrested or had faced university discipline — and yet we had had an impact. Our demonstrations had been covered in the newspapers and news magazines. South African black activists would not have gotten a comparable amount of publicity in the Western media unless dozens of them had been shot. We had a tremendous privilege in this respect because we were at Harvard, and I didn't see any reason to feel sorry for ourselves. I thought that the self-pitying attitudes of many SASC people were immature.

In addition, some of the most prominent South African anti-apartheid people had been closely following and taking heart from events at Harvard; and at the end of the year, largely in response to the campus movement, the university had conferred an honorary doctorate on Bishop Desmond Tutu. In 1979, which was five years before Tutu received the Nobel Peace Prize, he very much needed that type of recognition as a form of partial protection against government reprisals. Finally, Harvard modified its stand on South African investments somewhat — but only in the sense of participating in shareholder resolutions related to apartheid. This was far from divestiture, but some of the SASC people who were most upset that Harvard hadn't divested were the same ones who had initially been opposed to adopting the total divestiture demand because they thought it was too radical and unachievable.

Looking at it more broadly, the campus movement at Harvard and elsewhere built public awareness that ultimately led to the Anti-Apartheid Act of 1986, in which the U.S. Congress imposed limited economic sanctions over President Reagan's veto.

And for Ann and me personally, the anti-apartheid activities had been a tremendous learning experience. We had met and worked with some remarkable people — Bishop Tutu, Deborah Mabiletsa, Donald Woods, Mark Smith, and Luther. Seventeen years later — two years after the fall of apartheid and the election of Nelson Mandela as President of South Africa — we made our first trip to South Africa. Our visit to Cape Town was hosted by two colleagues at the University of the Western Cape — Renfrew Christie (who had been a political prisoner from 1979-1986) and Jan Persens — both of whom were veterans of the anti-apartheid resistance. When they took us to see the museums and university archives, we felt a special connection with the story that was told there.

PHOTO SECTION

3

1 My brother Donnie (two years old), my sister Ellen (five), me (going on seven), and my parents in India in late 1955.

2 Summer 1956 visit with the extended family of the Danish fisherman who found my uncle Alvin's wallet in the North Atlantic. I am third from the left on the ground in front.

3 Life magazine photo of the occupation of University Hall at Harvard on April 9, 1969. I am the only one wearing a coat and tie. (Reprinted with permission of the photographer, Timothy Carlson.)

4

4 With fellow PLP member Steve Wenger at the time of our court-martial, September 1970.

5 Dating Ann Hibner, 1971.

6 In our Moscow dormitory in May 1975. From the left: me, Volodya Berkovich, Galim Mustafin, Hà Huy Khoái, Anas Nasybullin.

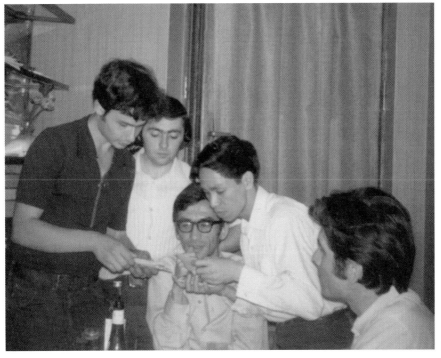

7

Советско-американский симпозиум по теории солитонов и Международная группа по нелинейным и турбулентным процессам в физике (Киев, 1979, сентябрь)

8

9

7 Talking to Harvard President Derek Bok following the faculty meeting on South Africa on May 1, 1979. (Photo from the front page of *The Harvard Crimson*, reprinted with the *Crimson*'s permission.)

8 Yuri Manin (left) with Mark Kac in September 1979.

9 Ann and I with Nguyễn Thị Đình at the Hanoi conference in January 1987. At the time she was President of the Vietnam Women's Union; twenty years earlier she had been General of the Army of the National Liberation Front of South Vietnam.

10

10 With Prime Minister Phạm Văn Đồng at the Presidential Palace in Hanoi, January 1987. He laughed when I mentioned that I supported Ann's views on not having children.

11 Ann and other Managua conference participants with Comandante Tomás Borge, August 1987.

12 Interviewing Hoàng Tụy in Hanoi, January 1989.

11

12

13

14

152 PHOTO SECTION

13 Along with Kovalevskaia Prize winner Rhyna Antonieta Toledo, I placed a wreath at the memorial to the martyrs of Central American University in El Salvador, March 1990.

14 Meeting with General Võ Nguyễn Giáp at the Women's Union headquarters in Hanoi, January 1991. His daughter, the physicist Võ Hồng Anh, is at the left.

15 With Tolya Fomenko on the Seattle waterfront, early 1990's.

16

16 With Mike Fellows in a third-grade class in Lima, Peru, June 1992.

17 Checking two children's solution to a graph theory problem in Cuzco, Peru, March 1993.

18 Ann helping a table of seventh-graders during a math enrichment lesson in Harare, Zimbabwe, September 1993.

17

18

19

20

19 With Ann and Sister Ellen Cunningham at Saint Mary-of-the-Woods College on the occasion of Ann's honorary degree, May 1995.

20 Vietnam Women's Union President Trương Mỹ Hoa pins the Women's Emancipation Medal on my jacket, August 1995.

21 A geometry lesson with geoboards in El Salvador, June 1996. Nora Vanegas is at the back.

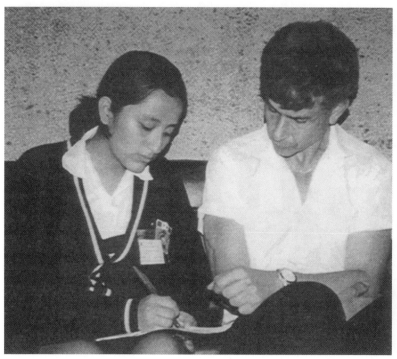

22

22 Liz Vivas explaining to me her solution to an Iberoamerican Mathematical Olympiad problem, June 1997 in Lima.

23 Discussions with Nguyễn Thị Bình in December 2002. She remained as chair of the Kovalevskaia Prize Committee even in the years 1992–2002 when she was Vice-President of Vietnam.

24 Trying to keep warm at the Great Wall near Beijing in December 2002.

23

24

25

26

27

25 Exploring Lake Powell near the Arizona–Utah border in March 2005.

26 The level of Lake Powell was over 100 feet (30 meters) below normal.

27 With Luther M. Ragin, Jr. on the Seattle-Bremerton ferry in May 2007.

28

28 With Alfred J. Menezes at the University of Waterloo in June 2007. Both of us are wearing our "I Love Chaldean Poetry" shirts in honor of the tenth anniversary of the short course on cryptography we gave at the MSRI in Berkeley.

CHAPTER 9 **VIETNAM PART I**

In the late 1960's, like most anti-war activists I viewed Vietnam almost as an abstraction. I admired the courage of the guerrillas and the fortitude of the people in standing up to the most powerful military machine that the world had ever seen. But I knew little about the country itself beyond what I had read in 1965 in the article by Kahin and Lewis in the *Bulletin of the Atomic Scientists*. To most of us in those years "Vietnam" was the name of a war, not a country.

That started to change when I was in graduate school at Princeton. During the year before I was drafted into the Army — this was 1969-1970 — someone had put up on the math department bulletin board a copy of a report by Alexander Grothendieck on his November 1967 visit to North Vietnam. In algebraic geometry, which was my field of study, Grothendieck was one of the giants of the 20th century; he was responsible for developing an abstract algebraic "machinery" that was powerful enough to handle the most complicated geometrical concepts and constructions. Like many French intellectuals, he was politically on the left and had long been an opponent of French colonialism and then American neocolonialism in Southeast Asia.

Grothendieck's report on his three-week visit, excerpts of which are given below in my translation from the French, was the first detailed account of mathematical life in Vietnam to reach the West. He lectured on abstract algebraic geometry for four hours a day and met with students and colleagues during the afternoons. Here is his description of lecturing in Hanoi during the bombing:

> Like most more or less public activities, the lectures were scheduled between about 6 and 10 a.m. because the bombing usually took place later in the day, rarely before 11 a.m. During most of my stay the sky was cloud-covered and consequently there were few bombing raids. The first serious bombardments had been anticipated; they took place on Friday 17 November, two days before we left for the countryside. Three times my talk was interrupted by alarms, during which we took refuge in shelters. Each alert lasted about ten minutes. Something which is at first very striking to the newcomer is the great calm,

almost indifference, with which the population reacts to the alarms, which have become a daily routine. I had the opportunity to observe many people during the alarms, both in the street and in the shelters, including children and old people, and I never encountered the least sign of nervousness among them.

It should be noted that things are extremely efficiently organized to reduce to a minimum the number of bombing victims: individual and group shelters everywhere in town, a very tight street-by-street and block-by-block organization of responsibility in case of an air raid, including first aid — a small red cross flag indicating the presence of a first aid station, which otherwise is carefully hidden beneath a protruding roof so as to avoid detection by enemy planes. One senses a great confidence in the populace — in the effectiveness of the air defense, for example — and a general interest in discussing the number of aircraft shot down (a topic of conversation which in North Vietnam seems to take the place of the weather) rather than the damage caused by the bombardments (about which, in any case, the radio tends to be rather discreet, for obvious reasons). As soon as the alarm is over, everyone (at least in the neighborhoods that were not hit) returns to their business as if nothing had happened.

During one of the air raids that Friday morning a delayed-action cluster bomb fell right in the courtyard of the Hanoi Polytechnic Institute, and (after the alert was over) it killed two mathematics instructors at the Institute. Tạ Quang Bửu, who is a mathematician as well as the Minister of Higher Education and Technology (and who attended the lectures that I gave while in Hanoi), was discreetly informed of this during the lecture. He left at once; the rest of the audience continued to follow the lecture while waiting for the next alert. The next day's lecture had to be rescheduled for the following week in the university in evacuation, so as not to have large groups of *cadres* in the city during the period of bombardment.

Interwoven among Grothendieck's descriptions of life in a war zone were comments about the organization of his lectures and other scientific activities:

It should be pointed out that for the past decade or so Vietnamese scientists have been in the process of creating a Vietnamese scientific language in its entirety — a task which, of course, is far from completed. (In mathematics, the first efforts in this direction go back to the mathematician Hoàng Xuân Hãn, who wrote the first French-Vietnamese mathematical dictionary in the 1940's.)

...The translator [of my lectures] at first varied according to the theme; but after a few days and by what seemed to be a general agreement on the part of the listeners, the choice devolved upon Đoàn Quỳnh, an instructor at the Pedagogical Institute and certainly one of the most competent and gifted mathematicians among our colleagues in North Vietnam.

The system of simultaneous translation seemed to me to work excellently, and on the whole suited equally well the lecturer and the audience. A sentence-by-sentence translation allows the speaker the luxury of collecting his ideas in an orderly way in the course of the lecture without an excessive effort at concentration, at the same time as it enables the listeners to follow at a pace which is more reasonable than that of an uninterrupted talk. Four hours talking at this pace (with two short breaks) seemed to me to be considerably less fatiguing than two hours at the usual pace. But it must be said that the interpreter's work is much more tiring, and at the end of my sojourn in North Vietnam I was in excellent form and well rested, while Quỳnh was visibly drained.

Notes were taken of all the lectures by Hoàng Xuân Sính, also of the Hanoi Pedagogical Institute, who is one of the few mathematicians (even more unusual, a woman mathematician) to have been educated in France (she received her first degree there in 1959).

Grothendieck made some comments about the scientific as well as practical difficulties that an aspiring mathematician has in such an isolated part of the world:

In a country which, by force of circumstance, has few relations with the outside (unless one counts the cluster bombs as a form of relations), it is particularly difficult for an inexperienced mathematician to orient himself among the multitude of possible directions, to distinguish what is interesting from what is not.

He explained that he was astonished to find an active community of research mathematicians in Hanoi:

The first statement to make — a rather extraordinary statement in view of the circumstances —is that *there is in fact a mathematical life worthy of the name in North Vietnam.* To properly appreciate this "existence theorem," first of all one must keep in mind that in 1954, after the eight-year war of liberation against French colonial occupation (i.e., thirteen years ago), higher education was practically nonexistent in North Vietnam. During the extremely brutal war of 1946-1954, the main effort in education was directed toward achieving literacy for the large masses of peasants, an effort which was carried through to its final goal in subsequent years, until about 1958, at which time illiteracy was practically eradicated in the lowlands.

... The method followed (undoubtedly the only one possible) was to send young people to universities in the socialist countries, especially the U.S.S.R. Among the hundred or so mathematics instructors at Hanoi University and the Pedagogical Institute, about thirty have gone abroad for four to six years of training. They have generally reached the level of a Soviet "Candidate's thesis," which, it seems to me, is slightly below the French degree (there is another, more demanding thesis requirement in order to be entitled to a university chair). This means that they have each published at least one or two original works, generally in a Soviet or East European journal. (In recent years they also publish directly in Vietnamese: in the packet of reprints I received when I departed, some were in Vietnamese.)

Nine days into Grothendieck's stay, because of intensified American attacks on the city, university classes were evacuated to the countryside. Here is his description of the conditions there:

Life is very primitive. Everyone — university administrators, teaching staff, and students — live in the same type of straw huts made of bamboo with mud walls, windows open to the wind, and the sun baking the earth. Some of them live with the peasants and others in communal dwellings, which they usually build themselves. Since there is no electric lighting, they use kerosene lamps; nor is there running water in the homes, so they take water from a well. As is the case in the populace as a whole, very few of the instructors live with their families: the husband works in one region, the wife and children are in evacuation in another, or else she works and the children are entrusted to relatives living in a third location. The family gets together when circumstances permit, perhaps one day a month, from which one must usually subtract about ten hours for the journey (by bicycle, of course). The trip is made chiefly at night, to avoid being strafed. Since the roads are continually being destroyed and rebuilt, the best form of transportation for a single person is a bicycle, which one can easily carry on one's back to detour around the rubble where the road is torn up. In both the village and the city one lives with the constant possibility of an air attack. Very often when the weather is clear enemy planes fly over the university, occasionally dropping their bombs — haphazardly, so as to get rid of them before returning to base — sometimes wounding or killing some civilians. In the month before my arrival two peasant children had been killed in this way.

One of the villages sheltering the evacuated university and one housing the Pedagogical Institute have thus far not been subjected to regular air attacks. Moreover, as everywhere else, a "self-defense" unit has been formed among the instructors in order to return fire in the event of an air attack. Everyone is required to wear a special hat for protection against fragments from cluster bombs; however, because of the relative calm in the countryside, the safety precautions are not always rigorously observed.

Next to almost all of the huts there are family bomb shelters, dug into the ground with a bamboo roof concealed under dirt; these are very effective against the projectiles and blast of a bomb. Special precautions are taken for lecture and meeting halls, as well as for children's classrooms. They have systems of trenches, usually extending from inside the room, which are hidden from the outside and allow a rapid evacuation of the room without detection by enemy aircraft. Generally the trenches run right next to the benches on both sides of the room, so that everyone can take shelter instantaneously in case of attack. The rooms are most often half buried in the ground, with the above ground part of the mud walls reenforced by a layer of dried earth about one meter thick to protect against bomb blasts. The part that remains vulnerable is the roof, which easily gives way to the blasts, and especially to the fragmentation bombs, which generally explode at a height of several meters in order to strike the populace with greater efficiency.

The problem of scientific equipment, a simple problem for mathematicians, gives rise to a multitude of difficulties for our colleagues in other departments. However, I saw a chemistry laboratory in action, with about twenty students engaged in practical experiments by the light of a kerosene lamp (which had been greatly modified so as to have the intensity of a powerful electric lightbulb). The chairman of the chemistry department, Nguyễn Hoàn, took me to his laboratory for me to admire the running water, which was stored in the gas tank of an American airplane that had been shot down nearby (this tank was carefully hidden from view by an overhanging bamboo roof).

His students took turns at "pump duty," refilling the tank by means of a hand pump from water coming from a reservoir farther down which was fed by a spring. In case of necessity, in the laboratories they could also obtain electricity from a gas motor.

Finally, Grothendieck concluded on an optimistic note:

> I can attest that both the political leaders and the senior academic people are convinced that scientific research — including theoretical research having no immediate practical applications — is not a luxury, and that it is necessary to promote theoretical scientific research (as well as the development of instruction and the applied sciences) starting now, without waiting for a better future.
> ...And through an effort undoubtedly without precedent in history, in spite of everything they are succeeding in increasing the cultural and professional level of their citizens, even as their country is to a great extent being devastated by the largest industrial power in the world. They know that, once the war ends, there will be people with the professional and moral qualities needed to reconstruct the country... They have confidence in themselves, and that is the best reason for us to have confidence in them and in their struggle on all fronts, cultural as well as economic and military.

Grothendieck's report caused me for the first time to think of Vietnam as a real country with people like myself who were studying and doing research in mathematics. I started to form the idea of some day visiting and working with colleagues there.

The first mathematician I met who had direct ties to Vietnam was the French-Vietnamese algebraic geometer Lê Dũng Tráng. He needed help with two projects to assist mathematicians in Vietnam — collecting books and journals, and raising funds for a delegation to attend the International Congress of Mathematicians (ICM) in Vancouver in 1974. I set up a table and solicited donations in the Princeton math department common room. When Tráng came to Princeton to speak at the algebra seminar, I told him that I'd like to visit North Vietnam after I got my Ph.D.

In August 1974, I attended the ICM in Vancouver, where I met the two Vietnamese mathematicians who had been able to come. One was Lê Văn Thiêm, who was the founder of the modern institutions of Vietnamese mathematics, particularly the Hanoi Mathematical Institute. The other was Hoàng Xuân Sính, who was the woman whom Grothendieck had spoken about in his report. Both Thiêm and Sính had received their doctorates in France. They encouraged me in my desire to visit Vietnam, but couldn't give me any assurances. We had some trouble communicating, since their European language was French and my spoken knowledge of that language was poor. Some of the time Chandler Davis of the University of Toronto, who a few years before had been the first North American mathematician to visit Hanoi, was with us and would translate between French and English.

Immediately after the ICM in Vancouver, Ann and I left for a year in Moscow. Our plan was to try to arrange a visit to Vietnam at the end of that year, taking advantage of the proximity of a Vietnamese embassy. We carried with us a "letter of recommendation" from Princeton Professor Richard Falk to the Vietnamese ambassador to the Soviet Union. Falk, the author of an important book detailing the violations of international law by the U.S. in Vietnam, had been part of a peace delegation that accompanied three repatriated American pilots back to the U.S. from Vietnam in 1972. He had met the Vietnamese ambassador when the group passed through Moscow. In his letter he recalled the meeting and told the ambassador that Ann and I "were active leaders in the American peace movement and were very effective here at Princeton."

Before I left for Moscow, Lê Dũng Tráng told me that a Vietnamese student named Hà Huy Khoái had just arrived in the Soviet Union to study number theory under Manin's direction. He requested that I find out from Manin how he was doing. Shortly after getting to Moscow, I asked Manin, who said that he was not able to communicate with Khoái because he could not yet speak Russian. Like other foreign students, Khoái had to spend a preliminary year getting acclimatized — mainly, this meant learning Russian — before beginning his graduate studies. Manin said that his students Volodya Berkovich and Anas Nasybullin, whom Ann and I had met the previous year, could introduce me to Khoái.

Although I met Khoái in the autumn of 1974, at first he was shy and, even though his Russian was rapidly improving, not very sociable. I later learned that in those days Vietnamese students were bound by strict rules governing their conduct when abroad. They were not supposed to have extensive informal ties even with their Soviet peers, let alone with Westerners.

There appear to have been two reasons for the restrictions on Vietnamese students. One was the insularity and paranoia in Vietnam that came from three decades of a war for survival against the French and then the Americans. The other explanation was the extreme poverty of Vietnam at that time — not only relative to the U.S. and Western Europe, but even relative to the Soviet Union. The Vietnamese authorities did not want the young people to become too accustomed to the student lifestyle in Moscow; did not want them, for example, to get married and remain abroad; and certainly did not want them to get drawn into the black market underworld.

Volodya, Anas, and I organized an informal seminar just among the students of Manin. When I learned that Khoái did not feel free to come unless it was part of his required program, I explained the situation to Manin, who promptly "ordered" Khoái to attend the seminar, which he happily did. Khoái still would not have felt comfortable coming to the room of an American couple, so we held the seminar in Zone B, where Volodya and Anas lived. Of course, if it weren't for the unnatural situation with Khoái, it would have made more sense to have met in our place in Zone V, where there would have been room to spread out and where Ann would have fixed us a wonderful meal.

In mid-April of 1975, I ran into Khoái at the university and commented to him that the liberation forces seemed to be making rapid progress in South Vietnam, and that victory should come soon. Khoái said that the tone of news broadcasts from Vietnam had become more optimistic, but he and most others were skeptical. After so many years, it was hard for the Vietnamese to believe that final victory was near. Saigon would be a difficult city to take militarily, Khoái told me, and there would be a protracted seige. I said that I didn't think so — at the U.S. Embassy I had seen the latest *Newsweek*, which was saying that there wouldn't be anyone who would want to defend Saigon, and the fall of the city would come in days, not months.

Just before 8 a.m. on Wednesday, April 30, 1975, the last of eighty-one helicopters left the American Embassy compound in Saigon, evacuating American personnel and South Vietnamese officials to aircraft carriers in the South China Sea. The Vietnam War — and the 21-year partition of Vietnam — was over. The next time I saw Khoái, I congratulated him on the victory and said that Ann and I were planning a party to celebrate both the Soviet holiday on May 10 marking the thirtieth anniversary of the defeat of Nazi Germany and also the defeat of U.S. imperialism by the Vietnamese. Under the circumstances, Khoái decided that he would be able to get permission to come, and he did. That was the only time in 1974-1975 that he came to our room.

For the Soviet graduate students our Victory Day party was, like all holidays, an excuse to get together with friends and have fun. And there was plenty of food, drink, and jokes. I took photographs that show our diverse group — coming from both Asian and European republics of the U.S.S.R., as well as from the U.S. and Vietnam — in animated conversations. All of us look very young. At one point in the evening I asked everyone to pause to honor the occasion, and I proposed a toast to those who have struggled, whether in Europe or in Asia, against fascism, racism, and superpower chauvinism.

Meanwhile, we hadn't heard anything in response to our request for a visa to visit Vietnam. We knew that the chances were slim, especially so soon after the end of the War. Finally I received a letter dated May 17 from Lê Văn Thiêm in which he explained that they wouldn't be able to accommodate us that summer. He concluded by asking me to "please let us know your desire to visit our country some other time and we hope we can welcome you then." So in the summer of 1975 the closest Ann and I got to Vietnam was a four-day trip to Vientiane, Laos.

When we next went to the Soviet Union in early 1978, things were different. In the first place, Khoái came to see us and immediately told us that there were no longer any restrictions on whom he could socialize with, and where. In the second place, we were hopeful that we would be able to visit Vietnam after our semester stay in Moscow.

When Khoái came to our apartment on Gubkin Street, we asked him to teach us some Vietnamese. He was from an illustrious family: his father had been a professor of literature, and his uncle had been a founding member of the Indochinese Communist Party. Khoái himself had won a prize in a literature competition in school before he decided to dedicate himself to mathematics. So what he taught us was not the usual Berlitz "can you tell me where's the bathroom?" type travel phrases, but rather some of the polite and literate conversation openers that sophisticated Vietnamese would use.

Eventually we learned some of the basics of the Vietnamese language and a rudimentary vocabulary of perhaps a couple hundred words. Ann was able to speak and understand a little, but I was hopeless, largely because Vietnamese is a tonal language and I'm tone-deaf. I simply could not hear the difference between certain of the six tones. (By way of comparison, Mandarin Chinese has only four tones.)

At one point Khoái was drilling me on a particular phrase that I repeated after him again and again. Finally he said, "Okay, that's very good. It can even be understood." Ann and I burst out laughing — he hadn't at all meant to be sarcastic. It's just that if you get the tones wrong, someone in Vietnam will simply not understand you. Well, that's not entirely true. I later found that I could be understood at the marketplace when I pointed to something and asked, *Giá bao nhiêu?* No matter how badly I mangled the pronunciation, the sellers somehow figured out what I was asking.

I wanted to start my lectures in Hanoi with a sentence in Vietnamese that paid tribute to the country's mathematical tradition. I wrote a long, elaborate sentence in Russian that Khoái translated — I'm sure into a beautiful literary vernacular. I memorized it and rehearsed it again and again. When I eventually delivered my brief speech in Vietnamese at the Hanoi Math Institute, most likely my pronunciation was so bad that no one except Khoái could figure out what I thought I was saying. But the mathematicians graciously applauded my effort anyway.

After Khoái left Moscow, his friend Nguyễn Đình Xuyên, who was studying geology, started visiting us in Zone V and teaching us a little Vietnamese. The language lessons were valuable not just as a way of picking up a few words and phrases. They led to interesting conversations and cultural insights, and so helped prepare us for our later visits.

We never learned enough Vietnamese to carry on a conversation in that language, so we talked with Khoái, Xuyên, and others in Russian. The fact that we were all foreigners in Moscow and were communicating in a third language meant that we had a common bond. For many years when we visited Hanoi our main language with our colleagues was Russian. We found that for them, just as for us, Russian was a language that was associated with student days, a language of jokes and stories, late-night parties, and shared experiences.

As the semester drew to an end, we still had not received a reply to our visa request. The people at the Vietnamese Embassy told us not to lose hope, since permission often would come at the last minute. But we had to leave at least a couple of days to buy air tickets and make arrangements either to go to a conference in France (if our request was denied) or else to Vietnam. Just about two days before we had to leave, I decided that it would be our last chance. As Ann, Volodya Berkovich and I were on our way to see the Tarkovski film *Andrei Rublyov*, I made one last call to the Vietnamese Embassy from a pay phone. As soon as I gave my name they said, "Yes, yes, permission has come! Come right in and receive your visas." I was ecstatic — and, thinking about our upcoming trip, I didn't pay much attention to the movie.

On June 28, 1978, Ann and I arrived in Hanoi on the twice-weekly Aeroflot flight from Moscow with a refueling stop in Bombay. The plane circled Nội Bài airport a couple of times before landing. The problem was not air traffic — Hanoi's international airport handled just one or two flights per day. Rather, the runway was far short of regulation length for large jets, and the pilot had to align the plane with great care. From the air, bomb craters, now full of water, were still visible near the runway and at various locations around Hanoi.

Because of a miscommunication, we were not expected and no one came to meet us. The guy who seemed to be in charge of the ramshackle little airport saw us waiting and looked for someone to take us into town so that he could close up for the day. The other foreigners had already been met by their hosts, and the Vietnamese students on the plane had all ridden into town in the back of an old truck. Finally, a representative from the Ministry of Light Industry, whose Swiss colleagues had failed to come on the flight, agreed to take us and put us up for the night in the hotel room that he'd reserved for the Swiss. After waiting for his driver to change a tire on his car, it was another two hours into Hanoi with delays caused by road construction and by rush-hour congestion at the one-lane bridge into the city.

All I had was the address of the Math Institute. The next morning I changed a few dollars into đồngs, bought a map of Hanoi, and set out walking in the direction of Đội Cấn, which turned out to be three or four kilometers from our hotel. After passing the Hồ Chí Minh mausoleum and the lotus-shaped One Column Pagoda, I headed into a part of the city where the wide avenues, French colonial mansions, and embassies and government buildings gave way to narrow residential streets full of potholes, bicycles, chickens, fruit- and vegetable-sellers, and children. After about a kilometer I turned into an alley at 208-Đ and found a group of one-story dilapidated stone buildings — more like sheds with their dirt floors — around a dusty courtyard. That was the Hanoi Mathematical Institute.

To say that I was not expected would be an understatement. The telegram I had sent a few days before had been either misdelivered or misplaced. The Institute director, Lê Văn Thiêm, was out of town, and Khoái wasn't there either. They quickly found someone who spoke Russian, and I explained who I was and where we were staying. They said not to worry, they'd find Khoái, and in the meantime Ann and I should wait in the hotel. By late morning the summer heat and humidity were oppressive, and they got someone to take me to the hotel on the back of his bicycle so that I wouldn't have to walk.

A few hours later Khoái came to our hotel, along with an applied mathematician by the name of Vương Ngọc Châu, who was the Institute's personnel director, general manager, and Communist Party representative. He was a dedicated, hard-working administrator, and seemed to specialize in handling awkward and impossible situations, such as knowing what to do when an American couple unexpectedly show up on one's doorstep for a three-week visit.

First, they moved us to the Thắng Lợi (meaning "victory"), which was located on the outskirts of town beside the Tây Hồ ("West Lake"). Although the setting was beautiful — rattan fishing boats, lush vegetation, fishermen and their homes by the lake, mountains in the distance — it was isolated from the rest of the city. Moreover, guards were posted to keep townspeople away — in this respect it reminded us of the compounds for foreign diplomats and journalists in Moscow.

One incident when we arrived left a bad impression. After Khoái and Châu got us checked in, we invited them up to our room to talk. They said no, they weren't allowed to go to the rooms; they'd let us settle in and then meet us later. We said good-bye and started to take our luggage toward the room. At that point the hotel personnel angrily shouted something to Khoái and Châu, and Khoái said that they wanted them to carry our luggage. Thus, our professional colleagues weren't allowed to come to our room as guests, but only as porters to drop off our things so that the delicate foreigners wouldn't have the burden. I was furious. What was the purpose of three decades of anti-colonial war if the result was to have the same mentality as in the French colonial days? Perhaps I overreacted slightly — I recall describing the incident in detail in a long letter to our geologist friend Xuyên back in Moscow (he had asked us to write him about our impressions).

Most of the people staying in the hotel were foreign businessmen or airlines personnel. One day when I was out lecturing and Ann was in the room, the maids came in and started talking with Ann in a mixture of Vietnamese, sign language, and a few words of English. Ann finally figured out that they were extremely curious about what I did, since my manner and dress were very different from that of the other men staying in the hotel. Ann knew the words *giáo sư toán học* ("math professor"). As soon as they heard that, the women clapped, hugged her, and congratulated her on having a husband in such an honored profession. Back home in the U.S., I would have had to be a famous actor or quarterback — not a mere math professor — to cause such a reaction.

I gave six three-hour lectures at the Math Institute on p-adic analysis, focusing on my recent work on Gauss sums. The talks were very technical and were not appropriate for the audience, of whom no one but Khoái was interested in this sort of number theory. He was able to dragoon about a half dozen others to politely listen to all my lectures, but I doubt that they benefited from them. Like many young mathematicians, I thought that I should always give the most advanced, high-brow lectures

I could. It wasn't until I got older that I learned to give broadly accessible talks that someone might actually learn something from. In addition, I was undoubtedly influenced by Grothendieck's report, which described the series of extremely abstruse lectures he had given on current developments in algebraic geometry. In 1980 the London Math Society published my Hanoi talks in their Lecture Note Series — it sold the fewest copies of any of my books and is the only one that was never either translated or reprinted.

Most of the time Ann and I were left by ourselves. In those days the Institute was not equipped to deal easily with guests. The director had part-time use of a car and driver, but he was away during the first part of our stay, and none of the people we knew had motor scooters, which did not become a common sight in Hanoi until about a decade later.

People who've visited Hanoi more recently see none of the primitive conditions and deprivation that we saw in the immediate post-war years. Only foreigners and VIP's had cars, and most of those were old American jeeps that had been jury-rigged to keep them going. Khoái told us that when Vietnamese diplomats went abroad, they would be issued a single suit, which they would return when they rotated back to Vietnam.

On the other hand, we saw no evidence of the extreme distress that one would find in most impoverished countries. There were no beggars on the streets (a small number started to appear in the late 1980's after the marketization reforms started), and there were no children with tiny limbs and distended stomachs from starvation. The level of hygiene was high for a poor country in the tropics: everyone boiled their water and recycled organic waste into fertilizer, and the city had a reasonable sewage system. Moreover, the large number of book stores, newspaper stands, and schools of various sorts all over the city attested to a high educational level among the populace. Despite the devastation of the War, which had ended only three years before, Vietnam compared well with other poor countries according to the most basic criteria — freedom from hunger, disease, illiteracy.

Ann and I spent a lot of time going on long walks around the neighborhoods. Once when we were hot and exhausted, we decided to take the tram, which would take us most of the way back to our hotel. The rickety old tram was very slow — it went roughly at the speed of someone walking — and it cost only 5 *xu*. Even at the official rate of exchange, which was at least an order of magnitude worse than what a free-market rate would have been, that was equivalent to only a penny-and-a-half. A few years later the tram lines were ripped up, and the tram — and also the *xu* — were phased out of existence. We should have been charged double because our butts occupied more space on the tiny benches than anybody else's. But the conductor decided not to charge us at all, perhaps because we provided much amusement to the other passengers as we ran to catch the tram and then tried to find a place to make ourselves comfortable inside without stepping on anyone's chickens or trays of vegetables.

In those days people tended to assume that the Westerners they saw must be Russians, and children would often call out *Chào các Liên-xô!* ("Hello Russians!"). Once when Ann and I were resting from the heat in a shaded park, a couple of children came up to us and showed off the Russian words they knew: *Eto mama! Pora!* ("That's Mom! Gotta go!") That must've been what the Soviet kids they played with would say when they had to leave.

After several days, at our request the Math Institute had us moved back to the Hotel Hòa Bình ("Peace") in the center of town where we had stayed the first night. Even though the Hòa Bình was a much less fancy hotel and had an abysmal restaurant, it had two big advantages for us — its location and its price. We had to watch our money carefully because we had had a mixup with air tickets, and because our American Express credit card could not be used in Vietnam.

The Hòa Bình was on a busy residential street. Like the Vietnamese, we would wake up with the sun at around 5 a.m. Usually the neighborhood would quiet down by about 9 p.m., but one night we were kept up until 11 p.m. because a group of children next door were singing *Guantanamera* over and over again. Perhaps they were rehearsing for a school event.

We noticed that at least two fist-sized, thick, hairy spiders were sharing our hotel room with us. Never having lived in the tropics, we found this disconcerting. I asked Khoái whether these spiders could bite, and he said that they were harmless, but there was a Vietnamese folk belief that it's bad luck if a black one falls on your eye.

When Lê Văn Thiêm returned to Hanoi, he used his car and driver to take us around town for one day. He and Khoái went with us first to Văn Miếu (Temple of Literature), founded in 1076. The remains of the ancient university feature an array of 82 large stone slabs, each containing a list of winners in the royal examinations that were held between 1442 and 1779 in order to determine who would become Doctors of Literature. The stele for 1463 includes the name of Lương Thế Vinh, who, in addition to his literary accomplishments, was a well-known geometer.

Thiêm and Khoái then took us to visit Đống Đa Hill, the location of a battle in 1789 that rid the capital of Chinese invaders for the last time. China conquered Vietnam in 111 B.C. and ruled for over a thousand years (until 938 A.D.). The Hai Bà Trưng (Two Trưng Sisters), who led insurrections against the Chinese in 40-43 A.D., are national heroes; a central street and a famous pagoda in Hanoi are named after them. Thiêm told us that, according to legend, Đồng Đa Hill was formed from the bones of the defeated Chinese soldiers. None of us would have expected that just seven months later China would again attack Vietnam, and thousands of Vietnamese and Chinese soldiers would die before China was pushed back across the northern border.

It was possible for foreigners to rent a car with a driver, and we did that one day. It was a little expensive, but there were a few places we particularly wanted to see that the Math Institute people had not shown us. We visited Bạch Mai Hospital and Khâm Thiên Street, both of which were destroyed by air attacks on December 26, 1972. The hospital had been completely reconstructed (thanks to Swedish aid), and the street had also been largely rebuilt. At one point on Khâm Thiên Street the remains of a bombed-out house were preserved as a memorial. The photographs there showed the utter devastation caused by the saturation bombing by the U.S. Air Force. The Christmas ceasefire was formally over, but most families had not yet dispersed, thinking that the next day would be soon enough. Early in the morning of the day after Christmas 283 people died on that single street.

When we returned the car, we had a dispute with the driver about some extra charge that should not have been there. The amount of money was not great — probably about ten dollars — but given our uncertainties about our tickets, we weren't going to pay it without a fight. The Russian-speaking young woman who worked in the rental agency tried to mediate. She had a hard time believing that we were Americans. It was not only our fluency in Russian but, more importantly, our tourist visas (which had been authorized by the foreign ministry at the special request of Lê Văn Thiêm) that seemed incongruous. In those years there was no such thing as an American tourist in Hanoi.

During the argument the driver claimed something related to the extra charge. Ann, who has an uncanny knack for interpolating what people are saying in a language where she knows only a few words, used her minimal Vietnamese vocabulary and her best guess to determine what the driver was saying. Without waiting for a translation, she answered in Russian, responding precisely to the driver's claim. At that point the woman gave in. This was just too weird — American tourists in a city that didn't have American tourists, knowing Russian, and now also understanding Vietnamese. We left without paying any extra charges.

As a general rule Americans did not visit Hanoi during the War and the years immediately after. There were some famous exceptions, such as

Henry Kissinger and Jane Fonda. Ann and I used to joke that we were the first *unimportant* Americans to visit Hanoi — and that made us feel important! (The reader might notice the similarity to the well-known proof by mathematical induction that there is no uninteresting number. Namely, let S denote the set of uninteresting natural numbers. If S is non-empty, it has a least element n. But being the smallest number in such a set is certainly an interesting property. Hence, n does not belong to S. This contradiction shows that S is empty.)

A major occupation while in Hanoi was trying to straighten out the mixup with our tickets. Our plan after Vietnam was to go first to Israel and visit our friend Ephraim Isaac, who was doing research in the Coptic Monastery in Jerusalem; then we'd visit Andrei and Betty Todorov in Bulgaria, and finally take the train through the Soviet Union to Helsinki in time for the International Congress of Mathematicians in August. Before we left Moscow, the Aeroflot office had sold us tickets on an Air France flight that supposedly went from Hanoi to Teheran before continuing to Paris; we then got El Al reservations from Teheran to Tel Aviv. The problem was that there was no such Air France flight; in fact, no Western airline flew into Hanoi at that time, and when we saw the conditions of the runway and airport, we knew why. For the socialist countries the air connection with Hanoi had strategic importance, and Aeroflot and Interflug (the East German airline) must have put their best pilots on the route. But the idea of Air France flying out of Hanoi clearly never got beyond the planning stage.

So we were stuck in Vietnam with useless tickets out of the country and little money to buy anything else. After it became clear that no one at the Math Institute knew where we could turn for help, I got the idea of stopping in at the Soviet commercial building, which, among other things, had administrative responsibility for Aeroflot's operations in Vietnam. After all, in some sense the whole mess was Aeroflot's fault. That building was in the center of town next to the Soviet Embassy.

There we met an amusing Russian guy who seemed to be the all-purpose trouble-shooter. He was jovial and laid-back, and wore a large bright tropical shirt over his big beer-belly. He reminded me somehow of Saul Bellow's *Henderson the Rain King*. When we explained the situation to him, he said that the Moscow people were crazy to think there was an Air France flight out of here. They just didn't realize what a remote outpost Vietnam was. Well, he said, he'd see if he could think of something.

The next day the Soviet commercial representative found us in our hotel and said that he had a solution. He could exchange our Air France tickets for tickets on an Interflug flight from Hanoi to Karachi that left three days later, followed by an Aeroflot flight from Karachi to Teheran that left six days after that. We did that, although we realized that there was no way we were going to wait around Karachi for the connecting flight almost a week later — to do so would mess up our plans for both Israel and Bulgaria. So in Karachi we ended up buying Pan Am tickets to Teheran; in the process we lost about $500, which was a lot of money for us in those days, all because of a non-existent Air France flight.

In Karachi I tried to get word to Ephraim Isaac that we had been delayed, and I also wanted to change our El Al reservations. But it was absolutely impossible to do either in Pakistan. No communication with Israel was permitted, and neither the Pan Am office nor the U.S. Embassy was willing to help. So when we arrived in Teheran at about 2 a.m. on Friday, July 21, all we could do was hope to get on the daily El Al flight standby.

That flight went without fail every day except Saturday, the Jewish sabbath (this continued until the Shah was overthrown a few months later, after which there was never again a direct air connection between Iran and Israel). It was a popular flight, not only because the Shah's regime had significant business ties with Israel, but also because for American travelers wanting to reach Israel from the east, rather than from Europe or North America, there were very few countries that had an air connection. So it wasn't at all clear that we could get on the flight standby.

The security was the most impressive I have ever seen before or since. A couple of hours before the incoming flight arrived, plainclothes men presumably from Savak (the Shah's security force) gathered in the check-in area. Six big tables were set up for examining luggage. When the El Al aircraft arrived, it parked at the opposite corner of the airfield and was surrounded by troops. The El Al personnel were brought in under guard, and proceeded to examine the checked luggage. All electronic devices were inspected by an El Al specialist. Other El Al security people questioned each passenger at length. The hand luggage was inspected later, just before boarding.

Finally they called the flight. When they came to our names on the standby list, there was only one seat left. We resigned ourselves to staying in Teheran for a couple of days and trying to be first on the standby list on Sunday or Monday. I went to another part of the airport to change money.

As Ann started to follow behind, she heard an El Al person on the other side of the partition tapping on the glass to get her attention. It turned out that one of the passengers had made a joke or remark about security and had been immediately removed from the flight. The El Al guy remembered that we were the next in line and signaled Ann that we should come through.

Ann motioned that we'd come and went running in my direction screaming "Neal! Neal!" Several Iranians then started running after her, trying to calm her down. Earlier that morning we had seen an instance when an Iranian woman, perhaps in public in such a strange place for the first time, had started screaming and had had to be quieted down; so the Iranians must've thought that Ann was having some kind of fit of hysteria. Fortunately, I was nearby, still in line to change money, and we went directly to the gate. Amusingly, in the rush the El Al people forgot to inspect our carry-ons, which included a big exercise bar that I was taking with me everywhere. This large metal object in its own case could have been a rifle for all El Al knew.

The last El Al person came with us in a van that sped across the airfield to the plane, which was being protected by the Shah's troops with machine guns drawn. We boarded the plane, they pulled away the stairs, and we were off. Despite all the drama and uncertainty, when all was said and done, we managed to get from Hanoi to Tel Aviv in less than 48 hours.

In August 1978 at the International Congress of Mathematicians in Helsinki I arranged a meeting between Hoàng Tụy, who was the senior member of the Vietnamese delegation (and soon after succeeded Lê Văn Thiêm as director of the Hanoi Math Institute), and Shiing-Shen Chern. Although Chern taught at Berkeley and was a U.S. citizen, even at that time he was known to be extremely influential in the Chinese mathematical world. My purpose in setting up the meeting was to try to improve relations between Chinese and Vietnamese mathematicians.

Before the Cultural Revolution, the Chinese had helped their Vietnamese colleagues considerably. For example, not belonging to international copyright conventions, they routinely photocopied Western math journals for their own use, and they sent a copy to Vietnam as well. However, contact largely stopped with the Cultural Revolution, and Sino-Vietnamese relations deteriorated during the 1970's. I thought that, despite the political tensions between the two neighboring countries (which erupted into war in 1979), it should be possible for mathematicians to get along.

The meeting between Hoàng Tụy and Shiing-Shen Chern was a friendly one, although I doubt that anything concrete resulted from it. Hoàng Tụy asked Chern about several Chinese mathematicians he had known in the 1950's and 1960's and had lost touch with during the Cultural Revolution. When he saw that his pronunciation of the names didn't ring a bell with Chern, he wrote the Chinese characters for their names, and immediately

Chern knew whom he was talking about. Vietnamese who have studied the Chinese characters, as Hoàng Tụy had, pronounce them in a way that is close to Cantonese, but very far from Mandarin.

In the 1980's, I contacted the editors of *Annals of Chinese Mathematics* and proposed an exchange with *Acta Mathematica Vietnamica* (AMV). I knew that at this time the Chinese government would not permit them to have direct relations with Vietnam. However, the journal exchange between Shanghai and Hanoi would be routed through Seattle, and the Chinese agreed to this plan. For many years I regularly received a copy of the Chinese *Annals*, which I passed on to the Vietnamese, and sent two copies of AMV (which was published half as frequently as the Chinese *Annals*) to Shanghai.

In 1978 Ann and I had a vague idea that we'd return to Vietnam some day, but we had no concrete plans. We were not collaborating with anyone in Vietnam, and in our frank moments we would have had to admit that little of benefit to Vietnam had come out of our first visit. I wrote an article about our trip for *The Mathematical Intelligencer*, a journal with a fairly large readership, and that presumably encouraged other mathematicians to think about developing ties with their colleagues in Vietnam. And Ann and I learned a lot and got some wonderful memories from the three weeks. But that was all.

Soon after we moved to Seattle in 1979, I received a letter from a physicist at Cal State Fullerton named Ed Cooperman. When working in France he had been impressed with scientists' activities in support of Vietnam and had decided to start a similar group in the United States, called the U.S. Committee for Scientific Cooperation with Vietnam (USCSCV). He had heard about me from Lê Dũng Tráng and wanted to know if I would head up the mathematics subcommittee of the USCSCV, which I agreed to do.

Between 1980 and 1984 what this meant in practice was that I advised Ed Cooperman on visits to the U.S. of Vietnamese mathematicians, and Ann and I hosted them in Seattle. I would introduce them to people in their field at the University of Washington, photocopy things from the math library for them, and talk with them about a range of subjects. We would rent a car (we didn't own one during our first ten years in Seattle) and take them to see some of our favorite attractions in Puget Sound. The mathematicians who came to Seattle in those years included Lê Văn Thiêm, Hoàng Tụy, Nguyễn Đình Trí (who headed the Polytechnic Institute in Hanoi), Nguyễn Văn Đạo (who later became chancellor of the Hanoi University system), and Phan Đình Diệu (the most prominent computer scientist in Vietnam). In addition, we helped with a visit to Seattle of two marine biologists from Nha Trang.

What seemed remarkable to a lot of people at the time was that throughout the Reagan years — a time of deep chill in relations with Vietnam — Cooperman was able to get U.S. visas for all these scientists. It turned out that the U.S. State Department sometimes liked to maintain small-scale low-visibility contacts with nations toward which the U.S. government was extremely hostile. As long as the scientific visits "flew under the radar," they would be allowed. Cooperman and his successor, Judith Ladinsky, maintained cordial and cooperative relations with the State Department's Vietnam Desk throughout the 1980's.

In 1983 when Ann and I were in Bangkok awaiting our visas for Vietnam, we met a Canadian woman from a Mennonite charity who was incredulous that we were allowed to bring Vietnamese scientists to the U.S. in apparent defiance of the embargo. She said that her organization had tried unsuccessfully to get permission to send badly needed soap and other supplies to Vietnam after the devastating typhoons of a few months before, and the shipment had gotten held up in the U.S.

One evening in 1981 a heavy-set man in a suit came to our door in Seattle, said he was from the FBI, and asked if he could talk with us. He wanted to know if any of our Vietnamese visitors had shown a special interest in anything or tried to get us to find out information for them. He said that he was concerned that people from countries such as Vietnam might spy for the Soviet Union or try to entice Americans into spying for them.

I politely explained that there was nothing the least bit suspicious about any of our guests, that the last thing they'd want to do was spy for anyone, and that I thought that it was out of line for the FBI to be inquiring about our scholarly contacts. Beyond the assurance that the visitors were above reproach, I wasn't going to talk to the FBI about the nature of conversations with them. However, I was happy to discuss the whole issue of whether or not the FBI should be questioning people about visiting scientists. Ann made tea for the guy, and we had a long discussion. I gave him a detailed argument explaining why he shouldn't be interfering with these visits. My tactic with the FBI was the opposite of the usual "don't say anything to them." Rather, I talked the guy's ear off.

At one point I told him that I had no connection to anything classified, and as far as I knew neither did my university. He asked me if I'd heard of the Applied Physics Lab on campus. At that time I hadn't, but apparently it was doing secret government research. I said that now that he'd mentioned it, that brought up one of the reasons why many professors had campaigned to prevent universities from engaging in classified research. Namely, the presence of that type of work on campus gave the FBI an excuse to go around questioning faculty about their foreign contacts. Well, this conversation (much of it more in the nature of a monologue) went on for over two hours. We were never again bothered by the FBI.

Ann and I met Ed Cooperman only once, when we were visiting Los Angeles in the spring of 1981. He took us to dinner in a Vietnamese restaurant that he said was one of the few where he would be served. Although the scientific exchanges had low visibility, some of Cooperman's other activities on behalf of Vietnam did not. For example, he organized a showing on his campus of films from Vietnam. Anti-communist Vietnamese refugees demonstrated angrily, and some threatened violence. There were many Vietnamese-American students at Fullerton, and for the most part Ed got along well with them. But he was *persona non grata* in "Little Saigon," and he had received death threats.

Although most Vietnamese refugees were anti-communist (or apolitical), there were a small number who either had been opponents of the South Vietnamese regime, had come to the U.S. before the War and been in the anti-war movement, or had been influenced by Cooperman or others who had a positive viewpoint toward socialist Vietnam. In the mid-1980's when Ann and I were in the Bay Area we sometimes stayed with a group we called the "Berkeley Vietnamese" who supported Cooperman's efforts. Another person we met who was in this category was the Vietnamese wife of Columbia University mathematician Pat Gallagher.

When I realized that Cooperman had funding to bring mathematicians to the U.S. and could get them visas, I started trying to arrange a visit by our friend Hà Huy Khoái. It was in 1978 in our apartment in Moscow when I'd first mentioned to Khoái the possibility of some day visiting the U.S. Ann said that that was a "stupid" thing to have said, since it seemed so unlikely, and Khoái laughed and said, "Yes, that really was stupid." Among other considerations, Khoái did not have any administrative status in Vietnam and was not a member of the Communist Party. He seemed to think that the Vietnamese authorities would never permit him to travel to the West. Undoubtedly Khoái was thinking by analogy with the Soviet Union, where the government tightly controlled exit visas for scientists.

However, in the case of Vietnam there were rarely any problems in getting permission for scientists to travel, provided that their trip was funded from the West. The Vietnamese government attached great importance to scientific ties with the technologically advanced countries. The potential problem in arranging Khoái's visit was rather at the U.S. end and arose because Khoái would be coming from Moscow, where he was studying for his Soviet doctorate (the "more demanding" degree referred to by Grothendieck that follows the Candidate's thesis) under Manin's direction. None of the other Vietnamese scientists to visit the U.S. had come from Moscow.

We set up Khoái's visit for April 1982. For most of the academic year 1981-1982 Ann was in Leningrad, but in the spring she went for a month to Moscow, where she was able to assist Khoái. Several times she went with him to the U.S. Embassy, translated for him when necessary (his English was still pretty shaky), and helped him understand the procedures. About three days before Khoái was scheduled to leave, he and Ann went to the U.S. Embassy again, but still nothing had come through. Ann called me from Khoái's apartment to say that there was no visa for him. I phoned Cooperman, who said he'd get back to me soon. After a while he called back to say that the person at the Vietnam Desk had just that moment sent a cable authorizing the U.S. Embassy in Moscow to give the visa. It was the middle of the night in Moscow. A few hours later, I called and woke Khoái up, said that he must go wake up Ann and return to the Embassy, must not take no for an answer, and must insist that they look for their latest cables. He later told me that he hadn't understood me well and had thought that I was raving; the version that he relayed to Ann made it sound as if I was saying that they should go back and demand a visa whether or not the authorization had come. This made no sense. But they went anyway, and the authorization was there.

Once Khoái had his U.S. visa, a number of urgent tasks remained before he could leave Moscow. Foremost was getting his ticket from Moscow to Seattle on Finnair. The people at Aeroflot were so impressed that a Vietnamese scientist had prepaid tickets from the West that they addressed him as *gospodin* ("mister" or "sir"), which he had never been called before (the Soviets always used the term *tovarishch* — "comrade" — with the Vietnamese). More importantly, the Aeroflot agent called in a repairman to tamper with the timestamp so that his ticket could be backdated; if they hadn't done this, his ticket would not have been available at the original fare. Khoái then made provisions for his wife Cúc to receive his paycheck in his absence. Ann saw him off at the airport and gave him a quarter for a pay phone in case I wasn't at the airport in Seattle (he wasn't permitted to carry dollars through Soviet customs).

Khoái spent two months in the U.S., visiting Seattle, Harvard, Princeton, and New York. He lectured on p-adic analysis, talked with people in his field, and worked in the libraries. In Seattle I borrowed a bicycle from a colleague for Khoái to use. However, the first time we went from home to campus he was unable to ride most of the way. Conditions were very different from what he was used to — hilly terrain and cars whizzing by at speeds that would've been unthinkable in Hanoi. After that we walked all the time, except when I rented a car for a few days.

While in New York, Khoái spent a day or two with Vietnam's U.N. delegates, who were the only Vietnamese officials on U.S. territory at the time. Once he was with them going to an event that had been organized by Americans who were friendly to Vietnam to mark the seventh anniversary of the end of the War. Suddenly a group of Vietnamese refugees attacked them, beating them and knocking them to the ground. Khoái was stunned. He minimized the physical injury — he compared it to falling off a bicycle — but he was clearly shaken. He had lived through the War, but for people in North Vietnam the enemy had been distant — American pilots dropping bombs. To be attacked by one's own countrymen was a new experience for him. Before he left, he told me that he would play down the incident in his report, because he didn't want to alarm people and discourage other scientists from visiting the U.S. But he wanted me to know that he thought that Cooperman was underestimating the danger from right-wing immigrant gangs in the U.S.

Ann and I made our second trip to Vietnam in April 1983. We went from Oakland to Bangkok on a special charter flight that was going to be returning with a plane-load of Vietnamese from the refugee camps in Thailand for resettlement in the U.S. The outgoing flight was almost empty — it had only fifteen passengers, mainly deadheading airlines personnel and United Nations people. Ed Cooperman had connections with the U.N. agencies that were involved with the refugee camps and was able to get us seats almost free of charge.

I had asked the Math Institute to arrange for us to pick up our visas in Bangkok, and we received them within a few days of arrival. We then took one of the thrice-weekly flights from Bangkok to Hanoi. We stayed in Vietnam for four weeks, which was the longest period we've ever been there.

The Institute had moved into a nice building on the grounds of the National Center for Scientific Research that had been constructed at the urging of Prime Minister Phạm Văn Đồng, who had visited the Institute in its earlier location and been horrified by the conditions there.

I lectured in Russian, with a Vietnamese translation, for three hours in the morning three times a week (ten lectures in all). Mainly I covered topics in elliptic curves and modular forms (essentially the content of my third book, which was published the following year). Ann and I gave talks in English at the Math Society — mine was on higher education in America and hers was on the life and work of Sofia Kovalevskaia — both of which were translated by the mathematician Nguyễn Đình Ngọc. And Ann also gave a talk in Russian on history of science at the Pedagogical Institute. Despite all this, our activities did not fill the four weeks, and there was a lot of dead time.

We later decided that the visit was too long. We must have been a burden on the Math Institute, which was paying for our hotel. We've found that we usually get as much done in a two-week as in a four-week visit. When people know you're there for only a short time, they schedule everything more efficiently and give you a higher priority.

On the other hand, we learned a tremendous amount in that four-week period. First of all, mathematicians everywhere love to gossip, tell jokes, and complain about bureaucracy. That was especially true in Hanoi, because our common language with most of the people at the Math Institute was Russian. That had been the language of their student years, and it felt perfectly natural to sit around trading amusing stories and sarcastic comments on various subjects. The people who spoke English with us had also spent time overseas — the algebraic topologist Huỳnh Mùi, who had studied in Japan, and the algebraists Hoàng Xuân Sính (whose English had greatly improved since the ICM in 1974) and Nguyễn Đình Ngọc, who had studied in France — and they were similarly uninhibited in conversation. Ngọc was a particularly unusual figure in the mathematical community. He had played a key role in the South during the War (see the postscript to this chapter), and had the habit of always wearing military fatigues. He had eclectic and sometimes bizarre interests — ranging from the occult to fuzzy set theory — but at the same time his comments on conditions in Vietnam and on organizational issues in mathematical life were thought-provoking and often surprising.

In addition, Ann gave English lessons that were popular with the Math Institute members. From reading and correcting their essays and listening to their oral presentations, she learned a great deal about their backgrounds and what was on their minds. One of her students, for example, had a brother in Cambodia in the Vietnamese army, which was fighting the remnants of Pol Pot's forces and helping with reconstruction.

We also had long conversations with Khoái and the Institute's general manager Châu, especially during a weekend with them at Hạ Long Bay. The bay was only about a hundred miles (160 kilometers) from Hanoi, but given the conditions of the roads in those years, it took four or five hours each way. There aren't many seascapes in the world that can compare in beauty with the islands and seastacks of Puget Sound and the Pacific Coast near Seattle, but Hạ Long Bay is one of them. Ann and I were thrilled to go out by boat among the giant rock monoliths and small islands in the South China Sea off the coast of Vietnam.

In our hotel most of the other guests were from Eastern Europe. Ann and I always tried to get hotel restaurants to bring us Vietnamese dishes rather than the greasy, terrible "Western" dishes that they insisted on preparing for non-Asian foreigners. We were rarely successful when we were on our own, but Châu was able to convince them to make us the Vietnamese breakfast, a type of chicken-noodle soup called *phở*. While Khoái, Châu, Ann and I, and two French women were happily eating our *phở*, several dozen other foreigners, almost all of them men from Eastern Europe, seemed equally happy eating greasy sausages, greasy eggs, and some awful-looking salami.

During our first six visits to Vietnam — in 1978, 1983, 1985, 1987, 1989, and 1991 — we always had problems with the hotel food. In those years, when there were few foreign tourists in Hanoi, the infrastructure and administration of hotels were in poor shape. The water and electricity were always going out, and the restaurant food was atrocious. During our seventh visit in 1993 we stayed in a tiny hotel that had been started by a former mathematician — this was one of the first privately owned hotels in Hanoi — and for the first time we did not get sick on the food. In subsequent visits we have always stayed at the Math Institute Guesthouse, which opened in the mid-1990's, and there the food and accommodations have been fine.

After we returned to Seattle in 1983, I wrote a detailed "Confidential Report" of ten single-spaced pages based on conversations and observations during our month-long visit. It dealt with a range of sensitive problems:

- Institutional rivalries in Hanoi among the Math Institute, Hanoi University, the Polytechnic Institute, and the Pedagogical Institute were standing in the way of efficient use of resources. There was an especially worrisome "psychological distance" between the Math Institute and Hanoi University.
- The Math Institute library, which was the best math library in the country, was often closed, and was underutilized.
- Research was divorced from teaching, and the best researchers, most of whom worked at the Math Institute, had little role in undergraduate education.
- Travel abroad, foreign degrees, and any type of foreign ties had too much prestige, and it didn't seem to matter whether or not anything of value resulted.

- Many young people tried to arrange trips abroad because of the money and status, even when they were unprepared to make scientific use of such a visit.
- Most young scientists lost touch with their former colleagues and professors after returning to Vietnam with their foreign degrees.
- Foreign visitors were received with too much formality; in particular, the large banquets were expensive and not appropriate for scholars and scientists.
- Too few women were being trained for careers in the physical and mathematical sciences and technology. For example, only 8% of the students at the Polytechnic Institute were women.

I gave Ed Cooperman copies of my Confidential Report to distribute to a list of leaders of the mathematical and scientific establishment in Hanoi. I later learned that my report reached the desk of Võ Nguyễn Giáp, who at that time was head of the State Committee for Science and Technology. During the French and American wars, General Giáp had become almost a legend in the West — the Pentagon had dubbed him "the Red Napoleon" — for having masterminded the humiliating defeat of the French at Điện Biên Phủ in 1954 and the Tet Offensive of 1968, often regarded as the turning point in the American war.

I doubt that my report had much influence, except perhaps in areas related to the administration of the Math Institute (especially the library, to which I'd donated many books). The main reaction of the other people who read it or had an aide read it probably was just to be surprised that a visiting American would have learned all these things — airing one's dirty linen in public (or to foreigners) was not the Vietnamese habit. We heard that General Giáp had called in the Math Institute director, Hoàng Tụy, and had commented to him about this.

Despite the harshly critical tone of some of my report, none of the Vietnamese reacted badly. They understood that I was trying to be constructive, and that I was a friend. Whether or not the report had any positive effect, at least it did no harm.

In the summer of 1984, a few months after Ann's biography of Sofia Kovalevskaia was published, she and I decided that we didn't really need the money from sales of the book for ourselves, and it would be nice to do something with it to support women in science and honor Kovalevskaia's memory. A project in the U.S. didn't seem to make much sense, since there were already many programs for women in science, and the amount of money we could come up with would be a drop in the bucket. On the other hand, after our visit to the Polytechnic Institute in Hanoi, we had become aware of an extreme gender imbalance in science and technology in Vietnam, and no one seemed to be doing anything about it. So we decided to start a project called the Kovalevskaia Fund that would address this problem.

Our initial idea was to work exclusively in Vietnam, and, with Ed Cooperman's help, we set up our fund under the rubric of the U.S. Committee for Scientific Cooperation with Vietnam. The very first initiative would be to bring two women scientists from Vietnam in 1985. (All of the scientists who had visited us in Seattle had been men.) And in fact, in August 1985 the mathematician Hoàng Xuân Sính and the medical researcher (and expert on the effects of Agent Orange) Dương Thị Cương came to the U.S., attended the International Congress of the History and Philosophy of Science at Berkeley, and visited several research centers. Ed Cooperman made Ann chair of a new subcommittee concerned with issues of women in science in Vietnam.

We realized that if we wanted to have ongoing projects in Vietnam, we had to start visiting more frequently than once every five years. We decided to go to Hanoi for two weeks during the spring of 1985, primarily for the purpose of meeting with the Vietnam Women's Union, of which Hoàng Xuân Sính was a vice-president, and setting up a program for women in science. We knew by then that we would be in Moscow for the first half of 1985, so we would once again go to Hanoi from there.

With our increasing involvement in Cooperman's committee, our relation to Vietnam was starting to look like a long-term commitment. But it was a horrifying and tragic event in October 1984 that would cause us to take on a much more active role in the functioning of the USCSCV.

On Saturday, October 13, 1984, I received a phone call from Pat Gallagher, whose wife had just been called by Vietnamese-American friends in California with shocking news. Ed Cooperman had been shot and killed by a Vietnamese refugee.

I was stunned. I tried to learn more, but the early information was sketchy. I called Ann, who was spending most of the year at the Institute for Advanced Study in Princeton.

At the time of the killing Judith Ladinsky, a professor of public health at the University of Wisconsin and the most active member of the USCSCV after Cooperman, was in Hanoi. It was already Sunday in Vietnam, she was out of touch with international news, and she didn't know what had happened until she was urgently summoned on Monday morning to the office of Foreign Minister Nguyễn Cơ Thạch. He informed her of the shooting.

On behalf of the government of Vietnam he offered his condolences to the USCSCV and asked Judith to convey his sympathies to Dr. Cooperman's family.

The reaction in Vietnam was extreme. It was a major news story, and initially the news service accused the CIA of having had Cooperman killed. I flew to Los Angeles the following weekend for a memorial service at Cal State Fullerton, and saw that the Vietnamese Mission to the U.N. had sent a large floral wreath.

It was not surprising that Vietnam suspected a CIA plot. Many former high-ranking officials in the South Vietnamese regime — men who had worked closely with the CIA during the War — had immigrated to the U.S. and settled in the Los Angeles area. For example, Nguyễn Cao Kỳ, the one-time prime minister of South Vietnam (who had acquired special notoriety after telling a reporter that he admired Adolf Hitler), had become powerful in the Vietnamese-American community. In 1984 a witness testified before the President's Commission on Organized Crime that Kỳ headed a Vietnamese "mafia" that engaged in extortion and politically-motivated killings.

However, it was doubtful that Kỳ and the others still had ties to the U.S. government. It is often hard for people in other countries to understand that someone might work closely with the U.S. for a while and then later become involved in terrorist activities completely on his own. The Cuban exile Luis Posada Carriles, whom Venezuela has been trying unsuccessfully to extradite from the U.S. for the terrorist attack on Cubana Flight 455 in 1976, worked closely with the CIA in the 1960's. And Osama Bin Laden was a valuable ally of the CIA in Afghanistan during the Soviet war there in the 1980's. But the terrorist attacks of Posada Carriles in 1976 and of Bin Laden in 2001 were presumably not CIA plots.

In the U.S. Committee for Scientific Cooperation with Vietnam we were concerned about the possibility that the Vietnamese government would overreact and cut off future visits. I telephoned Hoàng Tụy, who was in Paris, and stressed to him that, first of all, there was no reason to believe that anyone else was in any danger — it was most likely other activities of Cooperman, not the scientific exchange work, that had led to his being targeted — and, in the second place, the Vietnamese leaders should be told that the U.S. government almost certainly had no involvement in the murder.

Cooperman's killer, a Vietnamese-American student at the university, soon confessed, while claiming that it had been an accidental shooting. The Fullerton police declared from the beginning that the killing was a personal rather than political crime, and so there was no need for the FBI or for an extensive investigation. Even though the San Francisco police noted similarities between the crime and the shooting of a Vietnamese-American publisher in the Bay Area three years before, and the older brother of Cooperman's killer was known to be active in a right-wing extremist group in Hawaii, neither of those leads was followed up. The prosecutor tried to get a conviction for first-degree murder, but since the police investigation failed even to establish a motive, the jury rejected that charge and deadlocked on the charge of second-degree murder. Eventually the killer served a short jail sentence on a lesser charge.

To this day it is not known who ordered the killing of Ed Cooperman. No one can seriously think that the 20-year-old shooter, who had had a history of petty crime but nothing more, had done it on his own initiative.

Cooperman's death received much more attention than any of the earlier killings in the refugee community, because it was the first time that someone who was not Vietnamese-American had been targeted. The local press took a predictable blame-the-victim approach to investigating the circumstances of the crime. A reporter for the *Santa Ana Register* used records the police had found of Cooperman's last trip to Asia in order to "prove" that he had been a communist spy. For example, the newspaper noted that he had spent a night in Moscow on the way from Vietnam to France, but "shows no expenses, indicating lodging was provided for him," presumably by the KGB.

The attention in the media soon subsided, and Cooperman's friends and collaborators had to pick up the pieces and do our best to continue the work of the USCSCV. The obvious person to assume the role of chair was Judy Ladinsky, who had shortly before started making twice-yearly visits to Vietnam to carry out public health projects there.

Judy Ladinsky called a meeting of the surviving members of the Board of Directors of the USCSCV for Saturday, November 17 in Los Angeles. Ann flew in from New Jersey, and I came down from Seattle. About a dozen people attended the marathon meeting, the main purpose of which was to learn about the status of the various projects and reorganize the work of the USCSCV.

It was not an easy task. Cooperman had run the USCSCV as a one-man show, and other people on the Board of Directors knew only of their own projects. What was worse, Ed had had a cavalier attitude toward keeping the books. His records were in disarray, and he had been sloppy in his handling of funds, frequently using a grant for a different purpose from what it had been given for. Even though it seemed that the money had always been spent on something worthwhile, his financial practices had been improper and possibly illegal. One of the USCSCV's main granting agencies had already cut off funding after learning this, and Judy was desperately trying to convince them to reconsider.

In addition, not long after the November 17 meeting some simmering tensions within the USCSCV came to the surface. Three members of the Board started feuding with Judy Ladinsky, and it was never clear to us exactly what the issues were. They resented her leadership and eventually resigned from the group. On the other hand, a few of the Board members — including Ann and me — became more heavily involved in the affairs of the USCSCV after that meeting. When we visited Vietnam — almost always over Christmas vacation — we would often see Judy in Hanoi and work with her. On one occasion Ann assisted her in giving the TOEFL exam; and Judy consistently helped us administer the Kovalevskaia Fund projects.

Aside from the practical impact of the killing of Cooperman, there were psychological effects as well. None of the other people in the USCSCV had anything like the kind of visibility among refugees that Ed Cooperman had had, so we knew that we were not in danger. Yet the outbreak of violence had frightened and disoriented us.

For the Vietnamese, the killing probably increased their tendency to hold an exaggerated view of the sacrifices that people in the U.S. had to make during the War and afterwards in order to oppose our government's policies and support Vietnam. During a period of almost two decades of U.S. hostility, the USCSCV was one of the few channels for aid and cooperation from Americans. After normalization of U.S.-Vietnam relations in 1995, Ann and I at first feared that with the sudden influx of American groups the Vietnamese would lose interest in people like us with our relatively small projects. However, that did not happen. The Vietnamese have a keen sense of history, and they value old friendships. There is an expression *có tình, có nghiã*, which, roughly translated, says that longstanding friendships are, like family ties, of great importance.

Postscript

I wrote the following obituary for the July 2006 issue of the Newsletter *of the Kovalevskaia Fund.*

Maj. Gen. Nguyễn Đình Ngọc (1932 – 2006)

The Vietnamese mathematician and expert in military intelligence Nguyễn Đình Ngọc died on 3 May 2006 at the age of 74. Ann and I met Ngọc in 1983. In 1985 he was one of two mathematicians (the other being Hoàng Xuân Sính) with whom we had extensive informal discussions about initiating Kovalevskaia Fund projects in Vietnam. It was from these conversations that the proposal emerged to establish the Kovalevskaia Prizes.

In the 1980's Ngọc was a valuable source of information and candid insights into conditions in Vietnam, especially in the scientific and academic realm. In 1983 I gave my first public (non-mathematical) talk in Vietnam. Ngọc translated it from English into Vietnamese. Judging from the comments of our Vietnamese acquaintances, who described the talk in glowing terms, Ngọc's translation was much more eloquent than the original.

Despite his superb fluency in English, Ngọc's first European language was French. He studied in France for eleven years, returning to Saigon in 1966 with several advanced degrees in different branches of engineering. He was truly a polymath, finding abstract subjects (especially algebra and topology, fields in which he helped organize seminars in Hanoi in the 1980's) to be as fascinating as the more practical areas of math and science.

In the 1980's we heard stories and rumors about Ngọc's eccentricities. For example, he always wore combat fatigues. The reason, we were told, was that although he had been a high-ranking officer during the American war, he could not wear the uniform in those years because he was operating undercover as a college professor in Saigon. As a result, after liberation he made a point of wearing it always. When we asked about arranging a scientific visit for him to the U.S., we were told not to bother, since the U.S. government would know now of his role during the War and would never give him a visa.

We thought that the stories about Ngọc were probably exaggerated. But we were wrong. Shortly before his death from liver cancer, the Vietnamese government published tributes to his life, which revealed that he had played a crucial role as an intelligence agent during the War and had risen to the rank of Major General before his retirement. The son of a freedom-fighter who was executed by the French, Ngọc continued his father's tradition of struggle against French and then American imperialism.

Ann and I are privileged to have known this multifaceted intellectual and anti-imperialist. We are also grateful for his help and guidance during the early years of the Kovalevskaia Fund.

CHAPTER 10 VIETNAM PART II

Before Ann and I left Moscow for Hanoi in April 1985, we visited Khoái and his wife Cúc. After hearing about our plans to set up a project for women in science named after Kovalevskaia, Khoái commented that the idea would appeal so much to the Vietnamese that we'd get to meet a minister! And he was right. While we were in Hanoi, the mathematician Hoàng Xuân Sính set up a meeting with the Minister of Education Nguyễn Thị Bình. Both Sính and Bình were also vice-presidents of the Vietnam Women's Union (VWU), and it was largely in that capacity that they would collaborate with us on Kovalevskaia Fund projects.

Nguyễn Thị Bình had been a major figure in the history of the American war. After many years in the National Liberation Front in the south, in 1969 she was chosen as the chief delegate of the Provisional Revolutionary Government at the Paris peace talks. Western journalists were surprised to find a woman entrusted with negotiating with Henry Kissinger, and she attracted a lot of media attention. After the War she became Minister of Education, and in the 1990's she was Vice-President of Vietnam. Despite her prominence, during the past two decades she has taken a personal interest in the details of all of the Kovalevskaia Fund activities in Vietnam. She continues to be chair of the Kovalevskaia Prize Committee, a position that she did not relinquish even when she was Vice-President of the country.

Bình is also a warm, friendly woman with a sense of humor. When she asked about our lives and learned that Ann and I had academic positions on opposite sides of the U.S. (this was when Ann was at Hartwick College in New York state), she expressed concern and compassion. We later found out that she had been separated from her husband for long periods during the War.

Before we met with Bình we had had lengthy discussions with Sính and Nguyễn Đình Ngọc. Our original thought had been to start some kind of scholarship, but Sính and Ngọc believed that a prize would have much more visibility and broader impact. The whole idea of prizes for women scientists — which the Kovalevskaia Fund later carried over to other countries as well — originated with the Vietnamese, not with us.

We didn't want to impose a lot of conditions on the prizes, since we thought that the Vietnamese were in a better position than we were to judge what would work best there. We did insist that the sciences be defined as the natural sciences — including applied fields and medicine, but not the so-called "social sciences." Sính suggested that we put an age maximum on the recipients of the prizes, so that they would go to active, junior researchers. However, we felt that would be too intrusive. Sính anticipated that the natural inclination of the Vietnamese would be to give the prizes mostly to senior people as a type of lifetime achievement award — in some cases to people who were no longer active researchers. Like Sính, we would have preferred that more of the prizes go to women who were earlier in their careers. On the other hand, one of the reasons for the remarkable success of the Kovalevskaia Prizes in Vietnam has been that the selection of winners is entirely in the hands of the prize committee, with no interference from us.

The Kovalevskaia Prizes, which have now been given for over twenty years, are well known throughout the country. They are well publicized in the media — for example, in December 2002 a picture of Ann with the prizewinners was on the front page of *Lao Động*, which is the most popular newspaper in Vietnam.

Khoái once commented that if a name-recognition poll were conducted in Vietnam for all foundations, the Kovalevskaia Fund would easily beat out the Ford and Rockefeller Foundations. Our hope has been that as a result of the publicity about the prizes and other Kovalevskaia Fund activities, more young women have felt encouraged to go into careers in science and technology — but this is impossible to measure.

Besides the Kovalevskaia Prizes, a main topic of discussion with the VWU in 1985 was a regional conference on women in science that we agreed to cosponsor with them in Hanoi in January 1987. The proposed conference would address such questions as the following (taken from a list that was published in the first issue of the *Newsletter* of the Kovalevskaia Fund):

- What actions can be taken within the limited resources of developing countries to help women overcome the obstacles to their progress into scientific and technical work?
- What examples have there been in your country of women's leadership (cultural, political, scientific) that could be drawn upon to convince the public of the possibility of women's leadership in science and technology? What role did women play in your country in the development

of higher educational institutions, scientific and industrial research, and professional associations?
- Sometimes during a period of crisis, such as war or natural disaster, women are called upon to do scientific and technical work in greater numbers than ever before. Has this happened in your country? What occurred after the crisis passed? Were women able to continue playing a central role in scientific development?
- Are talented young women as likely as talented young men to go abroad for advanced study? What special obstacles prevent women science students from studying abroad and prevent women scientists from attending conferences abroad or visiting international centers of research? What can be done to increase the international contacts and reputations of women scientists in developing countries?

We discussed the specifics of the planned conference with Dương Thị Duyên, who was in charge of international relations for the VWU. Since we weren't going to be in Hanoi again before the time of the meeting, we knew that most of the details would have to be worked out before we left. Once we returned to the U.S., communication with Vietnam would be difficult. There was no such thing as e-mail in those days, and the postal service worked unreliably. Moreover, we found that the Vietnamese were much franker in person than in writing. As in other parts of the world (we've noticed the same in Latin America), a letter is regarded as a formal document — a part of the official records of the organization — that is not to be taken lightly. Especially when people are unsure of their knowledge of English, they are not likely to want to put their candid thoughts in the form of a letter to the United States. Thus, for many years we had difficulty getting useful information from the VWU by mail. We relied heavily on the messages that Judy Ladinsky gave us after her twice-yearly visits to Vietnam, but even that information was often sketchy. The situation improved somewhat after the advent of e-mail.

During our visit in April 1985 I also gave talks and spent time at the Hanoi Math Institute. One day the director, Hoàng Tụy, took Ann and me on an excursion to a pagoda called Chùa Hương that's located about 45 miles (70 kilometers) southwest of the city. At that time of year religious Buddhists make a pilgrimage to the site, which is at the end of a long trail up Perfume Mountain. Part of the way to the trailhead had to be traversed on a boat that went through several miles of rice paddies. The scenery was beautiful and evocative. Every now and then we'd glimpse a boy seeming to glide over the bullrushes; when we got closer we'd see the buffalo he was standing on barely visible above the water.

We were in a boat with some people who were making the pilgrimage. They asked Hoàng Tụy about us and looked startled when he said we were *Mỹ* (American). This was our first contact with rural people, and it was the closest we ever came in Vietnam to any type of negative response to our nationality. The reaction was not so much hostility, however, as a kind of fear. To peasants in the countryside the word "American" must have conjured up the image of some sort of alien creature that used to drop bombs on their villages. Hoàng Tụy seemed as surprised as we were at the reaction of the people in the boat. After that we noticed that he replied to curious queries with the term *Liên-xô* (Russian) rather than *Mỹ*.

It was during the long trip to Chùa Hương and back that we first heard some of Hoàng Tụy's stories about his past and the early development of mathematics in Vietnam. We were captivated by the tales of his struggle to study mathematics in the liberated zones during the French war (see the first postscript at the end of this chapter). Several years later I published a long interview with Hoàng Tụy in *The Mathematical Intelligencer*.

When Châu and Khoái took us to the airport for our flight back to Moscow, while in the check-in line we suddenly realized that we didn't have our Soviet visas. I had absentmindedly thrown them out when we packed up in our hotel room. This wasn't good — we remembered the often-repeated warning at our IREX orientation many years before *never* to get on a plane for the Soviet Union without a visa. In fact, most airlines headed for the U.S.S.R. insisted on seeing passengers' documents before allowing them to board. However, Aeroflot out of Hanoi didn't do this, and we decided to take the flight without our visas. There was no time to return to the hotel (a trip that in 1985 took over an hour each way), and the next flight to Moscow was at least two days later. I asked Châu to see if our visas were in our hotel room wastebasket and, if so, to send a telegram to the airport in Moscow informing them that our visas were in Hanoi.

The sixteen-hour flight (including a refueling stop in Tashkent) was awful. Many of the passengers were young Vietnamese who were leaving their families for the first time and seemed nervous and scared. Most of the men were smoking, and as the cabin filled with smoke the underweight young women started making extensive use of the air sickness bags. Finally, a British passenger decided that we'd had enough of this. She marched up to the cockpit and demanded that the pilots turn on the No Smoking signs. They did so, and the situation gradually improved.

We arrived in Moscow late on a Saturday night. I immediately walked up to a guard and said that we had a problem — we'd lost our Soviet visas. He said that they had received a telegram from Hanoi and were expecting us. I apologized for the trouble, but he said not to worry, "These things happen." The guards inspected our luggage and took us to a special part of the airport hotel called the "visaless zone." They explained that we'd have to stay there until Monday, when they'd contact the Academy of Sciences, which would bring new visas for us. They were very pleasant to us, even though we had delayed them from getting off their shift at midnight.

The zone for people without visas was a strange place. Partitions and shaded glass separated us from the rest of the airport hotel. The same kitchen served two cafeterias, one in each zone, and guards were stationed at the doorways from our restaurant to the kitchen, since in some sense it was a border. The convenience store sold postage stamps only in denominations for international correspondence, since internal letters were not permitted to be posted. Our room's windows to the outdoors were bolted shut, whereas the door couldn't be locked — in fact, we had to prop a chair against it to keep it from opening. During the night a guard would check that we were in our room. In short, it was a benign sort of house arrest.

We were the only ones who were awaiting visas so that we could go into the city. All the other guests in the visaless zone were transit passengers with overnight layovers. At that time some of the cheapest airfares between East Asia and Europe or Africa were on Aeroflot through Moscow. Passengers who had to overnight in Moscow were not required to go to the trouble of obtaining transit visas. Rather, they stayed at Aeroflot's expense in the visaless zone of the airport hotel while awaiting their connection.

Ann and I, on the other hand, were charged 25 rubles a night. Since international passengers were not permitted to take rubles out of the country, we had only dollars with us, and we didn't want to pay the hotel bill in hard currency at the official rate of exchange (a total of $70 for the two nights). Ann left her Lenin Library card as collateral, they let us check out, and she returned the next day with rubles to pay the bill and get her card back.

One thing we learned from the experience was why Ed Cooperman had had no expenses in Moscow when he traveled from Vietnam to France the previous summer. He had gone the cheapest way, and had been put up by Aeroflot while in transit in Moscow. So much for the *Santa Ana Register's* theory about Ed having been a spy for the KGB.

In my report to the National Academy of Sciences on my 1985 exchange visit to Moscow, I described all my scientific activities, including my time at the Hanoi Math Institute. To my surprise, the NAS bureaucrats tried to get me to refund half of a month's stipend because of my two weeks in Vietnam. They never would have done this if I had taken two weeks to go to France or Germany, or even to go on vacation. Their action was clearly political — most likely, they were afraid that some politician would learn that a scientist had visited Vietnam while receiving financial support from them, and this would cause them trouble.

Even though the amount at issue was small (the maximum stipend allowed me had been only $1500/month, which fortunately had been supplemented by sabbatical support), I refused to pay it. More precisely, what I did was offer to have the scientific advisory board for the exchange program arbitrate the matter; I said that I would happily pay back the $750 if their own scientific advisers thought that I should. The bureaucrats refused, since they knew as well as I did that the scientists would have taken my side. So I just didn't pay. Ann's colleague Margaret Rossiter, who had been a program director at the National Science Foundation for two years, said that if I just held back on paying until the end of the biennium, they'd probably drop the issue, since it would be a bookkeeping mess for them to record a repayment made in a different biennium from the original payment. She was right — the threatening letters from the NAS stopped at the end of the biennium.

In late December 1986 Ann and I went to Bangkok to wait for our visas for Vietnam. We had to be in Hanoi before January 8, when the Southeast Asian Seminar on Women and Science in Developing Countries was to begin. Relations between the U.S. and Vietnam were at a particularly low point, and some Americans who had plans to work in Vietnam had been having visa problems. Just a few weeks before, three Americans associated with the U.S. Committee for Scientific Cooperation with Vietnam had gone all the way to Bangkok and been unable to get visas.

After a few days it became clear that our visas were also being delayed. We explained that we were expected at a major international conference in Hanoi — among other reasons, because we had funds to reimburse the travel of most of the foreign delegates. We stressed that the Vietnam Women's Union was expecting us, and we asked for a meeting with the Vietnamese ambassador to Thailand. Because of the urgency of our request, we were given an appointment with Ambassador Lê Mai. We talked with him for an hour about the history of our activities in Vietnam and other subjects; he was extremely friendly and urbane, and spoke excellent English. (Later he rose to the rank of Deputy Foreign Minister, and in the early 1990's he played a key role in negotiating the normalization of relations with the U.S.; he died in 1996 at the age of 56.) Soon after the meeting with Lê Mai we received our visas.

When we came down the stairs from the plane in Hanoi, a VWU representative walked up to us on the tarmac and summoned us to a waiting van, which had special authorization to bypass passport control and customs. This red carpet reception set the tone for the treatment of foreign delegates by the Vietnamese throughout the conference. Reports on the seminar were carried on TV on the national evening news all three days of the meeting; the first day it was the lead item. Nhân Dân, the nation's leading daily newspaper, ran front page articles on January 10 and 11. Most of the foreign participants were interviewed by Vietnam's radio and/or TV networks. Teams of typists, technicians and translators were brought in to work full-time through the conference. In a country with acute shortages of paper, typewriters and audio equipment, each seminar participant had earphones to receive simultaneous translation of all reports and discussions (either Vietnamese-English or English-Vietnamese). Moreover, English language versions of the talks were typed, reproduced, and distributed to all participants before the end of the seminar.

A broad range of topics were covered in the talks. Some speakers from Vietnam, Laos, Cambodia, the Philippines, Thailand, Cuba, and Soviet Uzbekistan gave overviews of women's role in science, education, and the economy in their respective countries. Dr. Dương Thị Cương delivered a report on the effects on pregnant women of the dioxins used in the American war. She gave statistics showing the higher rates of ectopic and other abnormal pregnancies in the regions exposed to the toxic chemicals used by the U.S. military. She pointed out that stillbirths and deformed fetuses had also become a problem in the northern regions that had not been sprayed with Agent Orange or similar substances. In those cases the mothers had not been exposed, but apparently the chemicals had damaged the genes of the fathers who had fought in the south.

At the time of the conference the Kovalevskaia Fund had arranged for a mathematician named Arlene Ash to visit Vietnam. One of the Fund's projects had been a competition among women in applied mathematics for a travel grant. We got together a committee of mathematicians to judge the applications, and of the six we received they chose Arlene's as the strongest. Her specialty was medical statistics, and she later returned to Vietnam to work with colleagues in the south analyzing data on the long-term effects of the toxic chemicals used during the War.

Arlene wrote about her reaction to the seminar talks: "Many of the speeches were deeply moving. I was especially affected by the presentation of Võ Hồng Anh, who happens to be General Giáp's daughter, as well as a distinguished mathematical physicist. As she alluded briefly to the difficulties of completing her work on solid state physics amidst the American bombing of Hanoi in 1972, I recalled how, as a graduate student, I had protested that bombing at a meeting of the American Association for the Advancement of Science in Washington, D.C. But at the time I had no real sense of what it meant — no picture of a woman like myself, half a world away, struggling to complete her post-doctoral research project among the explosions."

Ann and I were given a small suite with a separate sitting area, and we took advantage of the opportunity to have informal parties there each evening of the seminar. On one occasion the Thai delegate, Ruankeo Kuyyakanon-Brandt, came to our door and saw that the two Cambodians, Ang Sarun and Ros Sivanna, were there. She later told Ann that for a moment she had thought that she wouldn't enter; at that time Thailand was almost at war with the Hun Sen government that came to power in Cambodia after Vietnamese troops ousted Pol Pot. But Ruankeo decided to join the group anyway, and was glad that she had. The women hit it off well.

At one point Ann mentioned that she and I rarely had evening parties, since I had the annoying habit of wanting to go to sleep early. She said that sometimes when we were in bed talking, I fell asleep at an interesting point of the conversation, and she felt like waking me up so we could continue talking. Suddenly Ros Sivanna grew quiet and thoughtful, as if reminded of something sad. After a pause she said that the same had been true of her and her husband when he was alive. Her husband had been one of the millions killed by Pol Pot's Khmer Rouge.

As often happens at conferences, the informal interactions could be more interesting than the formal presentations. Ann and I talked with Cương about the AIDS epidemic that was just starting to reach Asia. What prompted us to bring up the matter was that in Bangkok we had read an outrageous article in the *Nation*, one of the two main English-language newspapers of Thailand, that claimed that all the fuss about AIDS was exaggerated and that concern about the disease should not be allowed to interfere with Thailand's increasing success in attracting affluent foreign tourists (especially businessmen who came on "sex tours"). By the end of 1986 it was clear that Thailand was going to be hit by AIDS on a catastrophic scale, and we were incredulous that a newspaper could be so irresponsible as to print such an article.

I commented to Cương that Vietnam's isolation from the West would hopefully keep it from going through the horror of a major AIDS epidemic.

She said that, even though there were almost no reported cases in Vietnam then, the leadership of the Women's Union had started making plans to deal with the inevitable outbreak. She told me that they were not as optimistic as I was. In the first place, increasing numbers of foreign tourists and businessmen were coming to Vietnam. Moreover, the VWU had not been entirely successful in eradicating prostitution and drug addiction in the south. They knew that the social diseases left by the American troops would provide a breeding ground for the AIDS epidemic. So the leaders of the Women's Union and of Vietnam's medical establishment were developing public health measures as fast as they could, and they were soliciting international help.

Incidentally, one person who should be mentioned in this connection is Don Luce, whom Ann and I met in Vietnam and who told us fascinating stories of his travels during the War and afterwards. He was the journalist who, while in Saigon covering the War, had investigated and exposed the "tiger cages" that the South Vietnamese regime used for imprisonment and torture. After the War, he returned to Vietnam and initiated various projects to help the country rebuild. Don worked with the VWU to combat AIDS since the early days of the epidemic. His group, International Voluntary Services, was at the forefront of such efforts.

On the morning of the last day of the conference we all met with the Prime Minister of Vietnam, the legendary revolutionary Phạm Văn Đồng. There were many stories about Phạm Văn Đồng, who had been Hồ Chí Minh's right-hand man for several decades. One characteristic story was related to us by a Swedish participant in the conference. A few years before, a Swede who was working in Hanoi had fallen in love with a Vietnamese woman and wanted to marry her. In general the authorities discouraged marriages between Vietnamese and foreigners. But the Swede was persistent, and addressed an appeal to the top leaders of government. Finally he was granted an audience with Phạm Văn Đồng. A Vietnamese friend advised him not to bring up his request during the visit. So, as it happened, the hour-long meeting was devoted to discussing various works of Swedish literature which Phạm Văn Đồng had read in French translation and enjoyed. At the end the Prime Minister said, "Oh, and do not worry about your personal request — there will be no problem." Apparently Phạm Văn Đồng had satisfied himself that the Swede was a man of sufficient culture to be worthy of marrying a Vietnamese woman.

Early on January 10 a bus took the conference delegates to the Presidential Palace, where Phạm Văn Đồng, despite his advanced age, talked with us without aides. Two of the Vietnamese seminar participants served as translators between English and Vietnamese. After a few words of welcome, Phạm Văn Đồng asked for our questions and comments.

We had been told by our Vietnamese hosts to have no inhibitions about asking him whatever was on our minds. Esperanza Cabral, a prominent cardiologist and feminist from the Philippines, alluded to Corazon Aquino and Indira Gandhi and asked when there would be a socialist country with a female head of state. Phạm Văn Đồng said he hoped soon. (To this day there has never been one.)

Arlene Ash recounted that an interviewer from Vietnamese TV had asked her how she combined being a mathematician with being a wife and mother. She had replied that she was neither a wife nor a mother. This seemed not to please the Vietnamese, and the interview was not included in the evening news. Arlene now expanded on her answer, explaining to the Prime Minister that in her opinion, if a woman is successful in her career and is active in progressive social and political movements, then she is a suitable role model for the next generation of girls, irrespective of whether she has chosen to have a husband and children. Phạm Văn Đồng replied that he accepted her answer, which was a good one, but that he hoped that in a few years she would "have a better answer." Politely but unequivocally, he was making it clear that he did not share the view that a woman's goal in life need not include raising a family. A little later Ann spoke to him and reinforced Arlene's point that there was nothing wrong with a woman deciding not to raise a family; and the Prime Minister gave a similar reply to her. Here was one of the great revolutionaries of the 20th century, unwilling to part with a traditionalist view of women's fulfillment that many regard as conservative and retrograde.

Most people got into the spirit of the meeting and enjoyed the informal give-and-take with the Prime Minister. However, sometimes someone can have a strange reaction in such a setting. The scientist from Uzbekistan, Sanabar Khodzhibaeva, who did not know English (but who was a logical choice as a Soviet representative, since she had had several Vietnamese Ph.D. students), was accompanied by a Russian named Natalia Averianova who translated for her. But when Khodzhibaeva's turn came to speak to Phạm Văn Đồng, Averianova froze up. She whispered to Ann that she'd never translated for a head of state, she had butterflies, and all her English had left her. Could Ann please do it? Ann happily translated Khodzhibaeva's comments from Russian into English, after which Ngô Bá Thành (a member of parliament and VWU vice-president) translated from English into Vietnamese for the Prime Minister.

When my turn came to speak to him, I handed our camera to Ann, who got a photo of Phạm Văn Đồng laughing when I pointedly said that, as Ann's husband, I completely supported her opinion about not having kids. Ann remarked to Arlene, as the two of them were crawling around on the thick carpet in the Presidential Palace looking for good camera angles, "Isn't this a kick?"

The Southeast Asian Seminar on Women and Science in Developing Countries was a remarkable success under the circumstances. But there were a few negatives — for instance, the hotel. Almost all the foreign participants got sick from the food at some point during the three-day conference. On the day after the meeting with Phạm Văn Đồng, Ann and I found that a leak in the bathroom, which had supposedly been fixed after we reported it a few days earlier, had turned into a major calamity — water was now gushing from the damaged plumbing and spraying all over the place. I hurried to tell the people at the front desk, who informed me that, it being Sunday, no one would come to fix it until the next day. I argued with them, pointing out that, in addition to the tremendous waste of water over a 24-hour period, the uncontrolled flow was likely to do structural damage. But the hotel personnel seemed unconcerned. Finally, I appealed to a Foreign Ministry official who, presumably because of our VIP status, intervened on our behalf — and the leak was fixed in two hours.

Another feature of the conference that we didn't much like was certain obligatory formalities. For example, the group was taken to tour a school and a Pioneer Palace (activities center for children). The visits were highly ritualized, with tea and snacks in a big reception room and exchanges of pleasantries; there were no opportunities for meaningful discussions or observations. Moreover, to us it seemed patronizing and sexist to assume that women scientists would want to see institutions concerned with childraising. But apparently such excursions were *de rigeur* for all Vietnam Women's Union conferences in Hanoi.

Although the range of conference talks was excellent, it was disappointing that several people who had been expected ended up canceling, and a few countries of Southeast Asia — notably, Malaysia and Indonesia — were not represented. As a result, we had brought more money than was needed for travel reimbursement. At the request of the VWU we made a $500 donation to purchase sewing machines for a collective in Thái Bình province, an agricultural region to the southeast of Hanoi. This gift led to our being invited to visit Thái Bình as guests of the Women's Union; during our next trip to Vietnam in January 1989 we went there for a couple of days. Ann found the visit to Thái Bình particularly relevant to her teaching and research interests, which increasingly included questions of gender and rural development.

We also made a $1000 donation to a women's museum that the VWU was beginning to put together. While I was lecturing at the Math Institute, the VWU took Ann to see the makeshift holding facility for the material that they were starting to collect. Ann was impressed both with the artifacts they had gathered and with the person who showed her the exhibits. Ann's favorite item was a hollowed-out umbrella handle that had been used by Buddhist nuns during the American war to transmit messages for the National Liberation Front.

That was the beginning of our long collaboration with the Vietnam Women's Museum. Our modest contribution in 1987 was the first foreign donation they received, and the VWU chose the occasion of our Southeast Asian Seminar on Women and Science in Developing Countries to publicly announce that they would start constructing a major museum devoted to women in Vietnam. The museum expanded from the single room where material was being housed in 1987 to seven rooms in a two-story building, and finally to a four-story complex that was opened in 1995.

In the early 1990's Ann and I agreed to have a brochure they were writing translated into English and to edit it to remove any awkwardness of language. This led to what was perhaps the most remarkable meeting we ever had in Vietnam. When the brochure and an outline of the projected floor plan arrived and we found someone in Seattle to translate them, we were dismayed. The entire text was devoted to traditional women's roles — to Vietnamese as wives and mothers, to women's textiles and handicrafts, and to women of different regions in their costumes. There was nothing about the long history of women as leaders or about the role of women in the anti-colonial struggles against the French and Americans. Not a word was written about the tradition that started with the Trưng sisters, who led early uprisings against the Chinese, and continued through the late VWU president Nguyễn Thị Định, who had been General of the Army of the National Liberation Front of South Vietnam. It seemed that whoever had been entrusted with the plans for the museum had a very limited, conservative outlook.

This was an important issue. The Vietnam Women's Museum would become a major vehicle for outreach by the VWU, a showpiece for Vietnamese who come to the capital from around the country as well as for foreign visitors.

When we next were in Vietnam, we asked to meet with the new president of the VWU, Trương Mỹ Hoa (who herself had been a famous protest organizer in Saigon when she was a teenager and later had been a leader of the NLF), to discuss the direction the museum was taking. We knew that it was a sensitive matter to complain about the museum's leadership to the president of the Women's Union. We were fortunate in having an excellent translator, a young woman named Linh, and before the meeting we told her what we intended to talk about. We knew that the role of a good translator in Vietnam, as in many other parts of the world, is not only to translate words, but to explain and in some cases even subtly advocate for the foreigner's viewpoint.

When we met with Trương Mỹ Hoa, she had already been briefed by Linh, and she had called in the museum director. The five of us talked for about an hour. Ann and I started by explaining that the type of museum described in the brochure could be about the women of any country. Everywhere in the world there are women who are wives and mothers and who do nice embroidery. What is unique and interesting about Vietnam was missing — the legacy of women such as Hai Bà Trưng, General Định, Vice-President Nguyễn Thị Bình, and (although we didn't say it) Trương Mỹ Hoa herself. And what about the history of the VWU — probably the most powerful grassroots women's organization in any country of the world? This was also missing from the plans. Ann referred to some of the wonderful artifacts from the American war that she had seen in the one-room ramshackle women's museum in 1987 and contrasted that with the superficial exhibits in the floor plan that we had been given.

As Trương Mỹ Hoa listened to us, she was fingering a copy of the brochure and frowning. After we were done, she turned to the director and demanded an explanation. Of course, there was no explanation. What Trương Mỹ Hoa realized was that too much authority for the museum had been delegated to someone who had little sense of history, and that the VWU presidium would have to take a more direct interest in the detailed planning. She told us that the Women's Union would reexamine all the plans and the brochure would be put on hold until later. Afterwards there was a shakeup in the museum leadership, and the plans were changed. When the four-story building opened, the second floor was devoted to women in Vietnamese history, the third floor to the history and activities of the VWU, and only the fourth floor to handicrafts, national costumes, and the like. (The fourth floor also contains several exhibits about the work of the Kovalevskaia prizewinning scientists. Ann appears there with some of the winners; at one point the only Western women identified in photos in the museum were Ann and Jane Fonda.) The first floor contains a vestibule, meeting rooms, archives, and a gift shop.

Ann and I had violated two of the cardinal rules of what is supposed to be proper behavior of foreigners, especially in East Asia: *never* humiliate an official in front of a superior, and *never* interfere with the philosophical and cultural orientation of someone's project. In fact, just before leaving for Vietnam, in a bookstore I had browsed through a chapter about Vietnam in a book of "dos and don'ts" for businesspeople abroad; according to the author, one of the worst things one can do is to cause someone to lose face in public or in front of a superior. Despite this stereotype about the Vietnamese, what happened in our case was that, far from resenting us, the VWU leadership accepted our input, decided that we were correct, and proceeded accordingly.

This is not to say that the VWU always agreed with our suggestions for the museum. We objected strongly to a fourth-floor exhibit showing some of the beauty contests and fashion shows that had recently started to be held in Vietnam. Although some of the VWU officials didn't like those spectacles any more than we did (and there had been heated debates about how the VWU magazine for women should deal with them), they felt that the younger generation, which was increasingly coming under the influence of Western popular culture, would expect to see something about fashion models and beauty contests when they visited the museum. The VWU didn't want to appear out of touch with young people.

Ann and I have continued to collaborate in various ways with the Vietnam Women's Museum. The Kovalevskaia Fund donated $4000 for televisions and VCR's that the museum could use to show videos produced by the VWU, including a film called "The Sky and the Stars" that chronicled the first ten years of Kovalevskaia prizewinners. For several days during three different visits to Hanoi we worked with our translator Linh and a museum historian on correcting some of the English-language captions in the exhibits. This was interesting work, since we had to constantly ask for historical explanations and clarifications so that we would know how the captions should read, and we were often triangulating using the French captions as aids to translation.

After the conference in Hanoi in January 1987, I went to India for two months. I spent most of that time at the Institute of Mathematical Sciences in Madras (now Chennai), where the number theorist Balasubramanian hosted my visit. I gave a course on cryptography and had a great time with "Balu" and a group of students at the Institute.

Some of the graduate students had strong political views as well. One, for example, belonged to a "science for the people" organization in Kerala that had the goal of promoting science education and combatting superstition and Hindu fundamentalism. He asked me to write an article for their magazine about Vietnam, which he translated into Malayalam, the language of Kerala.

I also gave a talk about Vietnam at the Indian Institute of Technology in Madras. Some of the IIT faculty were concerned about the pro-American attitudes of their students — who a few weeks before had received a representative from the U.S. consulate with sycophantic adulation — and they wanted to provide a counterweight to that. Over a hundred students attended my talk, which put the U.S. role in Asia in a very different light from what they were used to hearing from Americans.

After my visit to IIT I was invited to give a similar talk in town. The sponsoring group was loosely affiliated with the Communist Party of India Marxist-Leninist (CPI M-L), which was a generally Maoist type of party, not to be confused with the pro-Soviet CPI. The CPI M-L had not, however, supported China in its attack on Vietnam in 1979. The party was pro-Vietnamese; when I visited the CPI M-L headquarters in Delhi a few weeks later, I saw that they had a portrait of Hồ Chí Minh in the lobby.

One of the people I met was N. Ram (he never seemed to use a first name). He belonged to a leftist branch of a very wealthy family that owned the *Hindu*, a national newspaper that was published in Madras. He was its editor, and after hearing my talk he asked me to write two articles about Vietnam for *Frontline*, the newspaper's weekly magazine supplement. I wrote one general piece that he titled "Vietnam: After the Agony" about the history and current conditions in Vietnam, and a second article, "Triumphs and Problems of Science in Vietnam," that described impressions from our work with scientific institutions and the Women's Union. The readership for those articles was much greater than the total number of people who've read everything else I've ever written.

After the 1987 conference in Hanoi, Ann and I received an invitation to visit Cambodia as guests of the Revolutionary Women's Association of Kampuchea (RWAK). We replied that we would very much like to do that after our next visit to Hanoi, which would be at the beginning of January 1989. They wrote that in that case we would be in Phnom Penh on January 7 for the tenth anniversary of the liberation of the country from the Pol Pot regime, and we could take part in the celebrations. We would be in Cambodia for a week, the first three days as guests of the Foreign Ministry. All of our expenses would be covered either by the Ministry or by the RWAK; all we had to pay for was our airfare from Hanoi.

At that time all Western governments still recognized the Khmer Rouge as the legitimate government of Cambodia, and Pol Pot's representative occupied the country's U.N. seat. The Hun Sen government had no embassy in Thailand (or anywhere else except for India and countries in the Soviet camp), so we had to wait until Hanoi to get our visas.

When our small Air Vietnam plane landed in Phnom Penh and stairs were wheeled up to the door, a protocol officer boarded and started to call out in English the different delegations that were expected to be on the flight. When each name was read, the head of delegation would stand and exit the plane, followed by the others in the group. After this happened two or three times, she called out "International Women's Organization," and I elbowed Ann, saying "That's you — get up!" Thinking of the game of "telephone," I figured that what the protocol person had said must have come from translating "Kovalevskaia Fund" from English into Vietnamese, then from Vietnamese into Khmer, and finally from Khmer back to English. Ann thought she must have meant someone else, but no one else was rising, so Ann went to the exit, followed by me carrying our bags. At the bottom of the stairs a man in a suit was greeting all the heads of delegation as they came off the plane. He shook her hand and said, "Hello, I'm the Prime Minister." It was Hun Sen.

We were ushered into the VIP lounge, where we sat with a jovial fellow who turned out to be the Indian ambassador. He had a rather cynical view of diplomatic life and joked that the basic maxim is "protocol equals alcohol." After a short while we were taken to our cars. Because Ann and I were a two-person delegation, they had two cars for us. This was absurd — so our first task was to convince the Foreign Ministry that Ann and I could ride in the same car.

That was the beginning of what can only be described as a surreal three days. We were housed in a small diplomatic hotel along with the delegations from Angola, Mozambique, and Nicaragua. In practice, the Cambodians were treating the Kovalevskaia Fund as if it were a small country. Our car always went in a convoy consisting of the minor diplomatic delegations — at the end of the procession, right in front of the ambulance.

We were the only official representatives of any U.S.-based group. In Bangkok I had stopped by the American Embassy to see if there was anything we should know about current U.S. policy or legalities before we went. A political officer met with me and said that there were no restrictions on private travel to Cambodia, but we must be sure not to give anyone the impression that we were representing the U.S. government. I assured him that doing that couldn't be farther from our minds. However, at one point when Ann and I were on the reviewing platform for the big parade with the TV cameras pointed our way, I couldn't resist joking to her that it was too bad that we hadn't brought along Ron-and-Nancy masks — it would've been fun (in a Monty Python sort of way) to pretend to be the President and First Lady.

On January 7 the parade started at 6 a.m. so as to avoid the oppressive midday heat. Float after float went by the reviewing stand, where they stopped to perform. Many of the floats illustrated the work of different occupations, and others were devoted to cultural themes. We had the best possible seats for the world-famous Khmer dancers. After the parade the delegates viewed boat races on the Mekong River.

The following day we were all taken by plane to Angkor, where we had a rare opportunity to tour Angkor Wat, Angkor Thom, and Bayon Temple. We were flown to Angkor early in the morning and brought back in the afternoon — it was too dangerous to stay overnight. In fact, since Pol Pot's forces were active nearby, the government had had to conduct a military operation in the area to prepare for our visit; it would have been a major embarrassment for the Hun Sen government and a symbolic victory for Pol Pot if there had been a successful attack during the tenth anniversary celebrations.

From our bus from the airport to the temples we saw soldiers lined up by the road to provide security. Ngô Việt Trung, a young mathematician we knew from Hanoi — whose father happened to be Vietnam's ambassador to Cambodia — was in the bus with us. He told us that, even though the soldiers were wearing Cambodian uniforms, they were really Vietnamese. There were some East German journalists sitting across from us, and their jaws dropped. What was he telling these Americans?! Let's face it — mathematicians are not good at keeping secrets.

During the trip we had some nice conversations with the East Germans. The subject of literacy rates came up, and Ann explained to them that U.N. statistics could not be taken at face value. According to official statistics, for example, Thailand had a higher literacy rate than Vietnam, but in reality the opposite was true. The problem was that different governments used different definitions. In Vietnam the measure was ability to read a newspaper, whereas in many countries it was just being able to sign one's name. The previous week when we had been in Thái Bình province, Ann had noticed several elderly women taking copious notes during a lecture we visited on pig-breeding. Such a level of literacy among elderly rural women is unusual in poor countries.

The day trip to Angkor went off without a hitch, and there was no attack by Pol Pot forces during the three-day commemoration. However, in some ways security was not very tight. When we attended a state dinner, my backpack was not even examined. I put it on the floor while I was taking pictures of Ann exchanging toasts with the heads of state of the three countries of Indochina. My uninspected backpack was just a few meters from them. Given the real risk of a terrorist attack on the festivities, I was surprised at the slack security measures. In the U.S. nowadays there's more rigorous security at an inconsequential airport in the middle of nowhere than there was at the state dinner we went to in Phnom Penh.

On Sunday our group also visited a crocodile farm. During the excursion Ann and I talked with the Mongolian ambassador to Indochina, who mentioned that it was unfortunate that the Mongolians had not been invited to send a delegation to the Hanoi conference of women scientists — the slight was still rankling him two years later. In her politest Russian, which was the common language she had with him, Ann assured him that, as far as the Kovalevskaia Fund was concerned, we had fully expected that they would participate. But invitations to the socialist countries had been the responsibility of the Vietnamese. So it was the Vietnamese, not the Kovalevskaia Fund, who had unintentionally snubbed the Mongolians. Ann had brought a spare copy of the Proceedings of the conference, which she gave to the ambassador as an expression of our good will. This exercise in Kovalevskaia Fund diplomacy took place as we were all making the rounds of the crocodiles.

Poor Mongolia seemed to be the Rodney Dangerfield of the socialist world. There was a tradition of exchanging gifts at diplomatic gatherings. At the big event in Phnom Penh the day before, Mongolia had been inadvertently left out of the schedule of speech-making and gift-giving. This *faux pas* had to be corrected at Angkor, where time during lunch was set aside so that the Mongolian ambassador could give his flowery speech and present his gifts.

Before our week in Cambodia I had had a rather naive view of what diplomats do. I had thought that after traveling a long distance for a special event, they would take the opportunity to negotiate trade deals and treaties — or something. But as far as we could determine, the various diplomatic delegations did nothing of any use; their presence was purely symbolic.

Angola had sent a relatively large delegation of four. They had terrible jet lag, and the head delegate collapsed on the stage during his speech, presumably from exhaustion and dehydration. It was a mystery to us why four had come — the poor fellows seemed to have little to do.

The two Nicaraguans knew only Spanish. Since Ann and I had traveled to Nicaragua several times by then and had made some progress in learning Spanish, we hung around with them quite a bit. On one occasion we were visiting an orphanage for the children of victims of Pol Pot. The Nicaraguan ambassador, a woman who had played a prominent role in the Sandinista revolution, wanted to tell the people running the orphanage how much it reminded her of similar facilities in Nicaragua for the orphans of the Contra war. I translated her Spanish into English, which our guide then translated into Khmer.

The only other Spanish-speaking people there were the Cubans, but the Nicaraguans seemed more relaxed with us than with them. The Cubans, who found it odd that the Nicaraguans were so friendly to us, seemed a little like fish out of water. On the reviewing platform during the parade the Cuban military attaché proudly greeted someone with the Vietnamese words *chào đồng thí* ("hello, comrade"). The only problem was that the woman he said that to was Laotian, not Vietnamese, and probably was not amused by the mistake.

The person we got to know best was Ambassador Gonçalve Sengo of Mozambique. He was posted to Moscow, had come to Phnom Penh for the occasion, and was staying in our hotel. The first evening we were there, he and I took a walk into town in search of bottled water and got into a long chat about life in Moscow. It was amazing how many impressions we had in common. Both of us were generally pro-Soviet (or certainly not anti-Soviet), but were exasperated by some of the cultural oddities of that country.

Sengo won his way into Ann's heart at Angkor Wat. At one point there was a long climb up steep stone stairs to the top of a tower, and Ann decided to wait and rest on a bench. While all the other men (including me) made the laborious ascent, Sengo said he'd wait with Ann since "I don't have to prove my masculinity by climbing the tower." Ann thought that was very cool.

Sengo was also the one who told the hotel restaurant that they really must stop serving us three different meat dishes at every meal. That was wasteful and inappropriate. Perhaps the diplomatic hotel was doing that as a way of making some sort of point about how far the country had come since the days of Pol Pot, when more than a million people had died of famine. But Cambodia was still a very poor country, and there was something grotesque about the way they were feeding us.

After the tenth anniversary festivities were over, the last four days of our visit were the responsibility of the Revolutionary Women's Association of Kampuchea. They took us on trips outside the city, and various RWAK activists talked with us about the difficult problems faced by Cambodian women, who at that time constituted 70% of the population because so many men had been killed during the 1970's. We expected to meet with the RWAK leadership and perhaps initiate a project or at least make a small donation (as we had to the sewing collective in Thái Bình two years before). However, we never had such a meeting; nor did we get any request to help with anything. This was odd — we knew that Cambodia was desperate for recognition and aid from the West, and we were among the few Westerners around. Our guess was that the RWAK must have been going through some kind of internal crisis — perhaps a leadership change or factional dispute. In any case, our visit to Cambodia, although a fascinating experience for us, didn't lead to anything concrete, and we have not visited again.

The Kovalevskaia Fund did finance a trip to India by Ang Sarun, who was director of Cambodia's Center for the Protection of Mother and Child, and two Vietnamese women physicians. The three doctors visited hospitals, rural and urban family planning clinics, technical training centers, medical research facilities, and village development projects in Madras, rural Tamil Nadu, Coimbatore, Baroda, and Bombay. That project of the Fund had already been arranged, however, before our visit to Phnom Penh.

The high point of our next visit to Vietnam in the winter of 1990-1991 was our meeting with the "Red Napoleon," General Võ Nguyễn Giáp. We had known his daughter Võ Hồng Anh since 1987, when she spoke at the Southeast Asian Seminar on Women and Science (hers was one of the talks that Arlene Ash described as "deeply moving"). In 1988 Võ Hồng Anh won the Kovalevskaia Prize, and in October 1990 she spent three weeks in the U.S. In addition to trips to research centers in her field, she visited Ann at Hartwick College, where she gave presentations on women scientists in Vietnam and her experiences as a physicist in a Third World country.

A colleague of Ann in the anthropology department objected to the choice of Võ Hồng Anh as a speaker on women in Vietnam on the grounds that she wasn't "authentic." This anthropologist seemed to think that unless a Vietnamese woman was wearing a conical hat and working in a rice paddy, she couldn't be "authentic." In reality, Võ Hồng Anh was about as authentically Vietnamese as anyone could be. Her whole family had been deeply involved in the Vietnamese struggle against colonialism. Exhibits in the Vietnam Women's Museum show photographs of her mother, who died in prison under the French, and her aunt, who was executed by the French. After their deaths Võ Hồng Anh as a baby and young child was raised clandestinely by Women's Union activists, who didn't want the French to know where she was out of fear that they'd kidnap her to use as leverage against her father. In a literal sense she was a child of the VWU.

General Giáp came with an aide to meet with us at the VWU headquarters. His daughter, another Kovalevskaia prizewinner, and several VWU officials took part in the discussion. As far as Ann and I were concerned, the main purpose of the meeting was to ask him to press the State Committee for Science and Technology to keep statistics about women's participation in scientific fields. We were particularly interested in knowing to what extent women had access to opportunities for study and research in other countries. We had the suspicion that few women made long-term scientific visits abroad, which were the most prestigious and useful kind.

The reason we had this impression was that we had been having great difficulty with one of the conditions for the Kovalevskaia Prize. The prize consisted of two parts: $1000 worth of scientific books which we would bring to Vietnam and a second $1000 that the winner would get when she went to a country in which she could receive dollars from us and use the money to buy scientific material. In those years it was hard to purchase scientific equipment in Vietnam (and most other Third World and Soviet bloc countries), and those countries also had complicated currency restrictions. So in order to receive the second half of the prize the winner had to go for a significant visit (more than a few days) to a Western country or a Third World country that had hard currency and easy access to scientific supplies. By the late 1980's we were hearing complaints that few of the winners had been able to collect the second part of the prize.

Võ Nguyễn Giáp wanted to talk about a completely different topic — the desirability of the Kovalevskaia Fund helping to finance significant ongoing science projects in Vietnam. Ann and I insisted on coming back to the issue of monitoring participation according to gender in science and technology generally and specifically in foreign travel. The women were all arguing on our side — no one seconded Giáp's comments — and at times the discussion became disorganized and animated, with different people talking at the same time in English and Vietnamese.

The VWU leaders knew us well and understood the financial limitations of the Kovalevskaia Fund, which precluded support of major ongoing projects; and they agreed with us that it would be helpful to keep statistics according to gender for various forms of scientific work and activity. General Giáp seemed taken aback by the united front presented by Ann and me, high officials of the VWU, and the two Kovalevskaia prizewinners, including his own daughter. His aide, who at the beginning had been taking careful notes, soon stopped writing and put away his notebook. The General was good-natured about the comments and criticisms that were coming from all directions, and at the end he said nice things about the Kovalevskaia Fund.

It's hard to say whether our meeting accomplished anything. General Giáp was almost 80 and probably not involved in the day-to-day running of things. Even if he had been swayed by our arguments, he would not necessarily have followed through effectively. To the best of our knowledge the Vietnamese ministries still don't keep the statistics we were asking for in 1991.

When we were next in Vietnam in 1992-1993 around Christmas time (which is not a vacation in Vietnam), one of our major tasks was to renegotiate the terms of the Kovalevskaia Prizes. We were going to propose that instead of two individual prizes, each consisting of $1000 in books plus a second equal amount in scientific supplies if the winner went abroad, one individual prize worth $1500 and one group prize worth $2500 would be given each year. The group prize would be project-oriented, and it would go a small way toward meeting the request of Võ Nguyễn Giáp and others that we try to support actual research activities in Vietnam. The individual prize would no longer be divided in two parts.

About half of the prizewinners during the first seven years never went abroad. A few of the winners were no longer actively involved in research, in which case presumably they didn't have need of the second half of the prize, at least not for their own work. In other cases a talented young researcher might not have gone abroad either because of sex discrimination in the programs for foreign collaboration or because of social pressures on her to stay at home. Whatever the reason, by 1992 it was clear that the second part of the prize was not having the desired effect of improving women's opportunities. Rather, it was creating resentment and misunderstanding.

Nguyễn Thị Bình, who just three months earlier had been chosen by the National Assembly to be Vice-President of Vietnam, met with us to discuss the detailed conditions for the Kovalevskaia Prizes. She quickly accepted the proposal for future prizes with only minor adjustments. However, the negotiations lasted over two hours because of what she regarded as the unresolved question of the second half of the prizes for the first seven years.

We insisted that the conditions for getting the second $1000 in supplies had been perfectly reasonable, as evidenced by the fact that several prizewinners had received it; and we did not want to change the conditions retroactively. At first I thought that Bình accepted this, and I passed a note to Ann saying that these negotiations might end quickly. Ann replied that we hadn't heard the last of this issue, and she was right. Every time we thought we'd made our case and the discussion had moved on, Bình would keep coming back to the question of the second part of the prize for the earlier winners. She wanted them to get the second $1000 in Vietnam even if they never traveled abroad. The total money involved was rather small, but we didn't think that we should change the ground rules governing the earlier prizes for no valid reason. After a while it seemed that she was trying to wear us down. I slipped another note to Ann saying, "I hate to say it, but I'm almost starting to feel sorry for Henry Kissinger right now."

In a way Bình seemed to be really enjoying the give-and-take. Here she was negotiating with Americans again, and both sides were hanging tough! When we finally reached a compromise (namely, we loosened the conditions slightly for the second half of the early prizes), we inserted the changes in the draft of the agreement, and the VWU made photocopies. Then Bình showed us the proper protocol for signing agreements: each copy gets signed in turn by each party and a different pen is used for each copy. It was all done with suitable pomp and circumstance, with a photographer recording the moment for posterity. I couldn't help thinking that for Bình it was like a sport: she had seemed truly to enjoy the afternoon's negotiations and the signing of the agreement, perhaps because it reminded her of her days as a much younger woman at the Paris talks.

In August 1995 Ann and I made a special trip to Vietnam to take part in a celebration of the tenth anniversary of Kovalevskaia Fund projects. The event was well attended and extensively covered by the press. Vice-President Nguyễn Thị Bình and VWU President Trương Mỹ Hoa presided. Many dignitaries were there, and eleven of the seventeen Kovalevskaia prizewinners of earlier years were able to come. The VWU showed the video "The Sky and the Stars" about the work of those scientists. They had translated the narration into English so that the Vietnamese delegation could show the film the following month at the World Conference on Women in Beijing.

The Women's Union had also organized a get-together of fifteen women students selected from universities throughout Vietnam. This was the first year of a joint project of the Kovalevskaia Fund (which gave about $1000, mainly to cover travel to Hanoi) and the VWU. In addition to attending the tenth anniversary ceremonies, the students were taken to visit various research centers around the capital. Ann and I joined the group at the Hanoi Math Institute, where the director Hoàng Tụy talked with them about why a young person should pursue a career that she/he truly loves — such as scientific research — rather than one that might be more lucrative or more favored by friends and relatives. The young women were spellbound; most likely none of them had ever heard an eminent scientist speak from the heart in such a manner. The get-togethers of talented undergraduate students have taken place once every two years since then; for the Kovalevskaia Fund it is a way to have a real impact for a small amount of money.

As part of the anniversary celebrations the Vietnamese conferred three different medals on each of us — a women's emancipation medal (pinned on us by Trương Mỹ Hoa), a science and technology medal (given to us by the Minister of Science and Technology), and the Friendship Medal (presented to us by Vice-President Nguyễn Thị Bình). We heard that they couldn't decide which of the three medals to give us, so they resolved the matter by giving us all of them. Before the ceremony they had inquired whether we would have any objection to receiving Vietnam's Friendship Medal. We replied that we'd be greatly honored (we knew that it was one of the most prestigious awards in Vietnam and had been given to relatively few people). Apparently most of the recipients had been communists, so the Vietnamese had wondered whether it would cause us any trouble or embarrassment to be associated with the Friendship Medal. On the contrary, we were flattered and touched.

During the 1990's Vietnam was changing rapidly. Starting in 1985 the government had introduced economic reforms that stimulated the rapid development of private enterprise and reduced the role of centralized government planning. As in other parts of the world, the effects of these changes were mixed. Most Vietnamese have become better off in material terms than they were in the immediate post-war years, and there would be little popular support for going back to a centralized socialist system. On the other hand, there is much more economic inequality than there used to be, and a class of "have-nots" has emerged. In addition, in many respects private industry is subjected to even fewer taxes and restrictions in Vietnam than in the U.S. and other capitalist countries.

For example, in 2003 we noticed a help wanted advertisement that the Caravelle Hotel had put in the English-language *Viet Nam News*. The ad invited males to apply for engineering jobs and females to apply to be floor attendants. Such a blatantly sexist job ad would be illegal in the U.S. and most European countries. Ann sent an e-mail about this to the editor of the newspaper, who wrote a polite but noncommittal response, and she also sent a copy to the VWU. The head of the VWU international relations department replied that she would forward Ann's letter to the VWU committee that was working on the draft of a new gender equality law.

Since the beginning of the economic reforms, the Women's Union has been striving to mitigate the negative effects on women's health and well-being that are the by-products of privatization in many parts of the world. In Chapter 7 I mentioned Ann's paper comparing the gender effects of marketization in the former Soviet Union and in Vietnam. In it she suggested that the situation of women was far worse in Russia and Eastern Europe than in Vietnam in part because those countries had no powerful women's organization that was effective on a grassroots level. Ann presented her paper at a conference of the International Studies Association in Acapulco in 1993 and published it in *Canadian Woman Studies* in 1995.

In higher education the first experiment in privatization occurred in the late 1980's. A group of academics in Hanoi decided that the public system was stagnant and resistant to change, so they devoted their energy to creating a private university that would serve as a model for the country. The result was Thăng Long University, Hanoi's first private college. The idea was to start with a top-notch program in the basic mathematical sciences, taught by some of Vietnam's leading mathematicians, and then expand into other fields.

The new university aroused interest in the Western press. Free market enthusiasts were delighted that prominent Vietnamese intellectuals seemed to be endorsing private higher education. *The New York Times* ran an article that was sharply critical of the public secondary and higher education systems in Vietnam and full of praise for the new private college. The *Times* heralded Thăng Long University as a break with socialism that held great promise for education in Vietnam.

I wrote a letter that was published in the *Times* taking issue with the negative remarks about Vietnam's educational system. I pointed out, among other things, that in a recent International Mathematical Olympiad the Vietnamese team had placed fifth, ahead of the U.S. team, which had placed sixth. Literacy and educational levels in Vietnam were high for a country that was still mired in poverty.

About five years after Thăng Long University was founded, Ann and I visited it. Here is what we reported in the May 1995 issue of the *Kovalevskaia Fund Newsletter*:

> On 26 December 1994, we visited Thăng Long to see what had happened to this dream. The institution's director, the mathematician Hoàng Xuân Sính, had at one time been a leader of the Women's Union and a member of the Kovalevskaia Prize Committee. But she had become totally absorbed by Thăng Long, had had to give up her earlier activities, and we had not seen her for four years.
>
> The so-called "university" bears no resemblance to what had been planned. It is now a trade school offering programs in only two areas: business/accounting and software management. No students are studying basic science, and the prominent mathematicians who initially supported the school have lost interest, presumably because of the low academic level... With the drying up of foreign donations to the school (which came mainly from the overseas Vietnamese community in Paris), tuition will go up considerably.

For us, the saddest part of the visit was not merely seeing that Thăng Long "University" had degenerated into a purely commercial venture of the sort that dots the higher education landscape in the U.S. and Latin America. The real tragedy was seeing the waste of the talents of two people we had known for over a decade — Sính and her managing director, Huỳnh Mùi. Sính, the first woman full professor in the sciences in Vietnam, had also been a member of parliament and for many years the coach of Vietnam's team in the International Math Olympiads. Mùi had been one of the country's most popular and respected math professors when he was at Hanoi University, and had been the thesis adviser of some of the first students to get their Ph.D. in mathematics in Vietnam.

Even though little of value remains of the Thăng Long project, it seems that Sính and Mùi are trapped. So much was invested — financially and emotionally — that it would be very hard to admit failure and abandon the effort.

The name "Thăng Long" comes from a Vietnamese word for "noble dragon." But unfortunately, the dragon has become an albatross for these two people.

A frequent lament among scientific researchers in Vietnam has been that few young people want to follow careers in the basic sciences. There has been a "graying" of scientific institutions. When I first visited the Hanoi Math Institute in 1978, the median age was about 30 — which was my age at the time. A quarter century later the median age was still roughly my age at the time. (Well, this is a slight exaggeration — the median age had gone up by about 15 years in the 25-year period.)

The Vietnamese people have traditionally had tremendous respect for scholarship — as Ann found out in 1978 when she told the maids in our hotel that she was married to a math professor. But with the increasing influence of value systems imported from the capitalist countries, intellectual pursuits have a hard time competing with more lucrative occupations.

This type of generational shift is common in one form or another to many countries, not just Vietnam. In December 2002, following our biennial visit to Hanoi, Ann and I spent a week in Beijing. My former Ph.D. student Daqing Wan (now a math professor at U. C. Irvine) arranged the visit for us. For both of us it was an intensive week of professional activities; we can't claim that we saw much of China, although we did fit in a trip to the Great Wall at Badaling, which in the snow looked like the setting for a fairy tale.

In our guesthouse we watched the English-language TV channel, which was showing a speech competition among Chinese students of English. In one of the events they had to choose the foreigner whom they most admired, and explain why. The students chose Bill Gates, because they were impressed with both his wealth and his influence on consumer technology (at this time he was not well known as a philanthropist). The two moderators of the competition were a Chinese woman and a Canadian man who were in their forties. The woman commented that when she was young, Chinese students would have had a different type of answer to the question. In earlier years, she said, the most admired foreigner among Chinese youth was the Canadian doctor Norman Bethune, who died in China in 1939 after selflessly performing battlefield surgery on several hundred Chinese communist soldiers wounded in the Sino-Japanese war. Thus, while she was impressed with the language ability of the new generation of Chinese students, she was uneasy about their ethical values.

When Ann and I went to Vietnam in April 2005 for the celebration of the twentieth anniversary of the Kovalevskaia Prizes, we thought that that would be a good time to interview Nguyễn Thị Bình about her life and work. We had known her for twenty years and admired her not only for her role in Vietnam's history, but also for her dedication to the Kovalevskaia projects for women in science. She had devoted countless hours to evaluating candidates for the Kovalevskaia Prizes, meeting with other committee members to choose the winners, and planning the publicity and the ceremonies. And we enjoyed meeting with her whenever we were in Hanoi.

Unfortunately, we did not get the in-depth interview we had hoped for. Initially we asked for eight to ten hours, but she agreed only to about an hour-and-a-half. Moreover, even though she was no longer Vice-President, she answered questions as an official speaking for the record. If there had ever been any tensions or conflicts among her colleagues when she was chosen for a role that in most societies was reserved for men, if there had been any mistakes made in the process of reunification with the south, if the Vietnamese of her generation had any concerns about the values of the younger generation — none of these themes were topics she felt it appropriate to talk about.

We had thought that after retirement a political leader would be likely to talk frankly about some of the difficulties that she had experienced many years before. However, in the first place, Vietnam has a tradition of calling upon elderly former leaders to take on new responsibilities — and as a result they never feel totally removed from current politics. In the second place, we had generalized too much from my experience with Hoàng Tụy, who had given me an unusually frank and provocative interview. He, however, is a mathematician, not a politician, and, as I remarked in connection with the bus ride to Angkor Wat, mathematicians do not like to keep secrets. Diplomats and political leaders, because of their professional background, do not share mathematicians' desire always to speak the unvarnished truth.

Ann and I returned from our 2005 visit without a usable interview with Bình. We had bought a small cassette recorder and had several empty tapes that it would have been a shame not to use. So we decided to tape our own recollections of the various controversies, travels, and interactions we'd been involved in over the years. We thought that all this would take six to eight hours, but it ended up stretching out to twelve two-hour tapes. Those recordings, which we made in the summer of 2005, were a big help to me in writing these memoirs.

Postscripts

The following are excerpts (slightly edited) from my interview with Hoàng Tụy, director of the Hanoi Mathematical Institute, which was published in The Mathematical Intelligencer, *Vol. 12, No. 3, 1990.*

Hoàng Tụy From 1947 to 1951, during the French war, I taught mathematics at the secondary school level in the province of Quảng Ngãi, which at that time had the best high school in our free region, which was called the Fifth Liberated Zone.

Koblitz Did education proceed normally during this period?

Hoàng Tụy To some extent, yes. Our free region was relatively stable, with a high level of economic and political organization and cultural life.

Koblitz It was at this time that you wrote a textbook?

Hoàng Tụy Yes, it was printed in 1949 by the Việt Minh [anti-French resistance] press. It was only an elementary geometry book for high schools, but perhaps it was the first mathematics book published by a guerrilla movement.

I was amused to see a reference to my geometry book in a recent popular novel. You know, in 1954 our country was partitioned at the 17th parallel. At that time some parts of the south had been liberated; but because of the Geneva Accords, the soldiers and many teachers from the liberated zones went north. Unfortunately, many of the schoolchildren could not leave with their teachers and were left to their own devices to continue their education. According to the popular novel, there were two books that were most prized by these youngsters: a book of poetry by the well known contemporary Vietnamese poet Tô Hữu [see the obituary below] and my geometry book....

Koblitz When did you decide to leave the south?

Hoàng Tụy In 1949, with Hanoi occupied and the university closed, some classes in university mathematics were established in the liberated zones in the mountains 200-300 km north of Hanoi, near the Chinese border. In addition, two other rudimentary universities had already been set up in the free regions: one under Professor Nguyễn Thúc Hào in the Fourth Liberated Zone, and one under Professor Nguyễn Xiển in the northwest.

Intellectuals of the Việt Minh organized an examination, which was administered by the Ministry of National Education. I was one of two candidates from my district in the south who took the exam. You must understand that the examination process was long and complicated, because the exam questions and our answers had to be carried over the mountain trail (later to be called the Hồ Chí Minh Trail) by guerrilla courier. Normally it took about three months to get a letter from the north...

After I sent off my answers, I had to wait eight months to hear the results. The exam tested general first-year university mathematics, mainly calculus and mechanics. Despite our primitive conditions, the exam was a rigorous one, and it was administered under strict conditions. The committee to administer the exam in our region was appointed directly by the Việt Minh governing council of the Fifth Liberated Zone. So the exam had a high prestige, and people were very impressed when the good news came of my success on the examination.

Koblitz Did you go north as soon as you heard?

Hoàng Tụy No, this was late 1949. I taught for two more years. In 1951 I learned definitively that Lê Văn Thiêm had returned to Vietnam and was working in the liberated zones of the north. I then asked for permission to go north, and it was granted.

Koblitz How did you travel north?

Hoàng Tụy There was only one way — on foot through the mountains. At this time the Hồ Chí Minh Trail was still a trail in the proper sense, a narrow footpath. But it was very well organized. Every 30 km there was a station where one could spend the night and a guide to take us to the next station. But, of course, there were many dangers.

Koblitz What were the main dangers?

Hoàng Tụy There were three — the French, malaria, and tigers.

Koblitz How long did your trek north take?

Hoàng Tụy The actual walk took three months. The beginning, in the region near Đà Nẵng, was relatively easy, since we could walk in the plains, taking the road by night. The cities and towns were abandoned because of the French bombing. But they were not occupied. Farther north, however, the French occupied all the lowlands, and we had to keep to the mountains. That was the hardest part of the walk — in the mountains of what is now Bình Trị Thiên province.

Of course, we did everything possible to lighten our load. We carried only rice and salt for food. Before I left, I had taken my math books, removed the covers, and cut out the margins on every page so that they would be lighter for the journey north.

I wrote the following obituary for the July 2003 issue of the Kovalevskaia Fund Newsletter.

Obituary: Tố Hữu

The great Vietnamese poet Tố Hữu died on 9 December 2002 at the age of 82. He was sometimes called the "poet laureate" of the anti-colonialist movement. A communist from age 17, Tố Hữu was a leader in the struggle to liberate his country from French and American imperialism. After the end of the war in 1975, Tố Hữu occupied high posts in the Vietnamese government.

Tố Hữu played a crucial role in the establishment of Kovalevskaia Fund projects in Vietnam. In 1985 Ann and I met with people in the Vietnam Women's Union (VWU), including Nguyễn Thị Bình (who at the time was Minister of Education) and the mathematician Hoàng Xuân Sính, both of whom were VWU vice-presidents. The Kovalevskaia Prizes grew out of these discussions.

Not all male officials liked the idea of a prize for scientists administered by the Women's Union. Powerful bureaucrats in the State Committee for Science and Technology objected for two reasons. First, they felt that any international collaboration in the sciences should be handled by the State Committee and not by the VWU. Second, they thought that the financial commitment by the foreign sponsors (between $2000 and $4000 per year) was too small.

Dr. Sính explained the situation to Tố Hữu, who at that time was the Deputy Prime Minister. He replied that the proposed prize for women in science, supported by American friends and named after a Russian woman mathematician and socialist, was such a "beautiful idea" that Vietnam should welcome it even if no money were being offered. And he didn't see anything wrong with the Women's Union coordinating such a prize. As a result of Tô Hữu's strong support, on 21 May 1985 the Council of Ministers of Vietnam passed a resolution (of which we have a photocopy) authorizing all the projects of the Kovalevskaia Fund in Vietnam, including the prizes administered by the Women's Union.

Sofia Kovalevskaia liked to explain her love for mathematics by quoting her thesis adviser Karl Weierstrass, who said that "a mathematician should have the soul of a poet." In this case it was a poet who had the soul of a scientist, and understood the value for Vietnam of an annual prize named after Kovalevskaia.

CHAPTER 11 NICARAGUA AND CUBA

When Ann and I moved to Seattle in September 1979, we initially assumed that our international interests would subside. Like most people who grow up in the northeastern part of the U.S., we thought of the Boston–New York–Washington corridor as the center of the world and almost every place else — certainly including Seattle — as the hinterlands. When we left New England to come to the Pacific Northwest, we assumed that as far as international politics was concerned, we had fallen off the edge of the earth.

To our surprise, shortly after our arrival we learned that Seattle and London were linked by daily polar flights that did not actually go through New York. In order to adjust to a Seattle perspective, one of the first steps we took was to buy a globe and throw out our Mercator-projection map of the world. It wasn't until after we left the big cities of the Northeast that we truly embraced the notion that the Earth is round and discarded once and for all the typical New Yorker's flat-earth point of view.

It wasn't only for geographical reasons that we expected to become less interested in international travel. I had just turned 30, I had my first tenure-track job, and we thought that we'd be putting down roots. We figured that we would settle down, develop an interest in local issues, and become involved in campus politics (perhaps the AAUP), school controversies (the PTA), and neighborhood grievances. We assumed that in a few years, as our peers started to have kids and we saw the wonderful effects that parenthood would have on their lives, we would inevitably want to do the same.

But exactly the opposite happened. The more we saw the impact of having children on the lives of our contemporaries — especially the women — the more we had second thoughts. By the mid-1980's, Ann's career started to show great promise, and our international activities — in the Soviet Union, Vietnam, and later Latin America — showed no sign of tapering off. After we started the Kovalevskaia Fund in 1984 — and established it as an independent tax-exempt organization in 1985 — we increasingly viewed it as our "child" that would grow and benefit society without the drawbacks of a flesh-and-blood child. (This was the gist of what Ann said to Prime Minister Phạm Văn Đồng in Hanoi in 1987.)

It's not that we didn't make a good-faith effort to adopt a more sedate lifestyle. Between 1979 and 1981 we traveled hardly at all. Moreover, inspired by the mantra that "all politics is local," we became members of our neighborhood community group. We devoted several long evenings to the monthly meetings of the Ravenna-Bryant Community Association. We tried without much success to convince ourselves that the issues being debated there were important and interesting. Perhaps the last straw came when a large chunk of one of the meetings was devoted to a discussion of whether the Puget Sound Consumer Cooperative (the "PCC") was good or bad for the neighborhood. A man got up and complained in graphic detail about PCC patrons coming with their dogs and permitting them to urinate on the flowers of people living nearby. This was a "life and death" issue, he wailed, since if something wasn't done soon, his roses would perish from the toxic onslaught. He was particularly irate that the police — many of whom shopped at the PCC — refused to act on his complaint. Soon after, we decided that local politics just wasn't for us.

One of our first impressions after moving to Seattle was that the racial climate was much better than in Boston. The Seattle public schools had been the largest school system in the nation voluntarily (i.e., without a court order) to institute busing for integration. Despite some controversy over the busing plan, there had been no violent protests — in contrast to Boston, where whites had stoned buses carrying African American children.

Near our house we would see multiracial groups of pre-schoolers walking together — something one would not see in Boston. Ann noticed that the jewelry ads in the newspaper sometimes featured interracial couples. This would have been extremely unusual, perhaps even impossible, in New England at that time.

Because we had come from Boston, Ann and I tended to see Seattle as a kind of utopia of racial tolerance by comparison. The truth of the matter was more complex, as we learned later. For example, historically there have been problems of racism in the University of Washington football program. In the 1960's and 1970's the head football coach Jim Owens was widely accused of systematically discriminating against blacks — an issue that came to a head with the suspension of four black football players in October 1969. Sensitivity on racial questions continues to elude the bureaucrats who run the university — as recently as 2003, at the urging of some white alumni who had fond memories of Coach Owens, they had a statue of him erected near the stadium. Local civil rights groups, such as the NAACP, objected strongly, but their protests fell on deaf ears. It's an embarrassment to many of us at UW that the only statue of someone other than George Washington anywhere on campus is a tribute to a football coach who was responsible for some of the worst racial incidents in the university's history.

Coach Owens' successor, who wanted to improve the university's image with black recruits, allowed a black player named Warren Moon to become quarterback. This was almost never done in those days, since the accepted wisdom among both college and professional football coaches was that a black player wouldn't have the intelligence to be a good quarterback. It turned out that Warren Moon was the best NFL quarterback ever to come out of the UW football program. But when interviewed in connection with his recent induction into the Pro Football Hall of Fame, he recalled that many of the UW fans in the 1970's would insult him and use racial epithets. After graduation, Moon chose to go to the Canadian Football League, which, unlike the NFL at that time, would permit him to continue in the position of quarterback. He said that it wasn't until he got to Edmonton that the football fans treated him with respect and did not insult him because of his race.

When we moved to Seattle, Ann and I also became aware of the long history of leftist political activism in the Pacific Northwest. The region was a center of union organizing by the "Wobblies" (Industrial Workers of the World) and later the Communist Party. The 1918 general strike caused many American officials to think of the region as a hotbed of Bolshevism, and this reputation continued at least into the 1930's. In 1937 one of Roosevelt's cabinet officers in jest proposed a toast to "the 47 states and the Soviet Republic of Washington." One famous early Seattle personality was Anna Louise Strong, a leader of the general strike and later a friend of Mao and supporter of the Chinese Revolution.

To some extent this tradition continues to the present. In 2002, about six months before the U.S. invasion of Iraq, Seattle congressman Jim McDermott traveled to Baghdad, where he condemned President Bush's aggressive stance toward that country. Some of his colleagues in Congress were outraged and accused him of treason. In contrast, his Seattle constituents showed their opinion of the matter by reelecting McDermott two months later by a landslide (about 75% of the vote). Seattle voters were overwhelmingly opposed to war with Iraq even before the war started.

In March 1982, while Ann was in Leningrad, I went to a demonstration in downtown Seattle against the Reagan administration policies in El Salvador. Between ten and twelve thousand protesters filled thirteen blocks. I was tremendously impressed with the turnout, the speakers, and the level of energy — and anger — among the demonstrators. This was not long after the assassination of Archbishop Oscar Romero and the killing of three American nuns and a lay worker in El Salvador, and the war of the Contras against Nicaragua was just getting under way. That was the first time that I realized that political activism in Seattle did not lag behind that in East Coast cities.

In the 1980's it turned out that proportionally more political action on Central America came out of the Pacific Northwest than any other region. When Ann and I started visiting Nicaragua and meeting other Americans who were there to support the Sandinista revolution, we were constantly bumping into people from Oregon and Washington. Seattle became sister cities with Managua, and when the mayor of Managua visited Seattle, he and our Mayor Royer issued a joint statement opposing Reagan's policy toward Nicaragua. Even *The Seattle Times*, which has always been a fairly conservative newspaper (often endorsing Republicans at election time), had its own independent and surprisingly good coverage of Central America. Their reporter, Emmett Murray, was fluent in Spanish and had spent a lot of time in Latin America. He knew the history of the region and was well aware that in the aftermath of the revolution, before the Contra war started, the Nicaraguans had made significant strides against poverty, disease, and illiteracy. The coverage of Nicaragua in the Seattle press compared very well with that, for example, in *The New York Times*, whose reporter in Nicaragua, Stephen Kinzer, had feelings of great hostility toward the Sandinista leaders and was biased in favor of the right-wing opposition.

In the early 1980's a young Oregonian named Benjamin Linder went to Nicaragua to support the Sandinista efforts to improve conditions in the countryside. A graduate of the University of Washington College of Engineering, he volunteered as a civil engineer in a rural community. One of his projects was to show the village how to construct an electric generator that would be powered by a nearby stream. Ben liked to ride a unicycle, and in this way he became popular with the children. He would lead them on his unicycle to the clinic and get their enthusiastic participation in the government vaccination program. On April 28, 1987 Linder was killed by a Contra hit squad. People throughout the Pacific Northwest were shocked, and there were massive demonstrations in both Seattle and Portland.

At the time I was a member of the Faculty Senate, which at UW is for the most part a powerless, pointless, and pompous assemblage of professors. I tried to get the group to observe a moment of silence after the killing of Linder. When the bureaucrat in charge refused to allow that, my colleague Tom Duchamp and I resigned in protest from the ridiculous organization. The UW student newspaper had a front-page article on our resignation.

Ann and I went to Nicaragua for the first time in March 1986 for two weeks. This was seven years after the Sandinista revolution, and the government was trying to achieve social goals while fending off military attacks from the Contras as well as political pressure by the internal opposition, both of which were largely financed by the U.S. government.

Through acquaintances in Seattle who had already been to Nicaragua we got some good contacts in Managua. However, no one told us — and, never having been to Latin America before, we weren't able to figure out on our own — that the week preceding *Semana Santa* (Holy Week) and the week of *Semana Santa* were the worst possible times to visit Nicaragua if one wanted to get anything done. By Thursday of our first week there, offices had all but emptied out and the vacation had begun. We managed to meet and talk with some interesting people during the first three days of our visit, but it was clear that we would have to return at a better time of year in order to plan anything.

Another lesson we learned from our first visit to Nicaragua was that we would have to learn Spanish. Ann had studied Spanish in school but had forgotten most of it, and I didn't know a word. In Managua we had to find people to translate for us from among the Americans who were working there and the few Nicaraguans who knew English. Actually, there were many Nicaraguans who knew English well, but they tended to be in the rightist opposition. Most Sandinistas had never been to the U.S. and knew little English. Until we learned Spanish, our visits would be a tremendous imposition on people who had better things to do with their time than serve as our translators.

Soon after returning from that first trip to Nicaragua, I began to study Spanish on my own. The following year Ann and I started getting weekly lessons from a Salvadoran-American who lived in Seattle; in addition, our friends Tom and Timmi Duchamp recorded Spanish cable news for us (at that time our region of Seattle had no access to any Spanish-language channel). "Noticias Univisión" was useful not only for language practice, but also for information about the culture and politics of Latin America.

Language is a political issue in many parts of the world, and nowhere more than in the U.S. Most Americans expect the rest of the world to communicate with them in English and see little reason to bother themselves with learning other languages. Culturally insular Americans are then affronted when they realize that a large and increasing proportion of the population right here in the U.S. have Spanish as their mother tongue. Hence the popularity of "English only" initiatives among Anglo voters.

After Ann got a position at Arizona State University in 1998 and we started spending a lot of time in the Southwest, we were struck by the extent of xenophobia and racism toward the Spanish-speaking population. When looking for roofers, we saw ads in the Phoenix Yellow Pages that included the words "no language barrier" as a code meaning "we don't have Spanish-speaking workers." Since companies with Spanish-speaking workers always have English-speaking supervisors, there was no practical reason why a customer should care. Rather, the point was that, as a Home Depot salesman said to us, "You don't want men all over your roof speaking a language you don't understand" — a typical example of Arizonans' ethnic bigotry. To me Arizona is reminiscent of Grenada, Mississippi in 1966. Much in the same way as white people talked about blacks in the deep south in the 1960's, Anglos in Arizona have no hesitation in making pejorative comments about Mexican-American immigrants in casual conversations with other whites whom they barely know.

Our impression has been that people throughout Latin America are keenly aware of the political dimension of language use by Americans. The quickest way to dissociate oneself from the negative image of Americans as arrogant imperialists is to speak Spanish. In the late 1980's, when Ann and I were still far from fluent in the language, we found that Latin American colleagues often preferred to speak Spanish with us even if their English was better than our Spanish.

We went back to Managua in September 1986 in order to plan for Kovalevskaia Fund activities in Nicaragua. We made contact with a group, known by the Spanish acronym CONAPRO, which was the umbrella organization for Sandinista-affiliated professional societies. Nicaraguan occupational groups were often polarized, with each profession having a pro-Sandinista association and a counterrevolutionary one. The full name of the organization we worked with was "CONAPRO Heroes and Martyrs," in which the epithet was a way to honor those who had died in the revolutionary struggle and at the same time leave no doubt what the political orientation was.

We used Vietnam as a model of what to propose. The people at CONAPRO — Silvia Narváez and Lilia Alfaro in the foreign relations department, and President Fredy Cruz — readily agreed to set up a Kovalevskaia Prize and also cosponsor with us a regional conference on women in science.

While in Nicaragua we became friendly with an American family that had moved there from Seattle in the early 1980's. Edgar Romero had received a Master's degree from my department at the University of Washington and had become a math professor at the National Autonomous University of Nicaragua (UNAN). His father Marco, who had taken early retirement at Boeing, worked for the Nicaraguan airline Aeronica, and his mother Frances ran a school for the disabled in Managua and maintained active ties with several pro-Nicaragua groups in the Pacific Northwest. During the five years when we were visiting Nicaragua frequently, the Romeros were a constant source of help and useful suggestions. I especially valued my long conversations with Edgar, who knew a lot about the intricacies of university and national politics, as well as about curricular issues in mathematics.

Edgar explained to me how good intentions in math education often went astray. In the hope of raising academic standards at the college level, university textbooks and curricula had been adopted from European countries with no changes. Since Nicaraguan students' preparation was far weaker than that of their European counterparts, in practice this meant that they would memorize parts of the textbook with no understanding whatsoever, and they'd usually be tested simply on their ability to repeat things from the text.

At the elementary level as well, tremendous damage was being done by the mindless importation of curricular materials. For example, the "new math" that had been developed mainly in the U.S. and France in the 1960's and had fallen out of favor in most countries by the 1980's, had somehow made its way into Nicaragua's K–12 program. The result, predictably, was disastrous. Once when Edgar and I were passing by a high school classroom we heard the students repeating in unison after their teacher, "An equivalence relation is a relation that is symmetric, reflexive, and transitive. An equivalence relation is a relation that is symmetric, reflexive, and transitive." I was reminded of the ritual of reciting the Lord's Prayer every day in the bus to the Catholic school I attended in second grade in India. "Forgive us our trespasses..." As a six-year-old Jewish boy I had had as little understanding of what that meant as the Nicaraguan youngsters had of what an equivalence relation was all about. To them the properties "symmetric, reflexive, and transitive" were something to be recited like the Holy Trinity: "... in the name of the Father, the Son, and the Holy Ghost. Amen."

The first few visits Ann and I made to Nicaragua occurred at a time when U.S. policy toward that country was at the center of national debate. The Iran-Contra scandal broke in November 1986, and a protest movement was growing among liberals, radicals, and church people. In fact, some said that Nicaragua would be the "Vietnam of the 1980's."

The Sandinista National Liberation Front (FSLN), which had been governing Nicaragua since 1979, fully appreciated the value of American sympathizers. During the Somoza dictatorship the country's educational system had remained backward, and the revolutionary government did not have a large number of highly trained Nicaraguans to carry out all of the professional tasks that were needed. So help from foreigners — especially Americans — was welcomed in all areas. For example, when we were interviewed on Sandinista TV we noticed that most of the people running the equipment were Americans.

A couple of years earlier the Reagan administration had closed the Nicaraguan consulates in most U.S. cities. At first it seemed that this would cause a bottleneck in processing visa requests from all the Americans who wanted to come down to help. The Sandinistas had an ingenious response — they simply abolished the visa requirement for U.S. citizens. They understood clearly that such a requirement never stopped the CIA from getting their people into Nicaragua; it was a bureaucratic obstacle only for their friends. So they did away with it.

By the mid- to late-1980's there were thousands of Americans involved in practically every part of the Nicaraguan economy and educational system. It got so that conservative pundits in the U.S. would snidely refer to Nicaragua as the "Disneyland of the left." In a sense they were right. Once when I was in the airport in Managua, I saw that the Bearded Lesbian Circus from Berkeley had arrived in town. Some Nicaraguans might have doubted whether that was what the country most needed during a time of crisis. However, the strategy of the FSLN was to welcome *all* forms of support from American sympathizers and to try to find a niche for any individual or group that flew in to help the revolution.

The Central American Conference on Women in Science, Technology, and Medicine was scheduled for August 1987, and in June of that year I traveled to five countries of the region in order to interview people who were interested in coming and encourage their participation. During this period the Central American governments that were the closest allies of the U.S. — especially Honduras and El Salvador — had extremely tense relations with Nicaragua, so it took some courage for scientists from those countries to go there.

My Spanish was still rudimentary, so the "interviews" were on a pretty superficial level, at least in the case of the people who didn't speak English. The Kovalevskaia Fund ended up paying travel expenses for a total of seventeen women from Costa Rica, Honduras, Guatemala, El Salvador, and Mexico.

In El Salvador I met Concepción ("Concha") Lemus de Bendix, who had received her Ph.D. from the University of Washington many years before and was a professor of microbiology at the University of El Salvador (UES). Her son Bill Boyle and his family had recently moved to Seattle, and later that summer Ann and I contacted him about giving us Spanish lessons. For a ten-year period until he left Seattle, Bill gave us lessons on a regular basis. Most of those sessions were devoted to laboriously correcting our Spanish in various talks that Ann and I were preparing to give in Latin America and in articles of ours that we were translating.

The UES campus where Concha Bendix worked was in the center of the capital. It was by far the largest and most important college in the country; the only public university (except for branch campuses in other cities), it was often referred to simply as the "National University." Concha also knew people at the other major university, Central American University (UCA), which was run by the Jesuits. She introduced me to the president of UCA, Father Ignacio Ellacuría, whose permission was needed in order for a chemistry professor Concha knew at UCA to be able to go to Managua for the conference. Concha thought it best if I explained directly to Ellacuría what the purpose was of the event in Nicaragua. Ellacuría decided that we would communicate best if I spoke slowly in English and he spoke slowly in Spanish, and this worked well. He agreed to approve the trip to Managua of the chemistry professor, Olga Esquivel.

After we left his office, Concha remarked to me that he was a very courageous person. Originally from Spain, Ellacuría was the leading philosopher and theologian in the country and was identified with the "liberation theology" movement. He had just written an article explaining why leftist guerrilla groups such as the FSLN in the 1970's in Nicaragua and the FMLN in the 1980's in El Salvador had a kind of legitimacy that rightist militias such as the U.S.-financed Contras did not have. Later someone at the U.S. Embassy described him to me as the leading opposition figure in the country. Of course, none of that prepared me for the horrific news of the atrocities that would take place on the UCA campus two years later.

When I was in Managua, I asked Edgar Romero to take me to the Interior Ministry to deliver a letter and some photographs for the Minister of the Interior, Comandante Tomás Borge. Ann and I had heard that Borge was an accomplished poet and thinker, as well as a key leader of government. He was also a living link with the very beginning of the Nicaraguan revolutionary movement; he was the only person still alive among the nine founding members of the Front. We thought that, just as Phạm Văn Đồng had helped attract a lot of publicity to our conference in Hanoi, it would similarly enhance the visibility of the Managua conference of women in science if Borge could be persuaded to come.

We had prepared a formal letter of invitation and copies of photographs of ourselves with Phạm Văn Đồng. Basically, we were using the precedent of Phạm Văn Đồng to convince Borge that a conference on women in science was the place for a true revolutionary to be. We were also using the Hanoi conference to establish our credibility with him.

The gambit worked. Not only did Borge come to address the conference, but he prepared an hour-long discourse especially for the meeting. After speaking for fifteen minutes about the political situation in the region — and praising the recently concluded Esquipulas peace accords signed by the Central American presidents (*The New York Times* had falsely reported that Borge had opposed the agreement) — he proceeded to give an overview of the history of misogyny by intellectuals and religious leaders. He then discussed the situation of women in Nicaragua, making astute observations that two decades later still seem fresh and topical. Recently Ann and I watched our video recording of Borge's address, which we hadn't done in many years. We were struck by the depth and sophistication of his analysis of the origins of feminism and its meaning for Nicaragua. The translated excerpts of his speech given as a postscript to this chapter do not fully capture the rich sounds and cadence of his words, which reflect his poetic sensibility.

On August 22, 1987, when Ann and I arrived in Miami to make our connection to Managua, we had a snafu. Although Nicaragua did not require visas, it did demand that we have U.S. passports that were valid for at least six months after the time of travel — but Ann's passport would expire in just three months. TACA airlines would not let her board the flight, because they had no assurance that Nicaragua would admit her. So I went ahead without her, and the following day (which was a Sunday) was able to arrange for the Nicaraguan government to telegram TACA guaranteeing that they'd admit Ann. Ann had spent the night in Miami, and as soon as she came to the airport the next morning the TACA guy waved to her and said that everything was okay and she could take the next flight. I had no way of contacting her, and was still extremely worried when I went to the airport with the CONAPRO people to meet the flight that we hoped she'd be on. We were all very relieved when she emerged from the plane.

Over fifty women, roughly half from Nicaragua and half from ten other countries, participated in the conference. The first pair of Kovalevskaia Prizes were conferred on two medical researchers from the university in León (Nicaragua's second city after Managua). Both of the pro-Sandinista daily newspapers carried front-page articles on the prizewinners.

The thirty talks fell into five categories — women and health, women in math and physical sciences, appropriate technology, education, and broad overviews. After the conference we published a bilingual volume of proceedings.

Participants were taken on excursions to laboratories and medical and educational facilities in Managua and León, and we also attended a "Face the People" encounter between government leaders and students. Finally, one evening a Nicaraguan feminist lawyer showed the group a video of a "Face the People" session the year before that dealt with the treatment of women's rights in the new draft of a national constitution.

Almost all of the Central American women who came to the conference had been politically sympathetic to Nicaragua before, and they reacted enthusiastically to what they were seeing of the country's progress in health, education, and women's welfare. They were also impressed with the level of internal democracy and government accountability. For example, at the three-hour "Face the People" session we saw student representatives harangue the government leaders, including President Ortega and Minister of Higher Education Joaquín Solis, with angry complaints and lists of grievances. The government leaders listened patiently, and from time to time an official would answer with explanations, promises, or apologies. Several of the foreigners commented that such a meeting would never take place in their own countries.

The only woman who was not politically on the left was Emilisa Callejas from Honduras. She was the sister of a candidate for the presidency of Honduras (who was later elected). At one point she objected to a draft statement from the group condemning "the actions which the U.S. government is conducting to finance and maintain the war in Central America." Emilisa said that in order to be evenhanded, the group should also condemn Soviet interventionism. Everyone reacted angrily to that, and she withdrew her amendment, after which the statement passed unanimously.

For Ann and me a long-term benefit of the conference was some of the friendships we formed there. Mary Glazman, the only participant from Mexico, was a mathematician at the National Autonomous University of Mexico (UNAM). In 1991 she organized an international meeting on the occasion of the centennial of the death of Sofia Kovalevskaia, and she invited Ann to speak on the life and work of Kovalevskaia. She was the nucleus of an active group of women in math and science at UNAM until she died in 2000.

I already mentioned Concha Bendix, who later helped tremendously with our visits and Kovalevskaia Fund projects in El Salvador. In 1991 we invited her to join the Board of Directors of the Kovalevskaia Fund. Another person who became a Board member was Sister Ellen Cunningham, who was chair of the department of science and mathematics at Saint Mary-of-the-Woods College in Indiana (the oldest Catholic women's college in the U.S.). When Ellen was in Nicaragua for the conference, she also met with church groups and with an organization called "Mothers of Heroes and Martyrs." She wrote an eloquent and moving article about her experiences for the college's alumnae magazine.

For several years Ellen lobbied to get Ann some special recognition for her work with the Kovalevskaia Fund. Finally, in May 1995 Saint Mary-of-the-Woods College conferred an Honorary Doctorate of Letters on Ann and had her give a Commencement address. I found it a fascinating experience to be with her during the festivities at the college. It appeared that Ann had been the faculty's choice; there was a second Commencement speaker (a corporate type and alumna of the college) who was the favorite of the Board of Trustees. Many of the nuns were, like Ellen Cunningham, politically progressive. They subscribed to the anti-Vatican newspaper *National Catholic Reporter*, and in the Convocation they incorporated animist and feminist elements into what was supposed to be a Catholic ritual. I thought that the ceremonial parts of the program were really well done, and I was extremely proud to see Ann at the podium receiving her honorary degree.

After the Managua conference in August 1987, Ann and I were pretty pleased — we felt as if we could walk on water. Never having organized any sort of conference before, we had put together two of them within an eight-month period in the capital cities of two countries that were considered enemies by the U.S. The regional meetings had obvious symbolic meaning, and they had also achieved the objective of giving support and encouragement to women scientists. Moreover, both conferences had been relatively inexpensive.

However, in certain respects we were extremely naive, and we made some mistakes. For example, we had several thousand copies made of the proceedings volumes for the two conferences. We thought that the content was interesting and at a high level — and we still think that when we look back at them — and foolishly assumed that that would mean that there'd be a big demand for copies. We learned the hard way that publications get distributed through existing networks — mostly, publishers and professional organizations — and not by sitting around waiting for people to ask for them. Ultimately we distributed only a few hundred copies of each proceedings volume; the rest sat in our garage for many years until we got around to carting them off to recycling. That was a waste of several thousand dollars.

We also underestimated the difficulty of following through with the Kovalevskaia Fund projects under the conditions in Nicaragua, which had no institutions that were comparable to the Vietnam Women's Union or the Hanoi Math Institute. The Kovalevskaia Prizes continued for two more years and then petered out in 1990. There was an attempt at a second Central American regional conference of women in science in Costa Rica in June 1989, but it was much less successful. Ann went, and the Kovalevskaia Fund financed the travel of several Nicaraguans and one Salvadoran. But most of the organizational work was in the hands of the Costa Ricans, and they were unable to pull it together. Ann found that Costa Rican feminists were split into several feuding groups. There was a saying that if you had five Costa Ricans in a room, you'd have six factions.

I continued to travel to Nicaragua twice a year from 1988 to 1991, and Ann would come with me on one or both trips. We learned a great deal from our conversations with people in CONAPRO and some of the foreigners working in Managua. We talked a lot with a German guy named Cornelius Hopmann at the National Engineering University and with a Russian named Elena who taught physics at UNAN. I especially liked talking with Edgar Romero and his parents.

In February 1990 the FSLN lost the national elections, and a conservative coalition took power. At first we thought that we could continue with our projects, since the government of Violeta Chamorro did not explicitly say that American volunteers would no longer be welcome.

The Kovalevskaia Prizes were not being given that year, supposedly because all the Sandinistas had been feverishly working on the election campaign and had had no time to organize the selection of winners. But after the election the people we had been working with appeared demoralized and out of energy. Nothing seemed to be getting off the ground. At this time we were also trying to support a women students' center at UNAN that, among other things, would combat sexual harassment, which was a tremendous problem on campus. After a promising beginning, that too fell through. In 1991 we talked with people in León about giving a Kovalevskaia Prize at the university there the following year. But when I called Edgar Romero from Seattle in December 1991, he said that as far as he knew no one was organizing anything. He also told me that his family was moving back to the U.S. If I came in March of 1992, as I had planned to, he knew some people who would try to help me. I thanked him, but Ann and I concluded that it was pointless to return to Nicaragua. We decided to visit Peru instead — something we'd been wanting to do for some time — and perhaps start Kovalevskaia Prizes there. We have not been to Nicaragua since June 1991.

In the mid- and late-1980's I joined various groups that were protesting against the Reagan administration's policies in Central America. In autumn of 1986, which I spent at the Math Sciences Research Institute in Berkeley, I took part in protests organized by students at U.C. Berkeley.

For the most part Ann and I found people in the solidarity movement to be congenial and politically astute. However, on occasion we were perplexed by some of their attitudes.

For example, when the Nicaraguan Minister of Higher Education Joaquín Solis visited the University of Washington, we went to a small meeting he had with faculty members and graduate students who had ties with Nicaragua. Ann and I asked Solis about women's participation in programs of study abroad. When it turned out that he had no data on that, we stressed the importance of monitoring that sort of statistic. Given the traditional attitudes toward daughters (that Tomás Borge had referred to in his speech), young women were probably underrepresented among those studying abroad, and it was the Sandinistas' responsibility to try to remedy this.

After the meeting, a couple of the other participants criticized us for putting Solis on the spot, making implicit criticisms of his ministry, and trying to "impose our agenda" on the Nicaraguans. What struck us as particularly bizarre about this objection to our comments was that one of the people chastising us was an activist in gay causes, and had questioned Solis about the Sandinista policy toward homosexuality. Now anyone who knows anything about Latin America should realize that an education minister would much rather answer questions about gender equity in higher education than about gay rights. *We* were not the ones who were trying to impose an alien agenda on someone.

A few months later we ran into Solis at a party CONAPRO had in Managua. He made a point of sitting with us and chatting. He clearly did not think that the questions we had asked him in Seattle had been out of line, and he had no objection to further conversations on related themes.

Although one cannot know for sure, it is likely that the Central America solidarity campaign of the 1980's played a role in preventing a direct U.S. invasion of Nicaragua to crush the Sandinista government. The memory of the Vietnam War was still fresh in the minds of most Americans, and the continual protests against the Contra war in Central America that were organized in cities and on college campuses reminded the Reagan administration of the bitter internal divisions that would occur if they sent U.S. troops to effect "regime change" in Nicaragua.

Many of the activists were too young to have participated in the anti-war movement of the 1960's and early 1970's. For example, Benjamin Linder was born in 1959, and the group I worked with in Berkeley in 1986 consisted primarily of students and recent graduates.

In our day many people have wondered about the absence of any significant organized opposition to the Iraq War among young people. In comparing the current decade with the 1960's, commentators often cite the absence of a military draft as the reason for the deafening silence of the younger generation on the war. However, this explanation is illogical. After all, in the 1980's there was no draft — and no American soldiers were dying — yet large numbers of young people actively campaigned against U.S. policy in Central America.

In early 2003, just before the invasion of Iraq, some professors at the University of Washington organized a "teach-in" that they hoped would jump-start a campus anti-war movement. Although they got a lot of publicity for the event, student interest was minimal, and nothing of significance resulted from it. In the years that followed, as the number of American deaths in Iraq climbed to over thirty-five hundred — with almost ten times that number returning to the U.S. from Iraq and Afghanistan with injuries, often serious ones, and with estimates of the number of Iraqi civilians killed ranging from 30,000 (President Bush's figure) to 655,000 (the conclusion of a team of American and Iraqi epidemiologists) — there has been little indication of any emerging activism at the nation's colleges and universities.

In March 2007 three demonstrations were held over a two-day period in Seattle to mark the fourth anniversary of the attack on Iraq; I went to two of them. According to local news reports, a total of roughly five thousand people participated in the marches — that is, less than half the number that had taken part in the protest over El Salvador that I had attended exactly a quarter century before. A dramatic contrast with the protests of the 1980's and earlier was that, although there was a sprinkling of young people in 2007, that age group was underrepresented among the marchers and especially among the organizers. My impression is that the main reason why the protest movement against the Iraq War has been sluggish and anemic is that American youth — which provided the critical mass for similar grassroots movements in the 1960's, 1970's, and 1980's — has been largely absent.

It might be unrealistic to hope for something on the scale of the anti-war movement of the 1960's. But the absence of anything comparable even to the anti-apartheid movement of the late 1970's or the Central America solidarity campaign of the 1980's cries out for an explanation. And one has to look deeper than just the absence of a draft to understand the apathy, complacency, and ignorance about world affairs of most young Americans today.

The "baby-boom" generation that was born in the late 1940's and early 1950's — and to some extent those born in the decade after that — had a moral sensibility formed in large part by two historical circumstances. In the first place, our parents had lived through the Great Depression and World War II, and we were raised in the aftermath of the Nazi atrocities. Many of us had a keen consciousness of what it had meant to be "good Germans" who sat idly by while one's government committed crimes against humanity. In the second place, the civil rights movement, which was largely led by young African Americans, reached its zenith during our formative years. By the end of the 20th century both of these influences had receded into distant historical memory and no longer had an impact on the social awareness of young people.

In superficial ways today's youth culture is remarkably similar to what it was when I was in college. Mick Jagger, now in his 60's, can still fire up a crowd of young people with "(I Can't Get No) Satisfaction," as he did in the 2006 Superbowl halftime show. And students' styles of clothing and appearance in 2007 are probably closer to those in 1967 than the styles in 1967 were to those in 1957. But beneath the surface much has changed, and college students now have very different priorities.

In the 1960's journalists often spoke of the "affluent society" with "permissive" childrearing practices, and noted that our generation was the first to have spent part of our childhood in front of the TV. Indeed, coming after a generation that had lived through a world war and economic depression, most "baby-boomers" seemed to have ample material comforts. However, compared to the University of Washington students I see every September unloading their vast possessions in front of the dormitories, students in the 1960's led spartan lives. We didn't have access to credit cards, and installment buying was also available only to those who were already in the workforce. It was almost unheard of to have a TV in one's dorm room; and even at elite universities, such as Harvard and Princeton, hardly any undergraduates owned cars. Whether one's family was poor or wealthy, it was considered uncool to spend a lot of money on clothes. In contrast, in our day young people are the choice demographic for upscale clothing stores; and if you see a beat-up old Honda next to a late-model SUV in the parking lot, chances are the Honda belongs to a professor and the SUV to a student.

This orientation toward material possessions has far-reaching consequences for today's students. For example, typically they are saddled with problems of paying off credit card debt, and feel under pressure to graduate and get a high-paying job as soon as possible. With these immediate worries, it is no surprise that issues of social justice and U.S. foreign policy are not on their radar screens.

I don't want to overstate the matter, however. From time to time one hears of counterexamples to the generalizations one can make about the younger generation today. A famous case is that of Rachel Corrie (born in 1979), who in March 2003 was killed by an Israeli Defense Forces bulldozer while protesting against mistreatment of Palestinians in the Gaza Strip. Like Ben Linder, she grew up in the Pacific Northwest — her childhood and student years were spent in Olympia, Washington, where she attended Evergreen State College.

In March 1988 when Ann and I were in Managua, Edgar Romero asked us if we'd be interested in talking with someone his family knew from the Cuban Embassy about possibly visiting Cuba. We said sure, although we had never seriously considered going to Cuba.

We were politically sympathetic to the Cuban revolution and were well aware of its influence as an inspiration for liberation movements throughout Latin America, but at that time our most recent impression of Cuba had been negative. At the Central American Conference on Women in Science, Technology and Medicine, many of the participants had wondered why no one was there from Cuba — and a few had even thought that the Kovalevskaia Fund, based as we were in the U.S., had stood in the way. But Ann and I had specifically asked CONAPRO to invite the Cubans. They later told us that they had, but the Cubans had declined to send anyone, perhaps because of the cost. This was strange, since Cuba had extensive ties with Nicaragua, having sent hundreds of physicians and technical advisers to support the Sandinista government.

We then asked the CONAPRO people why they couldn't just invite Cubans who were already in Nicaragua. Surely there must be some women doctors or engineers who could easily come and speak on women in science and technology in Cuba. What CONAPRO told us was shocking. It turned out that Fidel Castro had declared that, because of the danger of a Yankee invasion of Nicaragua, no women would be permitted to go there. Any Cuban who went to Nicaragua must be prepared to fight side-by-side with the Nicaraguans against the feared invasion, and this precluded any women. We later learned that Cuban women had been insulted by this, and few Cubans (or Nicaraguans) had thought that such a policy made any sense. But no one could argue with Fidel.

Some time after that Ann and I were talking with some Mexican academics (who were sympathetic to Cuba) about this episode and our dismay at Fidel Castro's attitude. They said, yes, well in Cuba the ideology is *machista-leninista* — a delightful play on the term Marxist-Leninist in Spanish. Even in Mexico, which was not known for being a country free of *machismo*, Cuba had the reputation of having a big problem with sexism.

When Edgar drove us to his parents' house, the man from the Cuban Embassy was sitting with them in the backyard. His name was Aníbal, and his card said in English "Vice-Technical Adviser," whatever that meant. It was apparent that he really worked for Cuban intelligence. He knew that the Americans who were coming to Nicaragua in large numbers to provide technical assistance to the Sandinistas would also be likely to be sympathetic to Cuba, and through their embassy in Nicaragua the Cubans could develop ties with Americans who opposed the embargo.

Aníbal offered to arrange a week's visit for us at the expense of the Cuban Academy of Sciences. We were planning another trip to Nicaragua for June of that year, and we'd go to Havana directly from Managua on a Cubana flight. Even our air tickets would be provided free of charge. This meant that we would not be violating the U.S. embargo, which is a prohibition on spending money in Cuba, not on traveling to Cuba. We gave Aníbal detailed information about our interests and whom we would like to meet with in Havana.

When we returned to Managua in June, we brought a computer for one of the Kovalevskaia prizewinners, a student of Edgar Romero at UNAN named Neyssa Calderón. Our friend Tom Duchamp, who's also a mathematician at the University of Washington and a member of the Board of Directors of the Kovalevskaia Fund, was spending part of his sabbatical in Managua. He'd be able to help Edgar and Neyssa set up her computer, which was no trivial task in those days.

The prize awards ceremony was impressive: UNAN organized a large convocation, attended by about 250 students and faculty. Two out of the three daily newspapers carried articles about the prizes (the right-wing paper *La Prensa* did not). The newspaper *El Nuevo Diario* featured a detailed interview with Neyssa, whose picture appeared on the front page. The interview ended with the observation that Neyssa, with her love of mathematics, "refutes the myth that of the Nicaraguan population a million and a half are poets, and the remaining million and a half also want to be poets."

After several days in Managua, we flew to Havana for our first visit to Cuba. The week was pleasant — as promised, all of our needs and expenses were covered by the Academy of Sciences. We stayed at the Riviera, a rather fancy hotel on the Malecón (the coastline road). An Academy official was our host; he spent a large part of the week taking us to tourist sites around Havana, the University of Havana, and the world-famous biotechnology research center.

However, most of the people we had asked to see were unavailable, and nothing substantive had been organized. Near the end of the week a talk was finally arranged for me at the Math Institute; it turned out that they had been told of my visit only at the last minute. My talk on elliptic curve cryptography, which I gave in a mixture of Spanish (which I still didn't know well), English, and Russian (which some of the mathematicians knew), was well received. But other than that our trip was largely just a tourist junket. As the week progressed, Ann and I increasingly wondered what the value was to Cuba of such a visit.

Our host never made any direct requests, although he said that Cuba was interested in getting information in two areas of technology that were priorities for them — microelectronics and bioengineering. We didn't understand exactly what he wanted us to do, so we countered with a concrete proposal: we would organize a visit by a leading woman researcher in bioengineering. Our model was Arlene Ash's successful trip to Vietnam just one year before that had been supported by the Kovalevskaia Fund. We said that the Fund would get together a committee of prominent specialists that would judge a widely publicized competition for a travel grant. The Fund would pay the scientist's travel as far as Mexico City, and the Cuban Academy of Sciences would fly her on Cubana to Havana and pay her local expenses, exactly as they had for us coming from Managua. We had no doubt that this would work for Cuba, just as it had for Vietnam.

However, our host was unenthusiastic. His main objection was that the Cubans didn't think that it was a good idea to be so open. He seemed fixated on the idea that cooperation with Cuba had to be done in secret. He spoke of the horrible reprisals that would befall anyone in the U.S. who was openly friendly to Cuba. Ann and I thought that was ridiculous. It was certainly true that people in Miami — especially Cuban-Americans — might be viciously attacked if they were sympathetic to socialist Cuba. But we tried to explain that the rest of the country, and especially Seattle, had nothing in common with Miami. Few people in Seattle were hostile toward Cuba, and someone could write a letter to a local newspaper praising Cuba without any fear of reprisals.

We also thought that we — and the prospective biotechnology expert — could be trusted to judge for ourselves whether or not it would be risky to participate in a well-publicized program with Cuba. We knew the conditions in our own country better than the Cubans did.

Moreover, what Cuba needed was not secrets. Cuban science needed to overcome the effects of the isolation that was caused by the embargo, and the way to do that was to have as many widely publicized scientific visits as possible. Science advances by the unfettered exchange of ideas and information, not by getting hold of secrets.

Ann and I tried to figure out why the Cubans had such a paranoid impression of the atmosphere in the U.S. regarding ties with Cuba. Unlike the Nicaraguans and Vietnamese, they seemed to view the U.S. as a monolith and had a hard time distinguishing between what was going on in Miami and in the rest of the country.

With time, we came to suspect that part of the blame lay with American travelers to Cuba who in some cases had conveyed an exaggerated impression of the sacrifices they were making and the risks they were running. When young leftists joined the *"venceremos* brigades" to help with the sugar harvest, there was an element of "forbidden fruit" in their trip to Cuba. Thinking of the famous posters of Che Guevara against the sunset, they might have found it hard to resist a little self-dramatization. For the most part this was harmless enough. However, when Americans gave an overblown description of the reprisals they expected to encounter at home, an unfortunate consequence might have been to reinforce Cuban paranoia about the possibilities for open collaboration with colleagues in the U.S.

After we returned to the U.S., I followed up on the Kovalevskaia Fund proposal by writing to the Treasury Department office that was in charge of the embargo. I described our project and asked if we were correct in thinking that, since the scientist would spend no money in Cuba, it was not in violation of U.S. government restrictions. I received a cordial reply, somewhat bureaucratically worded, that essentially said that we were right. However, training programs were not permitted to be given in Cuba, so we must be sure that the scientist's activities there were described as an exchange of scientific information, not as a type of training. Okay, that was fine with us.

We sent a formal version of our proposal translated into Spanish, along with a copy of the correspondence confirming that the project would not be in violation of U.S. law, to both the Cuban Interest Section in Washington and the Cuban Embassy in Mexico. In addition, we asked our friend Lee Lorch at York University in Canada (who personally knew several officials of the Cuban government) to help us get a response. However, we never got an answer.

After our visit in June 1988, we talked with several others — from Costa Rica, El Salvador, and the U.S. — who had had similar experiences when they went to Cuba. The people in their field were busy or unavailable, nothing was organized in time, and little of value for Cuba came out of their visits. In contrast with Nicaragua, which during the twelve years when the Sandinistas were in power was able to seamlessly integrate thousands of North American volunteers into every aspect of professional life, Cuba in that time period seemed not to know how to make use of ties with sympathetic professionals in the U.S. and Latin America.

When we arrived in Cuba the person who checked Ann's passport absentmindedly put a Cuban stamp in it. She wasn't supposed to do this — the Cuban policy was to honor requests not to stamp Americans' passports so as not to cause them trouble for having visited Cuba. Ann and I were concerned about the stamp, since we were going to be returning to Seattle through Miami. It wasn't that anything terrible would happen — but there was a slight chance that an inspector would take her aside for an unpleasant interrogation and make us miss our connecting flight.

On the plane back to Managua Ann made a brilliant observation: the Nicaaguan stamp was of the same size and colors as the Cuban one, but upside down. In other words, if the Nicaraguans put their stamp upside down on top of the Cuban one, it would look like nothing but a voided Nicaraguan stamp. When we arrived at Nicaraguan passport control, she asked the guard to do that. He said that he couldn't, because they respected the Cuban stamp and were not authorized to void it. Ann had noticed our old friend Aníbal seeing some people off at another part of the airport, and she asked whether the guard could cancel the Cuban stamp if Aníbal said it was okay. He said that in that case he could. He went and got Aníbal, and Ann explained what she wanted. Aníbal was impressed — I later said to Ann that probably a momentary thought had crossed through his mind that he should recruit her to become the Mata Hari of Cuba. Aníbal went with the guard into a back room and together they did an excellent job of disguising the Cuban stamp as a voided Nicaraguan one. Only someone who looked very carefully and knew what to look for would be able to tell what was really there.

As it happened, Ann didn't have to go through a detailed inspection in Miami, so all the fuss was probably unnecessary. But Miami can be a strange place, and it was best to take precautions.

After our Cuba visit of 1988 I had intermittent correspondence with a number theorist named Jorge Estrada who had talked with me at the Math Institute in Havana. He encouraged me to visit again, although for a long time I had no inclination to do so. Then I spent the academic year 1998-1999 at Waterloo, only an hour or so away from the Toronto airport, from which inexpensive flights leave for Havana all the time. In fact, Cuba is a favorite vacation destination for working-class Canadians, particularly during the long cold winters. In addition, while on sabbatical I didn't have a teaching schedule and so could easily leave for a week. In April 1999 I visited Cuba for the second time.

Much had changed in the decade since my first visit. After the collapse of the Soviet Union, Cuba had diversified its international ties. The place seemed more relaxed and less bureaucratic. This time my visit was productive, thanks to the planning of Jorge Estrada and his wife Victoria Hernández, who is also a mathematician. I gave talks at the Math Institute and the University of Havana, taught a lesson on cryptography at an elementary school, and spent a couple of hours with a group of high school students who were training for the International Mathematical Olympiads. I also talked with a mathematician named Lilliam Alvarez, who had been at my talk in 1988 and was interested in activities to support women in math and science. Encouraged by my visit in 1999, Ann and I decided to make another trip to Cuba two years later to set up a Kovalevskaia Prize there.

In 2001 we again went to Havana for a week and had detailed discussions with an ad hoc group of women scientists convened by Lilliam. Under the auspices of the Cuban Academy of Sciences they formed a Kovalevskaia Prize Committee. Every two years we would support three prizes — one in the mathematical, one in the physical, and one in the chemical sciences. The Cubans wanted to exclude the biological sciences because, in the first place, they were already being funded better than other fields of research by the Cuban government and international organizations; and, in the second place, women were not as underrepresented there as they were in the other sciences. However, in practice some of the women who were awarded prizes in 2003, 2005, and 2007 worked on applications of math, statistics, physics, or chemistry to the life sciences — so biology had not really been excluded.

One afternoon we took a stroll around the Plaza de Armas in Old Havana browsing the bookstalls. Private sellers would display old books that they had bought from collections of people who had died or left the country. This tourist excursion turned out to have a major impact on Ann's scholarly work. By chance we found two unusual books, one from the 1920's and the other from the late 1800's, that tied in directly with Ann's work on her third book, which was going to be a historical and cross-cultural study of birth control and abortion. One of the books was a Spanish translation of a French treatise by a famous physician named Ambroise Tardieu on abortion and forensic pathology. Ann was fascinated by the monograph, and it led her to read related material from the same period. A few years later Ann wrote a chapter for her third book on the close relationship between the early history of forensic medicine and the abortion controversy. So our trip to Cuba paid an unexpected dividend for her research.

Since 2001 our plan has been to travel to Cuba once every two years to give the Kovalevskaia Prizes and talk with colleagues. In 2003 the first prizes were given. Leaders of the Academy of Sciences were present, and a detailed article about the awards appeared in *Granma*, the newspaper of the Cuban Communist Party.

Travel to Cuba gives us a valuable perspective. A country that has stood up to a half century of bullying by the colossus ninety miles to the north, it has developed institutions that compare well to those of other countries in the region. Despite the bureaucracy and insularity we encountered in our first visit and despite the sexism that is still a problem there, it must be acknowledged that in many areas — literacy, education, science, health, reproductive rights, and emergency preparedness during hurricane season — Cuba continues to represent a beacon of hope for the Americas.

Postscripts

The following are translated excerpts of an address by Comandante Tomás Borge Martínez at the Central American Conference on Women in Science, Technology and Medicine, held in Managua from August 24-28, 1987. Borge was the Minister of the Interior during the 1980's. Well known as a poet as well as a political leader, he had tremendous prestige as the only surviving founding member (out of nine) of the Sandinista Front that overthrew the Somoza dictatorship in 1979.

...Aristotle — differing from his teacher Plato, who recognized that discrimination against women meant the loss of the resources of half the population — claimed that a woman was an incomplete and defective man... Today it is unthinkable to accept the stand of Saint Augustine, who gave woman the role of "temptress" and man the role of her owner.

Some of those who physically abused women in that era felt that they were authorized by the wise Saint Augustine, who had given men the right to physically punish women.... In like manner, Saint Thomas Aquinas — my namesake — joined the camp of Aristotle and not of Jesus Christ, in the sense that man supplies the essential human characteristics in reproduction, while the mother provides only the nutritive matter.

More explicit in their savagery were the Dominican Inquisitors Heinrich Kramer and Jakob Sprenger, who in their encyclopedia of demonology titled *The Hammer of Witches* stated in all seriousness that women were particularly susceptible to witchery because of their shallowness, their weak intelligence (there would be no women scientists), and their being cursed by an insatiable sexual appetite, even to the extent that their lechery could lead them to copulate with the devil.

The witch hunts started roughly with these Dominican Inquisitors, and scholars say that between 1500 and 1700 a million women were burned. Of course, the witch hunts have not ended, although nowadays they have taken on a subtler form, as we all know well, as the illustrious sexists and apologists of *machismo* know well...

To be sure, it was soon after that feminism arose, and in the midst of the revolutionary fervor of the 18th century a sort of first manifesto of the women's movement was written. Known as the *Vindication of the Rights of Woman*, it was written by Mary Wollstonecraft.

In the middle of the following century Schopenhauer, an archetypical misogynist, stated that the virtues and vices of women come from nature, and that it is in their very nature to be inferior to men. Women are weak, mistresses of foolishness and deception, unfaithful, ungrateful, hypocritical, and they love only by instinct. How many things have these people said about you women!...

I do not wish to conceal the fact that Marx once said that the most salient virtue of woman is weakness. However, in the work of Engels, Marxism took the position that discrimination against and exploitation of women could in some sense be compared to the class struggle, in which man represents the bourgeoisie and woman the proletariat...

A little later, Freud expounded his theory, according to which during a certain physical stage of childhood the girl's interest centers around her clitoris, but when she accidentally discovers men's genitals, she finds them superior to her clitoris, and this causes her to develop a castration complex and a ferocious penis-envy...

As some of the conference participants have pointed out (I had the opportunity to read the expositions, they were short, and I read them all: interesting, intelligent, concise), no one is in a better position than women

— and you in particular — to challenge these pseudo-scientific theories and their underlying assumptions...

The formation of a revolutionary power and a state based on a participatory process for all the people accelerated the transformation of the lives of thousands of women, who... have been ministers, political cadres, police chiefs — you may already know that the only woman national police chief in the history of the world is a Nicaraguan woman [Doris Tijerino]. The Revolution has been determined to explore new horizons by opening the windows to ventilate the darkened chambers of *machismo*...

However,... Nicaraguan women are victims of the tensions generated by their participation in the tasks demanded of them by the Revolution, which are tasks which are in contradiction to the historical role of housewife, and which give rise to the injustice of the double shift. Our confrontation with this problem has moved with the slowness, patience and stubbornness of an ant...

The intensive work in the home constitutes a source of inequality between the members of the household. A typical day includes 18 hours of work for a rural woman and 15 to 16 hours for an industrial working woman, including factory and housework. The unequal distribution of work limits women's access to the opportunities that have been opened by the Revolution and restricts the number of women who can join in scientific work, as has been noted by the participants in this seminar...

An object of our concern is the increasing proliferation of aggressive and backward values in the mass media, health, the educational system, the sciences, the home... For example, if a family is given the opportunity to have a daughter study on scholarship either nuclear physics in Europe or pedagogy in Managua, in most cases the parents will opt for the career which they think is more suitable to the "feminine mentality," unaware that what they are doing is mechanically transferring the image of gentle domesticity to the complex world of professional life...

We must condemn to oblivion the sexual blackmail so often present in various places, and the dishonesty; we must look at ourselves in the mirror with the blinders removed from our eyes; we must replace chastity belts with trust; we must convert our children into the true darlings of the Revolution, no longer the exclusive responsibility of their mothers...
As men, we must not hold back our feelings of sensitivity and tenderness, or, to put it another way, we must not dress our hearts in camouflage. As women, we must not keep hidden our strength, our capacity to be tough and firm. That is, like the strings of a violin, like breasts sculpted in marble, we must be strong and gentle, cry when our heart demands it, laugh when taken with joy, give ourselves to love when it seizes us, share

our bread and our thoughts, our vestments and our faith, irrespective of whether we are men or women.

In the May 1990 issue of the Kovalevskaia Fund Newsletter *I wrote a report on the electoral defeat of the Sandinistas the previous February. Here is an excerpt from my analysis*:

...The public loss of confidence in the FSLN was due not only to U.S. pressure, but also to internal weaknesses... a certain departure from the values of social justice, a tendency to use the Contra war as an excuse for everything. I saw an example of this last year when I attended the International Women's Day meeting between women from different occupational sectors and the leaders of government. Many women complained about the discriminatory way the economic cut-backs were being implemented, especially in agriculture: it was mainly women who were losing their jobs. But President Ortega's response was noncommittal and discouraging. He essentially said: women should stop complaining and concentrate on supporting the war effort; social issues will have to wait until after the war. Sandinista women with whom I talked afterwards were furious, and characterized the tone and content of his speech as "insulting." Even if Ortega had limited commitment to women's objectives, as an astute politician he should have taken into account that the majority of voters in Nicaragua were women and that within a year he would be running against a woman candidate of the opposition.

In many countries, rich and poor, an inability to overcome *machismo* has been a source of internal weakness in progressive movements. In the case of the FSLN, one manifestation of *machismo* has been an over-emphasis on the military aspect of their struggle. In the Museum of the Revolution in Managua, I was disappointed to see that virtually all of the exhibits concern the fighting to overthrow Somoza; hardly any treat the social or economic context of the Sandinista revolution.

Now, with the advantage of hindsight we can see that if the FSLN had acknowledged in 1987 or 1988 that the Contra war had lost much of its force, that some resources could be shifted back to social needs, and that the highly unpopular draft could be suspended — then in 1990 the outcome of the election may very well have been different. In any case, it is clearer than ever that feminist and anti-militarist values are a strategic necessity for the success and consolidation of a revolution.

CHAPTER 12 EL SALVADOR AND PERU

In the 1980's the two Latin American countries most in the news in the U.S. were Nicaragua and El Salvador, representing as they did twin pillars of the Reagan administration's anti-communist foreign policy. In the case of Nicaragua, the U.S. strategy was to undermine the Sandinista government by a combination of economic sanctions, financial support for the internal political opposition, and military attacks by a proxy army of "Contras" operating from Honduras. In El Salvador the policy was to give massive military support to the pro-U.S. regime in its fight against the forces of the Farabundo Martí National Liberation Front (FMLN), which was a coalition of five revolutionary parties, including the Salvadoran Communist Party.

In September 1988 I visited El Salvador in order to set up small projects of the Kovalevskaia Fund at the country's two main universities, the National University (UES) and the Catholic one (UCA). The value of what the Kovalevskaia Fund could offer was not great, but in both cases the universities needed moral support and contacts as much as material aid.

Of the two universities UCA was in better shape. Unlike UES, which had never received funds to rebuild after the big 1986 earthquake, the UCA campus, which had not been damaged, had attractive buildings and nice trees, lawns, and plantings. The National University had been occupied for over three years (1980-1984) by soldiers who ransacked and looted the library, which had once been the best in Central America. So now it was UCA that had the largest library in El Salvador.

While UES had been a hotbed of leftist politics for several decades, the Jesuit university UCA had been started by wealthy Salvadorans in 1965 as a conservative alternative to UES. However, under the leadership of Ignacio Ellacuría it had become a center of liberation theology, adopting the idea of "preferential option for the poor" that had been put forward at the 1968 Catholic bishops' conference in Medellín.

At UCA I met with the librarian, Mélida Arteaga, and the academic dean, a Jesuit sociology professor named Ignacio Martín-Baró who had received his Ph.D. at the University of Chicago. I explained that the Kovalevskaia Fund had very limited resources, but we could start a book donation program that would aim to get about fifty current technical books to the library each year. Martín-Baró said that a program of continuing small donations was actually more valuable to the library than a one-shot large donation. I also talked with him about UCA participating in events we would be sponsoring at the National University. There had been a history of tension and mistrust between the two institutions, and I wanted to understand what could realistically be planned. On several occasions I had frank and helpful conversations with Martín-Baró, who was our key contact at UCA. He had a professional interest in what we were doing; he had recently written an article analyzing the stereotyping of women in Salvadoran newspapers and in the TV series imported from the U.S.

At the National University we set up annual Kovalevskaia Prizes. The first prize would be awarded in 1989, and the project would be administered by a newly created Women's Commission. For most of the ten-year period when the prizes were given the person running the UES Women's Commission was a psychologist named Nora Vanegas, who became one of our closest collaborators in Latin America. Ann and I had long discussions with her about debates within feminism, strategies for supporting reproductive rights, and math education (which became an interest of mine and a focus of activities in El Salvador in the 1990's). Nora put a tremendous amount of effort into the Kovalevskaia Prizes — meeting with people in different departments in order to get more women to submit applications, organizing students to make posters, and setting up annual award ceremonies and seminars on women in science.

Before our trip to El Salvador in June 1989, Ann and I had great difficulty getting visas. In contrast with Nicaragua, which no longer required visas of U.S. citizens, the right-wing government of El Salvador was mistrustful of unofficial Americans who wanted to enter the country. Some people affiliated with progressive religious groups in the U.S. had gone to rural areas of El Salvador in order to act as witnesses to the government's repression of peasant communities that were suspected of harboring guerrillas. The government often forcibly removed peasants to areas under military control — a policy that was reminiscent of the "strategic hamlet" program in Vietnam and was presumably suggested to the Salvadorans by their American advisers. When Americans showed up to bear witness by accompanying the peasants when they attempted to return to their villages, the military felt constrained because any violence against the peasants and their American supporters would be reported in the U.S. press. For this reason the government made it much harder for private Americans to get visas. In all our travels Ann and I never had as much trouble getting a visa to any country as we did before our trip to El Salvador in June 1989.

Anticipating the difficulty, we decided that the week before our flight from Nicaragua to El Salvador Ann would apply for visas for both of us in Costa Rica, where she was attending the second Central American conference on women in science. A prominent Costa Rican anthropologist, Eugenia López-Casas, took Ann to the Salvadoran Embassy, where the consular officer said that she'd have to check first with the U.S. Embassy. Prof. López-Casas was incredulous that the Salvadorans would be so openly servile to the U.S. that they wouldn't even make their own decisions on issuing visas. The Salvadoran woman consulted with Charles Harrington, a political officer at the U.S. Embassy in San José, who instructed her *not* to give us visas.

Ann was not the only one at the Costa Rican conference who was alarmed by this. The Salvadoran participant, Misaella Molina, thought that the implications were ominous. The ultra-right ARENA party (whose founder, Roberto d'Arbuisson, is believed to have planned the assassination of Archbishop Romero in 1980) had won the presidential election in March, the new president had just been sworn in a few days before, and there was tremendous tension in the capital. Many were worried that the military and paramilitary forces would feel emboldened by the election. In these circumstances it would be especially dangerous to be cut off from contact with foreign colleagues. When Misaella returned to El Salvador, she told the UES people who were preparing for the first Kovalevskaia Prize ceremony that Ann had been unable to get visas in Costa Rica, but would make a final attempt at the Salvadoran Embassy in Managua.

Initially we had assumed that because relations were tense between the Salvadoran and Nicaraguan governments — and because most Americans in Nicaragua were known to be progressive activists — we would be less likely to get the visas there than in Costa Rica. But after Ann returned to Managua and told me about the visa denial in Costa Rica, we had no alternative but to make a last-ditch effort in Nicaragua. We asked Edgar Romero to take us to the Salvadoran Embassy, which turned out to be just someone's house. The ambassador was running it almost as a business, charging a significant fee for each visa. We happily paid the fees and got our visas with no questions asked.

When we got off the plane in El Salvador, a group from UES was there to meet us. They were overjoyed to see us — not primarily because we would be giving the Kovalevskaia Prizes, but rather because they took our success at getting visas as a type of victory for them.

The two-day seminar at UES went well. The two prizewinners, Ana Cáceres and Rhyna Antonieta Toledo, had systematically analyzed and catalogued the medicinal properties of native plants. Part of their work consisted of interviews with traditional healers, who had explained how they used various herbs. Their adviser, Prof. María Gladys de Mena Guerrero, and the dean of the UES School of Pharmacology, Prof. Salvador Castillo, took part in the ceremony.

Ann gave a talk (which we had translated into Spanish with Bill Boyle's help) titled "Social Commitment of Professionals Inside and Outside the U.S. Sphere of Domination." She spoke about our impressions of Vietnam and mentioned that "in the International Mathematical Olympiad last year, the Vietnamese team won fifth place (the U.S. team came in sixth)." The Salvadorans clapped enthusiastically upon hearing that the Vietnamese had bested the U.S. team in mathematics. Like many intellectuals throughout Latin America, they had tremendous admiration for the Vietnamese for their victory in 1975, but hadn't known anything about their achievements in other areas.

The Kovalevskaia Fund had also paid airfare from Managua for two academics from UNAN who spoke on women's participation in scientific training and university life in Nicaragua. In addition, we had brought a video that contained Tomás Borge's address at the 1987 conference as well as the "Face the People" meeting between women and national leaders about the new national constitution. The Salvadorans were extremely interested in the material from Nicaragua, a country that had had a successful revolution that they hoped to emulate. After they brought in a TV monitor, they closed and bolted all the doors to the auditorium and posted someone to watch outside each door; only then did they turn on the VCR. The hundred or so people in the room were known to one another, but they didn't want someone they didn't know to observe them watching the Nicaraguan video. Activities that would have been unremarkable on a U.S. college campus — such as showing a film of a speech by a Nicaraguan leader — were dangerous in the repressive conditions of El Salvador.

El Salvador was the only country where Ann and I ever maintained regular contact with people at the U.S. Embassy. Our hope was to convince them to reduce their hostility toward the National University. The ambassador, who viewed the university as a bastion of pro-guerrilla sympathy, had declared the main campus of UES to be off-limits to all official personnel. I believe that it was the only university anywhere in the world whose campus had been declared hostile territory by the U.S. Embassy. What we — and our friends at UES — were worried about was that the military and death squads would take such a policy as a "hunting license" to commit whatever violence they wanted against UES with tacit approval of the U.S. government.

The first Embassy person I met in my 1987 visit to El Salvador was Eugene Santoro, the Cultural Attaché. He was a pleasant enough guy, and he knew of Ann because he had read her biography of Sofia Kovalevskaia. Before he joined the Foreign Service he had taught math at a women's college (the College of New Rochelle) and had used Ann's book. He had a stereotype about the Salvadorans, though, and attributed all the violence in the country to national personality traits (not to U.S. policy, of course). When I saw him in March 1989 at the end of his three-year posting, he commented that he'd be happy to leave this violence-prone nation and take up his next position in a European country, where he thought people were more civilized. I asked what European country, and he said he was going to Yugoslavia. Speak of irony! He would be there for the outbreak of one of the most barbaric civil wars in modern times.

In April 1989, in preparation for the Kovalevskaia Prize ceremony in June, we wrote to U.S. Ambassador William Walker inviting him to send a representative of the Embassy, as would be normal in most countries when a U.S.-based foundation sponsors a major event at the national university. After receiving a polite refusal from him, we asked to meet with him to try to change his mind. To our surprise he agreed, and on June 5 met with us for over an hour.

Ambassador Walker started by saying that he completely approved of what we were doing for women in El Salvador (I restrained myself from asking in that case why the flunky in the Embassy in Costa Rica had tried to block our Salvadoran visas), and then he attempted to justify the U.S. policy of noncooperation with UES by claiming that it was a stronghold of anti-American and pro-guerrilla activity. We replied that public universities throughout Latin America were often centers of student and faculty opposition to U.S. policy, but in other countries that didn't result in a complete cut-off of official U.S. contact. Moreover, the National University campus was completely safe for Americans. We were certain that the university authorities would guarantee the security of anyone he might send as a representative to our awards ceremony. Walker insisted that he had sources of information that indicated that pro-guerrilla activity at UES went far beyond the type of peaceful meetings and protests that would take place at national universities in other countries. We questioned the credibility of his sources, pointing out that the Salvadoran military and the ARENA politicians had a long history of bitter hostility toward the National University — in fact, the military had closed and occupied the campus in 1980-1984, and Concha Bendix had a scrapbook of articles about university people who had been killed or driven into exile — and he shouldn't believe what his right-wing sources told him. Our conversation did not, of course, lead to any change in policy by Ambassador Walker. However, I thought that it was important for him to hear our arguments and to know that American academics had ongoing relations with the National University and were concerned about the danger of attacks against it.

In June 1989 I didn't think that we had to worry about UCA. With its beautifully landscaped hilltop campus, UCA seemed to be a peaceful sanctuary during the war and had never been occupied or bombarded. The U.S. Embassy did not consider the UCA campus to be a guerrilla stronghold, even though Santoro had told me that the university president, Father Ellacuría, was the country's most prominent opposition figure (among those who weren't either dead, in hiding, or in exile). Because Ellacuría had led UCA out of the conservative camp and had become a prominent exponent of liberation theology, Salvadoran rightists considered him, much as they had considered Archbishop Romero a decade earlier, to be a traitor.

The morning of November 16, 1989 I received a phone call from an acquaintance in Seattle's Central America solidarity movement saying that he'd just heard a report of killings at UCA but didn't have the details. He knew of my ties with the universities in El Salvador, and so he called me right away. I thought that he'd probably misspoken when he said UCA and had really meant UES, since that was the most likely target of rightist violence. I called *The Seattle Times* news desk, and they told me that it was indeed UCA and read me the names that had just come in over the AP wires: the priests Ignacio Ellacuría, Ignacio Martín-Baró, Segundo Montes, Arnando López, Joaquín López y López, and Juan Ramón Moreno; and the housekeeper Julia Elba Ramos and her 15-year-old daughter Cecilia Ramos. At 2 a.m. that morning an elite unit of the U.S.-trained Salvadoran army had entered the grounds of UCA and brutally killed several of the country's leading intellectuals.

I was shocked and numbed by the news. Two of the eight victims I knew personally; Martín-Baró had been the person I had talked with the most at UCA. This was the second time in my life that people I knew and worked with had been murdered for political reasons. The massacre at UCA occurred almost exactly five years after the killing of Ed Cooperman, founder of the U.S. Committee for Scientific Cooperation with Vietnam.

My first task was to call Bill Boyle. He had taught at UCA, and his wife Lastenia had been a student there and then worked for several years in the library and knew the faculty and staff well. The last words I had heard from Martín-Baró in June had been to give his regards to Bill and Lastenia. After getting off the phone with *The Seattle Times*, I immediately called them. As Bill repeated the names of the victims out loud, in the background I heard Lastenia scream.

A few days later I went with Bill, Lastenia and their little daughter Mayra to a memorial service held at Seattle's St. Joseph Church. The cathedral was overflowing with people who had come from all over Seattle to pay tribute to the martyrs of UCA.

This was Ann's first semester teaching at Hartwick College in New York state. The college's publicity office had efficiently put together a list of Ann's areas of professional expertise — including universities in El Salvador — for a central media data base. Both *The New York Times* and *USA Today* called her for information about the victims of the massacre in El Salvador. Of course, there were people who knew the priests better than we did, but the journalists were under deadline pressure, and Ann was one of the few sources they could locate on short notice. The *Times* ran brief profiles of the victims, and most of what they had on Martín-Baró had come from Ann.

I was interviewed by two local TV stations in Seattle. I tried to convey what it was like for a small country to lose those people. Beyond the shock and personal tragedy, the murders represented a devastating loss to El Salvador. I asked the viewers to imagine a military hit-squad coming to the University of Washington and killing the most prominent professors on campus.

We later learned of others whom the Salvadoran military had targeted during the FMLN offensive of November 1989. Norma Virginia Guirola de Herrera had been the coordinator of the Salvadoran Institute for the Investigation, Training and Development of Women. On November 13, her bullet-ridden body was deposited in a San Salvador morgue. The army had arrested her the day before while she was giving first aid to civilian victims of the military bombardment of poor neighborhoods of the city. Ann and I had never met her, but we knew her husband, who was in charge of international relations at the National University. We published her obituary in the *Kovalevskaia Fund Newsletter*.

During my trip to El Salvador in March 1990, on behalf of the Kovalevskaia Fund I visited UCA to present a wreath at the tomb of the martyrs. Their remains had been interred behind the campus chapel, and plaques commemorating them were hanging on a wall of the chapel beneath a portrait of Archbishop Romero. Several scientists and administrators from both universities, including the previous year's Kovalevskaia prizewinners at UES, joined me in placing the wreath.

The wreath had cost only a small fraction of what the Fund had set aside for the purchase. I left the remaining money to be given to a young woman student leader who had played a prominent role in the Kovalevskaia Prize event at UES the previous June. She was being sought by the military, and was in hiding. I was gratified that some of the money that we had allocated to honor the dead could instead be used to help the living.

During the November 1989 FMLN offensive, the military seized and occupied the National University. UES was closed for seven months, so of course the Kovalevskaia Prize could not be given in 1990. During my March visit I learned that the university was expected to reopen soon. When Ann and I arrived in El Salvador in June, the campus was still closed, but within days the military officially returned control of UES to the university authorities. On June 6, 1990 we were with the governing council of the School of Medicine (which consisted of senior administrators and professors, including our friend Concha Lemus de Bendix) when they entered the main building of the School for the first time.

The UES School of Medicine was the only public medical school in El Salvador — that is, the only one accessible to students from less affluent families. In addition, it was the only one that had a specific orientation toward improving medical care for the most needy sections of the population.

Our group examined the building thoroughly, going from floor to floor and lab to lab. Ann and I took extensive photographs. Virtually all the rooms were accessible without keys because panels in the doors had been smashed through.

Our first impression was of a large amount of broken glass everywhere — coming from smashed window panes, cabinet doors, microscope slides, and lab vessels. Also obvious was the extensive damage to administrative offices. File drawers had been ripped off their tracks and dumped on the ground; the contents of desk drawers, closets and shelves had been thrown onto the floor and in many cases trodden into a grimy mess of paper, books, office supplies, glass, and splintered wood.

The main chemical storage room and the adjacent lab had been destroyed by a rocket that entered through a window during the military bombardment of the campus. Ovens, scales, and chemical equipment had been removed from the nutrition lab, and its food supplies had been stolen or contaminated. The glass plates, transparencies, and microscope slides of the Department of Clinical Medicine had been trampled upon, and many had been pulverized, probably with a rifle butt. The formaldehyde was drained from the cases in the anatomy lab, and one of the skeletons was dismembered, its skull bashed in with its thigh bone. The anatomically correct mannequins in the nursing lab had been contorted into obscene positions and partially mutilated.

Ann and I found the vandalism by the troops to be disgusting and depressing. However, despite all the damage, the morale of the (mostly female) professors and administrators was surprisingly high. The destruction was not as complete as it had been in 1984. The soldiers had ignored certain pieces of expensive equipment, and much of what was vandalized would not be hard to replace. The attitude of the professors was summarized by the comment that "At least we're not starting from absolute zero this time." After a few weeks of cleaning up, sorting through everything, and taking an inventory of what had survived the military occupation, they would be able to more or less resume normal activities.

Later I wrote a detailed report on the damage to the UES School of Medicine for the University of Washington student newspaper, and of course we also published an account in the *Kovalevskaia Fund Newsletter*, which by then had readers in roughly a hundred countries. Finally, Ann and I met with Seattle's Congressman Jim McDermott. Besides being one of the most liberal members of Congress, McDermott himself is an M.D., and has had a longstanding interest in health care and medical education. We talked with him for about an hour and gave him a copy of our photographs and report.

By the following year things were back to normal at UES, and the Kovalevskaia Prizes resumed. The winners were an interdisciplinary team of eleven women scientists and physicians who had developed a plan for the ecological restoration of Cuscachapa Lagoon, which was the primary water source for a city of 115,000 inhabitants. In recent years the lagoon had become badly contaminated by waste from the town, especially motor oil from auto repair shops and the strong detergents used by launderers who washed clothes there. In June the Kovalevskaia Prize ceremony was followed by a two-day seminar on women and science and then a day-long excursion to Cuscachapa Lagoon.

The women scientists wanted to arrange a meeting with the municipal authorities to discuss a program for cleaning up the lagoon. In order to enhance their credibility in the eyes of the town politicians, who were rightists, we thought it would be helpful for the Cultural Attaché at the U.S. Embassy to come along. Santoro's replacement was an African American named Bob Dance who was unusual not only because of his race, but because he prided himself in having a positive attitude toward everyone in El Salvador, including the UES people. The ambassador's prohibition on entering the UES campus was still in force, and Bob was truly sorry that he couldn't come to the awarding of the prize. But he would be happy to join us for the excursion to Cuscachapa Lagoon and the meeting with the city leaders.

About thirty of us took the university's large new bus to the lake located forty miles (65 kilometers) northwest of San Salvador near the Guatemalan border. Even from a distance we could see the soap and oil films. As we walked around the lagoon, roughly a mile ($1\frac{1}{2}$ km) in circumference, we stopped to talk with some women who were washing clothes and some men who were fishing.

At one point a man asked us if we were Jehovah's Witnesses. At first I wasn't sure that I'd heard him correctly. It seems that groups of Salvadorans accompanied by North Americans had been coming there to perform their adult baptisms in the polluted waters. The local people saw us with a heavy-set African American in a wide-brimmed hat, and they thought that he must be our pastor. Bob Dance got a hearty laugh out of that.

After the visit to the lagoon we had a meeting with local politicians at which Bob, representing the U.S. Embassy, told them emphatically and repeatedly that they should work closely with the professors from UES. This was still a time when the U.S. Embassy officially regarded UES as a guerrilla front. The irony was not lost on the university people. If Ambassador Walker had heard what Bob said at this meeting, he would not have been pleased. In any case, Bob left El Salvador shortly afterwards.

Coincidentally, we ran into Bob Dance several years later at a party held by some acquaintances in Lilongwe, Malawi. He was still as free-spirited as ever and, somewhat surprisingly, still in the U.S. Foreign Service.

One of the most unexpected and eye-opening experiences Ann and I had in Latin America was the discussions and conferences on abortion that the Kovalevskaia Fund cosponsored with the Women's Commission of the University of El Salvador and the Salvadoran Women Doctors' Association. As in virtually all of Latin America except Cuba, abortion has long been illegal in El Salvador. In fact, the Salvadoran representatives at the 1995 World Conference on Women in Beijing were among the most strident of any delegation in their anti-abortion stance. Nevertheless, we found practically everyone we met to be remarkably open-minded about the issue.

The first event we helped organize was a one-day forum at UES on March 30, 1993. The topic was the medical and social consequences of illegal abortion, and the speakers included doctors from the Ministry of Health and the UES School of Medicine. About 140 people attended, roughly half of whom were medical students. Of the speakers only Nora

Vanegas, head of the UES Women's Commission, openly advocated legalization of abortion; the other speakers stated for the record that they opposed abortion. However, they all behaved respectfully toward Nora, they agreed that the current illegal abortion situation was a major public health issue, and they wanted to break the taboo on the subject and have more open discussions. The Minister of Health profusely thanked us for supporting the event. (The Fund had spent a grand total of $330 to cover lunches, coffee, photocopying, a newspaper ad, and a sound system.)

During the question period an American woman in the audience stood up and objected to the tenor of the discussion. She thought that the speakers, under the guise of being concerned with public health, had just been using the meeting as an excuse to voice their opposition to abortion. And she accused the Kovalevskaia Fund of being anti-abortion. Ann was also bothered (although not as much as this woman) that most of the speakers had started their comments with a statement of opposition to legalized abortion.

I saw the meeting in a different light. First, I found it remarkable that the anti-abortion speakers had been so friendly toward Nora, who was openly pro-choice, and toward Ann and me (who, as North American feminists, could have been assumed to be pro-choice). That would never have been the attitude of anti-abortion people in the U.S. In the second place, it seemed to me that the medical people were deeply troubled by the public health consequences of illegality, and were coming close to suggesting that there should be a way for women to obtain safe abortions.

Ann and I discussed the forum with one another at length and turned over our impressions in our minds. We concluded that the framing of the abortion issue in El Salvador and other Latin American countries was often quite different from that in the U.S. In North America the central question was legality. For the most part people on both sides assumed that when abortion is legal it is readily available, whereas the frequency of abortion is greatly diminished by restrictive laws. The Salvadorans didn't seem to be making either assumption.

It occurred to us that in Latin America legal codes sometimes have a different meaning than in the Anglo-American tradition. In the U.S. one assumes that laws are meant to be enforced. In Latin America, on the other hand, laws are often intended to express values and articulate a vision of society. When watching the video of the "Face the People" meeting in which Nicaraguan women voiced their ideas for the draft of the new constitution, we had been surprised by some of the suggestions. For example, one woman said that the constitution should mandate that husbands and wives share the housework equally. To an American this sounds foolish, since such a law would be impractical and unenforceable. However, to many Nicaraguans such an article (which never did make it into the constitution) would have been useful as an expression of a goal for society.

For the first time we understood why most women's rights advocates in Latin America were not focusing their efforts on legalization of abortion. They tended to see that as a losing proposition, because they would be forced to convey the impression that they saw abortion as an ideal option for a woman, which none of them did. Rather than framing the issue in terms of abstract rights, they preferred a pragmatic public-health approach and would have been happy with a *de facto* decriminalization with doctors providing safe abortions and the authorities basically looking the other way. This, in fact, had been the situation in Nicaragua under the Sandinistas, when police commander Doris Tijerino announced publicly that she had no intention of enforcing the country's anti-abortion laws.

Encouraged by the response to the forum at UES, we and the Salvadoran Women Doctors' Association (SWDA) made plans for a more ambitious regional conference on abortion at their biennial congress the next year. The congresses of the SWDA were devoted to medical reports, but before those sessions started they always held a "pre-congress" dedicated to a social theme. For example, in 1992 the topic of the pre-congress had been aging. The leaders of the SWDA decided to devote the 1994 pre-congress to abortion. The Kovalevskaia Fund would invite most of the speakers and would pay airfare for the ones who came from other Latin American countries.

The SWDA, which had been founded in the mid-1970's, was led by some of the country's most prominent doctors. For example, Leonor (Bessy) Linares, who was Concha Bendix's closest friend, was head of one of the best known clinics and medical labs in El Salvador. SWDA congresses and pre-congresses took place at ritzy hotels and had financial sponsorship from major pharmaceutical and medical equipment companies.

On June 9, 1994 the SWDA held its pre-congress on the theme "Abortion: Its Medical and Social Impact in Central America." Besides El Salvador and the U.S., speakers had come from Nicaragua, Puerto Rico, and one country in South America. Ann gave a historical and cross-cultural panorama of views on abortion, in which she explained that for most of its history (until 1869) the Catholic Church permitted first-trimester abortions. Yamila Azize of the University of Puerto Rico in Cayey spoke of Puerto Rico's experience since legalization in 1936; she pointed out that over 70% of the patients in abortion clinics identified themselves as Catholics and another 16% identified themselves as Protestant evangelicals.

The medical director of the most important abortion provider in South America explained that between 200,000 and 300,000 abortions were performed annually in his country, despite their illegality. Since its founding seventeen years before, his organization had "completed" the abortions of 231,052 patients without a single fatality. (The term "completed" was used to fit in with a religious justification of their activities that a Jesuit priest had provided to them — namely, that the "sin" of abortion started when the woman decided to have one, so that the clinic was only completing an abortion that in the theological sense had already been started.) The organization also trains professionals from other parts of Latin America. Ann and I had already known about the influence they were having in other countries: in Lima we had met a midwife who had received training from them. (Note: Ann and I visited this organization's main clinic during our South American trip in April 2007, and spoke again with the medical director. Because of the legal issues involved we were asked not to mention the name of the medical director, his organization, or even his country in any publication; that is why I'm being deliberately vague.)

Prof. María Gladys de Mena Guerrero, whose pharmacology students at UES had won the 1989 Kovalevskaia Prize, spoke of her work analyzing traditional folk remedies in rural El Salvador. She noted that fifty of the 283 plant species she had studied had abortifacient properties that had been well known to the indigenous peoples of Central America.

All of the speakers either implicitly or explicitly supported abortion rights. Concha Bendix, who spoke on RU-486, had a way of presenting her position that we found particularly amusing. She said that she was neither pro- nor anti-abortion. She further explained that her neutrality meant that she believed that no one should force a woman to have an abortion, and no one should forbid a woman from having one — it should be the woman's choice.

Interestingly, the anticipated vocal presence of anti-abortion groups did not materialize. There were, however, many people in the audience who strongly opposed abortion because of the stand of their church. In Ann's talk she cited the only Biblical passage that explicitly deals with punishment for killing a fetus (Exodus 21:22-25) — it says that if someone knocks into a pregnant woman and kills her fetus (i.e., causes a miscarriage), that is a property crime, not murder. The following day a woman came up to her and said that she was an evangelical Christian, and she thought that Ann's talk had been very scholarly. She and her religious friends had looked through the Bible in search of other passages on abortion, but had not been able to find any. Ann said that her mother was also an evangelical (having converted from Catholicism several years before), and had also tried without success to find a basis in the Bible for equating abortion to murder. The woman was clearly very bothered, but couldn't think of any argument to make. So she just ended by saying, "Please give my regards to your mother."

Between 150 and 300 people attended the different talks. This included 65 medical students from UES, whose registration was subsidized by the Kovalevskaia Fund. In Latin America it is unusual to have a conference that addresses the abortion issue in an open, public, comprehensive manner under the auspices of a medical association. That it happened in El Salvador was particularly remarkable.

We published a proceedings volume in Spanish, of which we distributed several hundred copies, mainly in El Salvador with the help of the SWDA. This time we knew better than to have several thousand copies printed — we now have only a few dozen left in our basement.

It is interesting to note that before the abortion conference I had contacted a U.S.-based solidarity organization called the Committee for Health Rights in Central America (CHRICA) to ask if they'd cosponsor the event and perhaps bring a speaker. The CHRICA person I talked with said that they would not want to be involved, because they thought that the theme of abortion was inappropriate. She said that we were imposing an alien agenda on the Central Americans.

As it turned out, she was completely wrong. Even people who were opposed to abortion — for example, the Salvadoran Minister of Health — thanked us for supporting the conference and arranging for the outside speakers. In fact, their collegiality and open-mindedness contrasted sharply with the attitudes of most anti-abortion people in the U.S. If anything, it was easier in El Salvador than it would have been in the U.S. to have the type of meeting that the SWDA and the Kovalevskaia Fund organized on abortion.

Not everything at the abortion conference lived up to our hopes. The medical director of South America's most important abortion provider had come to the conference with the idea of finding people in El Salvador who would want to set up a low-cost clinic modeled after the one he worked in. In El Salvador, as in most countries where abortion is illegal, wealthy women could usually procure a safe abortion, whereas poor women generally could not. He met with doctors and women's rights groups, but no one was prepared to make the necessary commitment, and no one was ready to go to his country for training.

Another disappointment was the visit of Yamila Azize from Puerto Rico. The previous year, while I had been attending a math conference in San

Juan, Ann had visited her at her university in Cayey. She was interesting to talk with, and Ann was impressed with her research on the history of abortion in Puerto Rico. Yamila was politically *independentista*, that is, opposed to U.S. colonial rule of the island. We thought that she would be an ideal fit with the Salvadorans, and we hoped that her trip (financed by the Kovalevskaia Fund) would result in collaboration between the women's studies program at Cayey and Nora Vanegas' Women's Commission at UES.

Ann and I couldn't have been more mistaken. To a woman the Salvadorans were put off by Yamila's manner, which they found to be arrogant, pretentious, and condescending. Yamila's talk was the only one that the moderator cut off as soon as her time was up, while other speakers were given great latitude to run over. Clearly it was not the content of Yamila's remarks — which actually was excellent — that offended the Salvadorans. Rather, it was the manner of delivery. Yamila sprinkled her talk with English words such as "weekend" and "issue" (as is the custom among Puerto Rican academics, but not elsewhere in Latin America). She tossed off a couple of sarcastic references to U.S. "imperialism" in Puerto Rico. But there was a disconnect between her *independentista* posture and her North American mannerisms that clearly disturbed the Salvadorans. Moreover, in El Salvador so soon after the war people we knew did not use leftist jargon lightly, and criticisms of U.S. imperialism were made carefully, usually in an indirect, aesopean language. Whatever the situation in Puerto Rico might have been, in El Salvador being anti-imperialist was not an empty gesture; it was a dangerous and difficult lifelong commitment.

In addition, Yamila was extremely patronizing toward Nora Vanegas. Nora might not have known English or been conversant with current U.S. feminist literature, but her extensive knowledge of women's circumstances in El Salvador and her longtime devotion to women's causes deserved far more respect than Yamila was willing to give.

Needless to say, no collaboration between UES and Puerto Rico resulted from Yamila's visit. Ann and I talked over and tried to analyze our miscalculation. We were reminded of observations we had made in Vietnam when so-called Việt Kiều (overseas Vietnamese) came to visit. Although official Vietnamese policy has been to strongly encourage return visits and investments by people of Vietnamese ancestry who grew up in the U.S., France, Canada, and Australia, the interactions of Việt Kiều with the Vietnamese in Vietnam often have had an undercurrent of tension.

Once Ann and I sat in on a lecture by a Canadian-Vietnamese mathematician in Hanoi. He had what to us seemed to be an exaggerated Western manner of dress and speech. His tone was brash and slightly patronizing.

We noticed that some of the Vietnamese listeners were exchanging glances with one another and quietly snickering. It was almost as if they saw him as "neither fish nor fowl," as the saying goes, and they resented his pretentions. This guy, they might have been thinking, claims to be one of us and at the same time better than us.

Ironically, it's sometimes easier for someone like Ann or me to be accepted precisely because we come from a different background and have no claim to any ethnic or cultural affinity. Although we often joked and gossiped with the Vietnamese mathematicians in a relaxed manner — especially when we were speaking Russian and reminiscing about shared student days in Moscow — this informal professional interaction did not violate anyone's sense of pride or cultural autonomy.

In 1993, a few months before the formal end of apartheid in South Africa, we were just across the border in Gaborone, Botswana listening to South African TV. Some African Americans visiting the country for the first time were lecturing the interviewer about what black South Africans should do and what they could learn from the American experience. They seemed to think that their racial background and knowledge of the U.S. civil rights movement gave them special expertise. Our impression was that the studio audience was not completely pleased with the visitors' manner of speaking to them.

Of course, many people whose background bridges two cultures are able to travel to the land of their ancestors without stepping on toes or offending anybody. Not all Việt Kiều in Vietnam or African Americans in Africa fall into the trap that these visitors did. Probably many Puerto Ricans would have known how to relate to the UES people with understanding and respect. However, we found it an unpleasant surprise that someone with the insights and outlook of Yamila Azize had such an unsuccessful visit to El Salvador.

Our activities in El Salvador continued through the 1990's. In 1996 we cosponsored a pre-congress on sex education with the Salvadoran Women Doctors' Association. In 1992 we visited a UES campus for teacher training in San Vicente (about two hours from the capital) and gave a demonstration math class to young children; in 1995 we held the Kovalevskaia Prize ceremony in the city of San Miguel, where the prizewinner, Irma Lucía Vides, did research on water contamination and purification techniques; and in 1999 we visited Santa Ana, where that year's prizewinner, Blanca Elba Berganza, worked on ways to use soy and amaryth to enrich the diet of rural families. Over Thanksgiving of 1998 we traveled to El Salvador to give workshops at a national teachers' convention.

For Ann and me sometimes the trip between the capital and a provincial town was as interesting as the actual activity once we got there. We liked observing the countryside from the bus, and we learned a great deal from the long conversations we had with the UES people. The ride back from San Vicente was especially memorable — it took us over four hours to travel the fifty miles (80 kilometers) because of the worst traffic jam we had ever been in. Food and drink vendors would come up to the barely moving vehicles, so people bought a few things and turned the back of the bus into a small party area. Some danced to the radio, while others were deep in conversation. Until that evening in the outskirts of San Salvador, we had never realized how enjoyable a traffic jam could be.

But throughout the 1990's the conditions for the Kovalevskaia Prize at UES deteriorated. The UES administration changed hands shortly after the civil war ended in 1992. The new university president, Fabio Castillo, had spent the war in exile in Costa Rica. Although he had a high reputation and the people we knew were initially happy that he was returning to resume his duties as university head, it soon became clear that his main priority was to get rid of the people who had played important roles at UES during the war and whom he perceived as a political threat. The excuse he used was that he was raising standards — most of the faculty we worked with at UES had not spent much time abroad, did not know English, and did not have Ph.D.'s. But in fact he purged some of the most dedicated, hard-working people — and a disproportionate number of women. Several of those who were ousted from key positions — for example, the vice-rector Catalina Machuca — had made tremendous personal sacrifices and at times had literally risked their lives in order to keep UES going during the difficult war years.

The UES president during our early visits, José Argueta Antillón, was a dedicated leader, as were several of the other administrators we met. The dean of pharmacology, Salvador Castillo, put great effort into creating a supportive environment for the (predominantly women) faculty and students of his school. He almost always came to the events organized by the Women's Commission and the Kovalevskaia Fund, and even joined us on our trip to the San Miguel campus in 1995. The UES people of this period emphasized community outreach, and they had excellent rapport with students. But unfortunately, in many countries of Latin America — and this became true of the UES under Fabio Castillo — formal titles were valued much more than experience, competence, dedication, and quality of teaching.

Since the Women's Commission under Nora Vanegas was associated with the earlier leadership, UES President Castillo eventually cut its support to zero. He never attended a Kovalevskaia Prize awards ceremony, and he repeatedly refused our requests to meet with him. In fact, I was never treated so rudely by any college administrator as by Fabio Castillo — that is, until University of Washington President Richard McCormick's insulting letter to me in August 1999 (a subject for Chapter 15).

By the late 1990's it had become impossible for Nora to organize the prizes. Morale among women scientists had declined, it had become harder to convince people to apply for the prizes, and Nora no longer even had an office out of which to organize anything. The Kovalevskaia Prize in El Salvador was given for the last time in 1999, and that was also the last year when Ann and I visited the country.

In June 1989, when the first Kovalevskaia Prizes were given at UES, we were accompanied to Nicaragua and El Salvador by a delegation of University of Washington faculty and other activists from the Pacific Northwest. In El Salvador the group did not remain with us, partly because all of our activities were in Spanish with no translation and partly because they wanted to meet a cross-section of society rather than spend most of their time at the universities. We enlisted the help of a North American resident of El Salvador named Serena Cosgrove who belonged to Peace Brigades International. That group tried to support and protect leaders of unions and popular movements by organizing people from North America and Europe to go with them and be witnesses if they were attacked. We paid her $200 to provide the Seattle visitors with guides and translators for a few days. The arrangement worked out well, and the Seattle people were impressed by the Salvadorans they met.

Serena wrote a report on the visit which criticized us for housing the group in Escalón, which was an affluent district of San Salvador. She felt that they should have stayed in a poor neighborhood. There was also an implicit criticism of the Kovalevskaia Fund, which was working with professional women and was not doing anything directly with the neediest segment of the population.

Yes, it was true. Ann and I always stayed in Escalón, which was where our closest colleague Concha Bendix lived, and we sprang for a $20 to $40 hotel room. We spent our time with people at the universities and in the Salvadoran Women Doctors' Association — people who, like us, were academics and professionals.

We had tremendous respect for the North American activists in El Salvador who worked in the poorest communities and gave direct support to the labor movement and the FMLN. It took courage and dedication to do that. But that was not what we did.

We have always seen our activities in Vietnam, Central America, and elsewhere as a type of mutually beneficial collaboration with colleagues. Ann's research interests have been greatly influenced by conversations and experiences during our travels and by publications that she has found that would have been difficult or impossible to get in the U.S. Many of her observations have also enriched her teaching, especially after she moved to the Women and Gender Studies Program at Arizona State University. Our travels have had a direct impact on my professional life as well — for example, they have given me a valuable perspective on issues of mathematics education. And much of what we do while abroad — visiting the Great Wall in the snow, seeing Machu Picchu ringed by early morning clouds, snorkeling in Lake Malawi — is simply fun.

Often people who see their activities in other countries in terms of aid or charity or self-sacrifice have unrealistic expectations. They want to hear expressions of gratitude from people, and they want to see dramatic success. Americans especially are subconsciously influenced by a Hollywood version of reality: the situation looks bleak, but then two hours later everyone is saved by the noble sacrifices of a few.

Our approach, which is based on a notion of "enlightened self-interest" rather than self-sacrifice, has the advantage that it's easier to sustain over many years. There's less danger of burn-out. Disappointments can be chalked up to experience, and we can learn from them. Our objectives are limited. The Kovalevskaia Fund is not out to save the world — but only to offer some encouragement and moral support to women scientists in a few places.

Once in Peru in the mid-1990's Ann and I were staying at a hotel in the resort town of Paracas, about four hours down the coast from Lima. The off-shore islands are a breeding ground for sea lions, and the northernmost of the famous Nazca lines are visible nearby. At around 6 p.m., the normal dinner hour in the U.S., we were in the hotel's bar, where we met another American couple who, like us, did not want to wait for supper until the restaurant opened at 8 p.m.; the bar served open-faced sandwiches as snacks, and that would do nicely for a meal. They taught at the American School in Lima. The woman had been hired from the U.S. and promised an American salary. Her husband had been given a job after they arrived at a salary that was much lower, but still at least twice that of any of the professors we knew at San Marcos National University. What surprised us was the way the woman spoke of their teaching as a type of self-sacrifice. She said that they were "giving of themselves" to help the Peruvian people.

In reality, the American School was one of the two most expensive and elite schools in Peru (the other being the Humboldt School). It catered to the very privileged, and for this reason could pay salaries that were many times higher than what teachers normally earned. From talking with the couple we got the impression that they had come to Peru in large part because they could not have gotten such good jobs in the U.S. Nevertheless, they had a self-image of noble altruism.

Unfortunately, this type of disconnect between self-image and reality is common among Americans abroad. On a national level the notion that Americans are a generous people whose presence on Earth brings great benefit to citizens of other countries is pervasive in the U.S. mass media. One of the few places where this idea is held up to the ridicule that it deserves is in an early scene in the movie "Ace Ventura: When Nature Calls." The Jim Carrey character has to leave the Buddhist monastery in the Himalayas where he's been living. He's sure that the monks, who he feels have much appreciated and benefited from his long stay with them, will be heartbroken when he tells them that he has to leave without having "achieved oneness." To his surprise, the head monk hastens to assure Jim that he really has achieved oneness. "You're more one than anyone," the monk says and then adds eagerly, "I'll help you pack." The monks pop champagne corks and have a rollicking celebration of Carrey's leaving. Puzzled, Jim says, "I've never seen them like this." Ann and I loved that scene, which we saw as a metaphor for the discrepancy between the way Americans see themselves and the way others see them.

Just as the U.S. media give a largely positive view of Americans' role in the Third World, they also convey an almost entirely negative picture of conditions and social values in those countries. For example, suppose that we ask what associations Americans would have had with Peru in the mid-1990's, when Ann and I were starting to collaborate with schools and universities there. (Here I'm referring only to relatively well-educated Americans who follow the international news, not to those who wouldn't have any idea where Peru is on a world map.) The words that would have come to mind most likely would be poverty, cholera, civil war, terrorism, corruption, and drug trafficking.

Yet there is another side to the country that rarely reaches the consciousness of Americans. Ann and I have encountered example after example of the incredible respect Peruvians have for education at all levels. Taxi drivers treat us differently when they take us to the universities than they would if we were just going to tourist destinations — in particular, they don't overcharge us. Parents make tremendous sacrifices to pay to send their children to a better school. Even impoverished communities in the Andes attach great importance to their local schools. There still seems to be a consciousness there of the intellectual and cultural achievements of the great civilizations of their ancestors, the Inca and the Wari, even though they were destroyed by the Spanish a half-millenium ago.

Peru has a rich history, and is multiethnic (Quechua, Aymara, Chinese, and Japanese, among other groups). There is some racism, of course, but there's a lot that compares favorably with the situation in the U.S. Once we noticed that at the end of the evening news the TV carried five minutes of reports of the day's events in the Quechua language. In Arizona the analogy would have been ending the news five minutes early to leave time for a summary in Navajo.

Even the corruption has another side to it. The mathematician Michel Helfgott told us a story of being stopped for a driving infraction in Lima. He knew that a bribe would be expected, but on his San Marcos University salary he didn't want to pay it. Michel asked the policeman if he had children, and it turned out that he did. Then he told him that he didn't have any money with him, but he was a professor at San Marcos and had just published an algebra book for children. If the policeman went with him to his house, which was nearby, he'd give him a copy of the book for his kids. The cop was delighted, and gratefully accepted the algebra book as a "bribe."

Just try to bribe a Miami cop with an algebra book!

In the early and mid-1990's most of the people Ann and I knew in Peru were at San Marcos National University. The oldest college in the western hemisphere (founded in 1551, eighty-five years before Harvard), San Marcos is the largest and most important university in the country. For five years the Kovalevskaia Fund gave annual prizes to women scientists at San Marcos and other public universities. However, conditions for faculty deteriorated under the so-called "neoliberal" policies of the Fujimori government, and several of the people we were collaborating with either had to take second jobs or else left the university entirely. By the late 1990's there was no longer a group for us to work with at San Marcos, and our university contacts shifted to the Pontifical Catholic University in Lima.

In recent years our closest colleague at the Catholic University has been the mathematician Uldarico Malaspina, who is also academic dean. Despite his many administrative responsibilities, Malaspina devotes much time and energy to helping organize the country's math olympiads and raise funds for the teams' airfare to various multinational competitions. He, his Catholic University colleagues Emilio Gonzaga, Cecilia Gaita, and Mariano González, and Felix Maldonado (founder of the Gauss School, which will be mentioned again in Chapter 15), for many years have worked closely with Peru's most talented youngsters and shouldered the burden of logistics and fundraising for Peru's olympiad teams.

Through colleagues at San Marcos and the Catholic University I have met some remarkable math students. My first such encounter was in March 1992, when I gave a talk at San Marcos on my joint paper with Mike Fellows on "self-witnessing" problems, in which we proved a small result about factoring and primality testing and applied it to a theoretical question in computer science. A brilliant boy in his mid-teens named Harald, the oldest son of mathematicians Edith Seier and Michel Helfgott, attended my talk and at the end asked an interesting question (which I don't remember) about the computer science part of the work. Unable to answer it, I made the lame excuse that that was in Mike's section of the paper. The moral of the story is that every author of a joint paper should be sure to read and understand all of it, not just the part that he or she wrote. Otherwise you might find yourself in the embarrassing position of being unable to answer a teenager's question about your own paper! Harald, by the way, is now a mathematician at Yale; he received his Ph.D. from Princeton in 2003.

In 1997 at a women in science conference in Lima I met a girl named Liz Vivas, who had been a star of Peru's olympiad teams since the age of 13. I asked her what the most interesting problem had been in a recent competition, and she gave me a nice, clear explanation of the problem and her solution. Because Liz came from a family of modest means, the Kovalevskaia Fund helped pay for her airfare to some of the regional meets, such as the Iberoamerican Math Olympiad in 1996. Several years later the Fund also paid the application and testing fees when Liz applied to graduate school in the U.S. At present she is working for a Ph.D. in mathematics at the University of Michigan.

In Peru the kids who do well in the math olympiads for the most part are not from affluent families of European ancestry, but rather are poor or lower-middle class and of indigenous (Indian) or mixed ethnicity. Malaspina once explained to us that those youngsters see excelling in mathematics as a route out of poverty for them and their families.

In April 2007 we met in Malaspina's office with four of these students (two girls and two boys). Among them they had won about a dozen medals in several math olympiads, including the International Mathematical Olympiads (IMO) in Mexico in 2005 and Slovenia in 2006. Two of them were already in college and so were ineligible for the 2007 IMO in Hanoi. The other two — Jossy Alva and Daniel Soncco ("soncco" means "heart" in Quechua) — were hoping to go, but knew that because of the exorbitant airfare between Peru and Vietnam it was unlikely that Malaspina and the other leaders of the mathematical community would be able to raise enough funds to send a full team.

After I returned to Seattle, I learned that Jossy and Daniel (and four others) had qualified to represent Peru at the IMO. I then obtained tickets to Hanoi for the two youngsters, using a combination of frequent flyer miles on American Airlines for the Lima/Los Angeles portion and Kovalevskaia Fund money for the Los Angeles/Hanoi part on China Airlines. Both Jossy, as a girl, and Daniel, coming from a Quechua family, belong to groups that have been severely underrepresented in the IMO's.

Postscript

In late summer of 1993 I spent ten days at the Electronics and Telecommunications Research Institute (ETRI) in Taejon, South Korea. While I was there the big international Taejon Expo '93 was taking place, and one of my tasks during my visit was to deliver the following letter in person to a representative of the Expo Organizing Committee. The letter was written by Ann in connection with an incident that had occurred during our most recent visit to El Salvador.

My host at ETRI at first refused to take me, saying that someone working at a government institute could not have any connection with such a matter, but then he reluctantly agreed to drop me off in the vicinity of the Expo offices and point me in the right direction. I found an Expo official who was willing to speak with me and receive the letter; however, we never got a reply.

Dear Sir/Madam:

I would like to respectfully call your attention to a possible source of misunderstanding related to Taejon Expo '93 that might have some unfortunate consequences in the area of international public relations.

I am the Director of a small foundation for the encouragement of women in science in developing countries. While I was visiting the School of Pharmacology at the University of El Salvador (UES), a letter from your organization was delivered to the offices of the Dean of the Faculty, Lic. Salvador Castillo, and the Secretary of the Faculty, Dr. Gladys de Mena. The letter announced a so-called "World Miss University Contest," and invited participation from UES women.

It is difficult for me to sufficiently describe the astonished reactions of Professors Castillo and de Mena as they read your letter. Well over half of the students in their School are female, and there are several distinguished women among the professors, including Dr. de Mena and two previous winners of the Kovalevskaia Prize (awarded for outstanding scientific research by women). Professor Castillo commented that he would hope that the women of their School would be honored for something more appropriate than mere physical appearance. Dr. de Mena remarked that it was indeed unfortunate that their only contact with your country was in the form of a request for women science students to participate in a beauty pageant.

The University of El Salvador is the foremost institution of higher education in El Salvador. For decades it has worked to produce dedicated scientists and professionals under conditions of extreme economic hardship. Women have participated fully in this educational process, and they are represented at all levels (both the Vice-Rector and the General Secretary of the University are women).

In fact, in most countries of the world, as in El Salvador, women participate heavily in academic and scholarly life and in scientific and technological development. Your own country has a distinguished institution of higher education for women (Ewha University), and Korean women scientists have taken part in international conferences that I have attended (most recently, the International Conference on 'Women's Vision of Science and Technology for Development' in Cairo, January 1993).

From newspaper articles and other publicity, I understand that the purpose of Taejon Expo '93 is to promote the appreciation of science and technology and at the same time contribute to peace and mutual understanding between different countries and cultures. It would be consistent with these objectives to offer some form of support and encouragement to women scientists, especially those who are working under the most difficult circumstances, as in El Salvador. However, the invitation to participate in the "World Miss University Contest" was perceived by many as an insult to their dignity and professional status, and can only lead to impressions that are contrary to the purpose of your Expo. On behalf of the Kovalevskaia Fund, I would like to respectfully request that you take appropriate measures to rectify this misunderstanding.

Sincerely yours,
Dr. Ann Hibner Koblitz, Director
The Kovalevskaia Fund

CHAPTER 13 TWO CULTURES

At the Harvard faculty meeting on South African investments in 1979, Karl Deutsch quoted the British author C. P. Snow commenting on the moral shift from their generation to the young people of the 1960's and 1970's, who were "very permissive about the sex life of their contemporaries but would not be caught dead serving South African sherry." But what C. P. Snow is best known for is not his remarks about changing moral priorities from one generation to another, but rather his analysis of a cultural gulf between people who work in the sciences and technology on the one hand and those who are trained in the humanities and social studies on the other. In his novels, lectures, and 1959 book *The Two Cultures and the Scientific Revolution* he discussed the vast differences that he observed between the two groups in background, knowledge, attitude, and outlook.

Although Snow was criticized by some, I have often found the "two culture" framework to be helpful in understanding what would otherwise seem paradoxical. Indeed, divisions along disciplinary lines — not only science versus humanities, but even between fields that are close together, such as math and computer science (this will be a theme of the next chapter) — sometimes play a greater role than political, ethnic, racial, or gender differences.

For example, in the 1990's at the University of Washington a committee of students and faculty, virtually all of whom were in the humanities or social studies, proposed adding an Ethnic Diversity Requirement (EDR) to the undergraduate curriculum. In order to graduate, students would be required to take a course on some U.S. ethnic or sexual minority (a course on women would count). After this proposal was voted down, supporters of the measure decried what they saw as an increasingly right-wing faculty that was unsympathetic to minorities.

However, in reality most UW professors were and are liberals, not conservatives. Support for affirmative action and programs for minorities is high among my colleagues in most departments, and in fact is higher than among the students. What had really happened in the EDR vote became clear if one looked at a breakdown of the vote by departments. Most non-science faculty saw the EDR proposal as a chance for the university to make a statement about the importance of diversity and for that reason favored the requirement. In contrast, the science faculty voted overwhelmingly against it because we saw the matter as a curricular issue, not a political one, and felt that adding the EDR would be unhelpful to science students, who have a hard enough time taking all the courses they need without the additional burden of a new requirement. Some of us in the sciences also thought that the EDR was a bad proposal for other reasons. Students could not meet the EDR by studying foreign languages or countries, but could meet it by taking a course about their own ethnic group (a Jewish student taking a course on Jewish history, a black student taking black studies, and so on). And given the slide in academic standards at UW in recent years, it was unlikely that many of the new EDR courses would be at a high level. So most science professors opposed the EDR because we thought it made no pedagogical sense.

Not only were there few science faculty who thought much of the EDR, but even among minority students it was largely the non-science majors who believed that it was a good idea. I saw no evidence of support for the new requirement among minority students in science and engineering. A proposal that might have worked if it had been drawn up carefully in consultation with science faculty and students went down to defeat because it failed to bridge C.P. Snow's two-culture divide.

The ethical norms in the non-sciences — for example, in faculty hiring — are sometimes hard for me to comprehend. Ann had terrible difficulties getting a job in the 1980's. During the five-year period after she got her Ph.D. she applied for practically every academic job in Russian history, history of science, or women's studies that was advertised. She did not restrict her search geographically, and she was open to working in practically any kind of institution (well, except for Bible colleges). Each year she made it to the shortlist at four or five places, but until 1988 she got no tenure-track offers.

For the first couple of years I kept telling her that by the law of averages she was almost certain to get a tenure-track job very soon. I was reasoning mathematically: if, say, you're on a shortlist of three people, then each time you have a 2/3 chance of not being the first choice. But if that happens five times, then there's only a $(2/3)^5$ chance — in other words, a 13% chance — that you won't get any offer. The math was flawless, but my assumptions — essentially, that we were dealing with unbiased rolls of the dice — were wrong. Time and again Ann would make it onto the shortlist, presumably because some people on the hiring committee were impressed with her work. But she would never be offered the job.

For instance, at the University of Dayton Ann was one of the top two choices for a tenure-track position. She had had her Ph.D. for four years, had published a book and several articles, had given dozens of invited talks at universities in the U.S. and Canada, had spent 1984-1985 at the Institute for Advanced Study in Princeton, and had taught for a year at Wellesley. But the department decided instead to hire a male ABD ("all but dissertation") because, as they said to Ann, he had tremendous "promise." In this case Ann knew the guy (and liked him personally) because he had been on the IREX graduate student program in the Soviet Union the same year that she was on it. But in the five years since then he had not managed to finish his dissertation. Other than his affiliation with a prestigious graduate school (Stanford) and his gender (male), it was unclear what basis there was to think that he had great promise. As so often happened in those years, a less qualified male was selected in preference to Ann.

It was at the State University of New York at Binghamton where we had the most information about the hiring process and the reasons for rejecting Ann. Her campus visit was so strange that she knew right away that the most powerful professors in the history department would block her. When she met with the chairman, he made disparaging comments to her about both tenured women in the department. When Ann told me this, I was incredulous — I'd never heard of a chair doing anything so unprofessional in a hiring interview.

Ann had a friend on the search committee from whom she learned what was going on behind the scenes. Lynne Viola had been on IREX the same year as Ann, was one of the leading scholars in the department, and soon after left for a better university (the University of Toronto). Lynne told Ann that the chair was upset at the women in the department because they were taking the side of graduate students who had complained about sexual harassment by a male faculty member. The department was polarized over this issue, and there was no way that the chair and his cronies would hire another woman, at least not while this dispute was going on.

When Ann told me about her depressing visit to SUNY Binghamton, my reaction was that we should sue the sexist pigs. My sister Ellen referred us to a lawyer we could talk to, and she dissuaded us. She said that discrimination in promotion and salary was one thing, but courts would almost never intervene in a university hiring decision. She added that if Ann had been trying to get a job in the fire department and they had treated her that way, then she might recommend suing. But not in academia. The lawyer also reminded us that a lawsuit takes a lot of time and money and can easily come to dominate one's life. Her suggestion was to forget about it and move on. After all, would Ann really want a job in a department that was dominated by misogynists?

We decided that Ellen's friend was correct. We had seen the corrosive effect of litigation on the life of someone we had met at Harvard in the 1970's. Ephraim Isaac was a scholar in the African American Studies Department who specialized in the history of the Ethiopian Coptic Church; it was he who hosted our visit to Israel in 1978 (the time we got from Hanoi to Tel Aviv in less than 48 hours via Karachi and Teheran). After Ephraim was denied tenure, he sued Harvard. The case dragged on for many years and became an obsession with him. Eventually Harvard settled out of court for a paltry sum that didn't nearly compensate for the damage to Ephraim's career caused by the lawsuit itself, let alone that caused by the tenure denial.

There were other reasons besides gender why Ann never got beyond the shortlist. It was clear from her work that she was a feminist and politically on the left. In addition, in those years neither history of science nor history of women was universally acknowledged as mainstream history. In both cases there was controversy in most history departments about whether scholars in those fields should even be hired at all. In the 1990's interdisciplinary research started to come into fashion, but a decade earlier people whose research bridged different fields of scholarship were often considered to "fall between stools." Ann perhaps would have had an easier time on the job market if instead of a biography of Sofia Kovalevskaia she had written a book about the liberal sociologist M. M. Kovalevskii (a distant relative of Kovalevskaia's late husband), who was a contemporary of Kovalevskaia and in fact her lover. He was a much less interesting person than the pioneering woman mathematician whom Ann chose to write her book about, but as a male and a non-scientist he would have been considered a more mainstream topic in intellectual history.

During this period Ann was often invited to speak to mathematicians and historians of science. For example, in January 1985 she gave the Kenneth O. May address on history of mathematics at the University of Toronto; they even paid a $500 honorarium, which was a lot of money to us at the time. On several occasions mathematicians who had seen her curriculum vitae commented that a mathematician with analogous credentials would have had a position at a major research university. People found it hard to understand why Ann was bouncing from one temporary job to another and was never able to get past the shortlist despite more than twenty on-campus interviews. Scientists generally believe that hiring should correlate with merit, and that someone with no record of accomplishments should not have been hired in preference to Ann simply because he came from Stanford and was male.

Finally, in the spring of 1988 Ann got two tenure-track job offers, one from Ursinus College and one from the University of Western Kentucky. She turned down Ursinus because the chair of the history department at Western had told her that she'd be hired at the associate professor level. However, when the written offer came it was only as an assistant professor, the same as at Ursinus. Because of the blatant dishonesty Ann had no choice but to turn down Western Kentucky and spend another year in temporary jobs. Then in 1989 she was offered and accepted an associate professor position at Hartwick College in Oneonta, New York.

The repeated rejections during the five-year period after her Ph.D. were of course discouraging, frustrating, and upsetting. But in retrospect everything turned out for the best. She ended up with more collegial, supportive, and stimulating colleagues during her nine years at Hartwick than she would have had at most of the other places where she had been interviewed. And her present position at Arizona State University is better from almost any point of view than any of the jobs for which she was turned down in the 1980's.

Ann and I have joked that Hallmark should put out thank-you notes of the type that a girl could send to a boyfriend who had dumped her after she ended up with someone far better than him. Ann could send such notes to the departments that treated her shabbily — at Dayton, SUNY Binghamton, Western Kentucky, and other places — expressing her gratitude that they didn't hire her. But unfortunately no one seems to produce this sort of sardonic thank-you note. Someone in the greeting-card industry could make a bundle by selling to this niche market.

A fundamental difference between the sciences and many of the humanities and social studies relates to reprisals against junior faculty for speaking out. I got a vivid illustration of this in May 1977 when Seymour Martin Lipset, who was Professor of Political Science and Sociology at Stanford's Hoover Institution, wrote a letter complaining about me to the Harvard math department chair Shlomo Sternberg.

The previous week Serge Lang — a prominent mathematician and the author of several dozen books, including the graduate algebra text that was the only book I had had with me when I was in solitary confinement in the Army stockade — had visited Harvard. On April 16 he had written a letter to Lipset in response to a questionnaire he had been sent called "The 1977 Survey of the American Professoriate." Lang's letter read in part:

You guys want us to take you seriously, answer your questionnaire, and cooperate, thereby transferring to you our academic respectability. No way! Did it occur to you that many of those who did not return the questionnaire think that Seymour Martin Lipset is a (characterization deleted)? Is there any place in your questionnaire giving us a chance to express this opinion? To express the opinion that the idea of the questionnaire is stupid? To express the opinion that Seymour Martin Lipset is not qualified to send out questionnaires and does not deserve being answered by the academic community? To express the opinion that the whole manner in which the questionnaire is put together already prejudices the issues to the point where one does not want to deal with them on your terms? To express the opinion that we have better things to do than answer questionnaires, especially those sent out by Seymour Martin Lipset?

You are parasites on the academic community. Lay off!

Several of us in the Harvard math department were delighted with Lang's letter. It was obvious to us that the questions in the survey were worded in a tendentious way that was likely to produce the results that the sociologists who designed the questionnaire wanted — basically, the conclusion that professors were more conservative than many people thought. The younger members of the department enjoyed Lang's irreverent tone and use of slang, but even some senior people liked the letter. John Tate, for example, wrote to Lipset's coauthor, Everett Ladd, saying, "I consider the whole project ridiculous... By and large, my reaction to this is the same as Serge Lang's."

Lang gave a seminar talk, and several of us went out to lunch. I asked whether it would be okay to give the letter to *The Harvard Crimson*, and no one could think of any reason why not. I did that, and an article in the April 30 *Crimson* quoted me to the effect that several members of the math department agreed with Lang's opinion of the questionnaire. Lipset found out about the *Crimson* article and was upset. On May 3 he wrote a letter to the department chair Sternberg in which he said:

The *Crimson* quoted a comment by Neal Koblitz, saying that many members of your department agreed with Lang. Since I do not know how many members of your department could possibly have been acquainted with the nature of the research, I was somewhat surprised and annoyed by his comment...

I will probably be in Israel this summer. I hope to see you and the family there.

After Sternberg showed me the letter, he told me that he was surprised that Lipset was so thin-skinned about the matter and he did not really see what "Marty" expected him to do. Even though Sternberg was friends with Lipset and was much closer politically to Lipset's neoconservatism than to my left-wing radicalism, it would never have occurred to Sternberg to try to do anything against me. I told Sternberg that if he was worried about Lipset having the impression that the math department had taken a position on the issue — which it obviously hadn't — I'd be willing to write a note to the *Crimson* clarifying that I was not speaking on behalf of the department. I wouldn't ask the *Crimson* to publish it, but I'd send a copy to Lipset so that it would look like Sternberg had taken some action against me. This I was happy to do.

Later I showed a copy of Lipset's letter to a few acquaintances who were junior faculty in non-science departments. They found it frightening, because in their fields such a letter could be extremely detrimental to the career of a young scholar. Although as a mathematician I could shrug off Lipset's note to the chair of my department, someone in sociology or political science would see it as a poison-pen letter. Thane Gustafson (who is now a well-known expert on Soviet and Russian political history at Georgetown University but then was just a junior instructor like me) said that it reminded him of a *sovietskii donos*, meaning a Stalinist denunciation letter.

In retrospect, Lipset had even more reason to be furious at me than he realized when he attempted to get Sternberg to take reprisals against me. Lang had written his original letter only to "give vent to my feelings," as he later explained. Initially he had had no intention of carrying the matter any further. It was my egging him on when he visited Harvard — along with publication of the *Crimson* article — that led to a flurry of further correspondence and ultimately to a three-year campaign by Lang against the Lipset-Ladd survey. Lang kept a complete file of all documents and correspondence related to the controversy, and in 1981 Springer-Verlag published it in book form. Titled simply *The File*, the volume goes to more than 700 pages.

In addition to my role in encouraging Lang, on May 11, 1977 I sent my own letter to Lipset giving a detailed analysis of the logical flaws in one of the questions in the survey and also informing him of two glaring factual errors in his account of the Harvard student movement of the late 1960's in his book *Education and Politics at Harvard*. I didn't receive a reply until I sent him a reminder note, and then I finally got a brief letter from Lipset telling me that he "did not find your criticisms of sufficient technical relevancy to reply to in detail."

Even before Lang's letter of April 16, 1977, I had been skeptical of the increasing use of quantitative methods among sociologists and others. No matter how glaring the flaws in their question construction and sample selection, Lipset and Ladd could be counted on to trumpet their conclusions to the media and describe their results with plenty of quantitative jargon and numbers given to several significant figures of precision.

I have always regarded terms such as "political science" and "social science" as oxymorons. During the 1970's I collected examples of the bogus use of mathematics by non-scientists, and wrote an article called "Mathematics As Propaganda" in a collection of essays titled *Mathematics Tomorrow* that was published in 1981 by Springer-Verlag. The volume sold poorly, and few people read my article. However, one person who did read it (because I sent him a preprint) was Serge Lang. Because of that one reader it turned out to be one of the most important articles I ever wrote. That's another story, though, and it begins with Ann at Boston University.

In 1977 in a graduate seminar on historical methodology Ann was assigned an article by Samuel Huntington titled "The Change to Change: Modernization, Development and Politics." In the paper, which summarized his influential book *Political Order in Changing Societies*, he wrote that the main points could be conveyed in a series of three equations. They have the form $a/b = c$, $c/d = e$, $e/f = g$, where the letters denote (a) "social mobilization," (b) "economic development," (c) "social frustration," (d) "mobility opportunities," (e) "political participation," (f) "political institutionalization," and (g) "political instability." Ann thought that these equations made little sense, and she showed them to me. I agreed with her, and later criticized the equations as follows in my article "Mathematics As Propaganda":

> Huntington never bothers to inform the reader in what sense these are equations. It is doubtful that any of the terms (a)–(g) can be measured and assigned a single numerical value. What are the units of measurement? Will Huntington allow us to operate with these equations using the well-known techniques of ninth grade algebra? If so, we could infer, for instance, that
>
> $$a = b \cdot c = b \cdot d \cdot e = b \cdot d \cdot f \cdot g,$$
>
> i.e., that "social mobilization is equal to economic development times mobility opportunities times political institutionalization times political instability!"

Interestingly, when Ann objected to the equations in the seminar, neither her professor nor the other students could understand her reasoning; in particular, they couldn't follow her algebra.

In early 1986 Serge Lang, as a member of the National Academy of Sciences (NAS), was informed that the social sciences section had nominated Samuel Huntington for membership. Normally a section's nominations are rubber-stamped by the rest of the membership — a chemist, for example, wouldn't presume to know better than the physicists whether or not a certain physicist should be elected. However, Lang decided to oppose Huntington at the annual meeting. He distributed a packet of material to all of the members that included the Huntington section from my "Mathematics As Propaganda" article and some excerpts from Huntington's works.

Huntington's supporters reacted with virulence. In a letter to Lang the chairman of the social sciences section, Princeton Professor Julian Wolpert, wrote:

> Koblitz merits censure for this irresponsible piece of scurrulous [sic] "journalism" which somehow got past a peer review process of "scientists." Koblitz has distorted severely the arguments made by Huntington and himself created a phony semblance of spurious algebra to provide a better "strawman" to support his preconceived bias. Koblitz's piece in contrast is a trivial and superficial foray of anecdotal capsules taken out of context and no doubt intended to provide amusement about the "rape" of pure mathematics.

After a lively debate at the NAS annual meeting in April, Huntington failed to receive the two-thirds vote needed to override Lang's challenge.

The social sciences section tried again the following year. This time they lobbied intensively on his behalf. In March 1987 they wrote a letter to all NAS members in order to "explain why we have renominated Samuel Huntington and why we believe he should be elected to the Academy." In defense of Huntington's three equations, the letter cited Nobel laureate and NAS member Herbert Simon, who "explained in the debate that...his mathematical reasoning was entirely correct." They went on to mention that Simon had "expanded these remarks into a short paper on the use of ordinal relationships."

Despite the lobbying efforts and the strong support from Herbert Simon, Huntington again failed to get the required number of votes. Although the membership elections in 1986 had received little public attention, the debates and final vote in 1987 burst into the media. On April 29 *The New York Times* carried an article on the front page with the headline

"'Pseudoscience' Charge Bars Honor for a Harvard Scholar." The article started out: "The National Academy of Sciences, the nation's leading honor society for scientists, rejected a prominent Harvard political scientist today after a bruising internal struggle over whether his work amounted to 'pseudoscience' in which equations were used to dress up mere political opinions." *The New Republic* weighed in on Huntington's side and accused Serge Lang of "blood lust" in having conducted a "crude, unsubstantiated witch hunt" in the NAS. And syndicated columnist George Will wrote that the government should withhold funding from the NAS in retaliation for the rejection of Huntington. Both *The New Republic* and George Will cited Herbert Simon's endorsement of Huntington's mathematical methodology.

I wrote to Simon and got a copy of his article, which was titled "Some Trivial But Useful Mathematics." As soon as I read it, I saw that it was full of poorly written definitions, sloppy reasoning, and self-contradiction. I had suggested to Sheldon Axler, the editor of *The Mathematical Intelligencer*, that he ask Lang to contribute an article about the dispute over Huntington. When Lang declined to do it, I agreed to write about the matter for the journal. My article, along with Simon's reply and my response to his reply, appeared in the first issue of 1988; another reply by Simon, my answer, and his "final reply to Koblitz" were published in the following issue.

Next to my article the *Intelligencer* printed a long excerpt from Simon's paper that included most of the mathematical content. I started my comments by noting that "the first difficulty in studying the paper is that one has to be a bit of a mind reader in order to understand Simon's definitions." After describing the ambiguities, I mentioned some ways that one could patch them up. I then pointed out the unclarities in the statement of his main result, and suggested how one might possibly fix some of Simon's confusing and contradictory assertions. Then I said:

> But at this stage we are more likely to lose patience with all the patchwork, and ask ourselves: What is the point? ... Simon fails to give any evidence that the use of mathematical jargon, mathematical notation, theorems about "sums modulo 2," and so on will ever lead to any insight that would not have been arrived at more quickly without all that.... Simon's paper illustrates... how mathematical formalism can lead to confusion and obfuscation.... What Simon does accomplish is mystification. Non-scientist readers are likely to be impressed and intimidated by sentences such as: "More precisely, the ordering of a set of elements by a variable associated with that ordering is invariant under any positive strictly monotonic transformation of the

variable." Mathematical verbiage is being used like a witch doctor's incantation, to instill a sense of awe and reverence in the gullible or poorly educated.

In Simon's reply to me he tried to dispute what I had said on technical grounds. That was a mistake. All he showed was that he did not have a firm grasp of elementary mathematics. For example, his reaction to my attempt to patch up one of his definitions was bizarre. He said that he had "no objections to limiting the discussion to strictly monotonic transformations that are bijective, [but] limiting ourselves to continuous transformations would be a much less desirable restriction of generality." He was apparently blissfully unaware that a bijective order-preserving map from the positive real numbers to the positive real numbers *must be continuous*. (That's an elementary fact that's easy to prove.)

In the next round Simon dug himself in deeper, revealing, for example, that he did not know the standard definition of a continuous map. In my last response to Simon, in reference to his shifting uses of words I quoted the following lines from Lewis Carroll, which Ann had found for me:

"That's a great deal to make one word mean," Alice said in a thoughtful tone.
"When I make a word do a lot of work like that," said Humpty Dumpty, "I always pay it extra."

Although Simon thought that he had gotten the better of me in our dispute (his autobiography refers to the outcome as a "clear knockout" for him), to a mathematician all he had succeeded in doing was making a fool of himself. Not by any stretch of the imagination had the mathematical gibberish in his article strengthened the case for electing Samuel Huntington to the National Academy of Sciences.

It was not only because of his misuse of mathematics that Lang and I were glad to see Huntington voted down by the NAS. During the Kennedy and Johnson administrations he had been one of the architects of American policy in Vietnam and had played a key role in establishing the "strategic hamlet" program. In a 1967 article in *Foreign Affairs* Huntington had expressed satisfaction that the influx of refugees to the cities (caused by the bombing of the countryside) would hasten what he called South Vietnam's "urbanization" and "modernization" at the same time as it undercut potential support for the guerrillas. The coercive actions of the U.S. and the South Vietnamese regime in removing people from their villages were typical of American contempt for international law. According to Article

7(d) of the Statute of the International Criminal Court, "forcible transfer of population" is a crime against humanity.

Four decades later Huntington has not changed his stripes. In 2004 he made a splash with his book *Who Are We? The Challenge to America's National Identity*. According to the review in *Publishers Weekly*, "This book is an aggressive polemic whose central argument — that America, at heart, has been and... should remain a Christian, Anglocentric country — wouldn't be out of place on many a conservative radio station." The main target of the polemic is Hispanic Americans. According to *The Washington Post*, "In Huntington's nightmare vision, we are headed for a 'bilingual, bicultural society,' where Latinos take over some states, Anglos retreat to others, most of our big cities look like the barrios of Los Angeles, and U.S. foreign policy is dictated in Mexico City." This manifesto based on racist paranoia is perhaps what one should expect from someone who forty years before was at the center of U.S. policy-making on Vietnam.

To end this story on a lighter note, Ann and I were amused to find that not all conservatives took Huntington's side in the dispute over the NAS election. In August 1987 a journalist named Charles Sykes contacted us in connection with a book he was writing. Published the following year, his *ProfScam* is a wide-ranging indictment of university professors for an assortment of real and imagined sins, such as our supposedly easy workloads, neglect of teaching, grade inflation, dumbing down of courses, and turgid jargon in academic journals. Even though Sykes is a conservative — for example, he laments and ridicules "the new ideology of diversity and anti-elitism" — he shares with Ann and me a distaste for the misuse of quantitative methods by social scientists. He devotes a chapter of his book to the Huntington affair, and I particularly enjoyed his dramatic way of describing the origin of the dispute (p. 208):

> Samuel Huntington's election to the National Academy of Sciences would probably have been little more than a formality if it had not been for a graduate student named Ann Koblitz. The dispute that would shake the social sciences to their quantitative foundations, that was featured on the front page of *The New York Times*, in articles in *The New Republic*, *Science*, and *Discover*, and that would convulse the normally insouciant National Academy of Sciences, can be traced back to a single assignment in a graduate seminar on historical methodology at Boston University in 1977.

Soon after Ann started teaching at Hartwick College, one of the senior administrators saw her and said that he'd just been reading Sykes' book.

He was not only impressed to read about Ann's role, but visibly intimidated as well. After all, if as a graduate student she was already capable of causing such upheaval — "shaking the social sciences to their quantitative foundations" — just imagine what she could do as a professor in a small college! Perhaps Ann's reputation among people who had read *ProfScam* was one reason why none of the administrators at Hartwick ever hassled her about her frequent absences for a long weekend during term time or caused her any other difficulties.

Many scholars and administrators who come out of the non-sciences seem to have an almost mystical faith in the power of numbers that is not shared by those of us who work in the mathematical sciences. Once information is quantified, it acquires a certain aura for these people, who assume it to have greater validity than a mere qualitative assessment could possibly have. An example of the tendency to be mesmerized by numerical data is the widespread reliance on student ratings as the most important component in evaluating someone's teaching.

In the late 1980's, when I was briefly on my department's personnel committee, I was surprised to see that a colleague whom I knew to be a dedicated and conscientious teacher had not received nearly as high student rating numbers as I would have expected. Her calculus students seemed not to be appreciative of the tremendous effort she put into teaching them. I wondered whether perhaps they were harsh on her in part because she was a woman.

I wrote a query to the *Newsletter* of the Association for Women in Mathematics (AWM) asking whether anyone was aware of studies of whether or not student ratings tend to discriminate against women. I received responses from several readers, a few of whom sent me reprints of papers on the subject. It turned out that a fair amount had been written on this question, but not in journals that mathematicians normally read. In one study, for example, fifty male and fifty female students had been given a set of descriptions of the teaching practices of professors in various specialties. In the forms received by half of the students the professors had been given names of the opposite gender from those in the forms received by the other half. It was found that the male students were biased against women, while the female students were not.

A lot seems to depend on the type of course. Sometimes students perceive (probably correctly) that the women instructors tend to be more sensitive to their needs, more concerned and caring, and more dedicated to teaching than the male instructors (it also helps if the women are thought to be easy graders) — and as a result they reward them with higher ratings.

Note that these traits match the sex-stereotyped expectations of women as "mother figures." According to a paper in the *Journal of Educational Psychology* (D. Kierstead et al, 1988), "[Our] results suggest that if female instructors want to obtain high student ratings, they must be not only highly competent... but also careful to act in accordance with traditional sex role expectations. In particular,... male and female instructors will earn equal student ratings for equal professional work only if the women also display stereotypically feminine behavior."

This suggests that the difficulty for women occurs when they teach courses such as calculus that are supposed to be tough. Indeed, according to an article in *Signs: Journal of Women in Culture and Society* (E. Martin, 1984), in certain classroom studies "male students were far more likely to give lower ratings to those female faculty perceived to be hard graders... This finding is consistent with a series of experiments... that indicated that college students of both sexes judged female authority figures who engaged in punitive behavior more harshly than they judged punitive males." Another study showed that women will be rated highly only if they are especially accessible to students and spend a lot of time with them, whereas men can receive equally high ratings while remaining more aloof.

When Ann was at Hartwick College, on a couple of occasions a male colleague's advisees in another department asked her if they could meet with her on a regular basis, since their adviser was "too busy" to see them often. These students assumed that their professor, being male, was occupied with important matters, whereas Ann was not. The fact of the matter was that Ann was hard at work on several research projects during this time, whereas the male professor who was supposed to be advising these students was not a productive scholar. He did, however, *look* like a busy professor to the students, whereas Ann, as a woman, did not.

I wrote an article for the September-October 1990 issue of the *AWM Newsletter* describing what I had learned from the responses to my query. I ended with four conclusions, of which the last two were the most important:

> 3. In certain teaching situations which are frequently encountered in math departments (especially in introductory-level courses), students tend to discriminate against women instructors on the rating forms.
>
> 4. Math departments and administrators have an ethical and legal obligation not to base promotion and salary decisions on data which are biased against women.

The short piece did not take me long to write, but it turned out to have a greater impact than I would have expected. Several mathematicians wrote me that they had used it to support their case for tenure. When the AWM published a collection titled *Complexities: Women in Mathematics* (2005) that contained the most important articles that had appeared in the AWM *Newsletter* over the years, my article on student ratings was included.

In May 2007 Susan Landau and Whit Diffie, who are cryptographers at Sun Microsystems and the authors of the definitive book about computer privacy (with the delightful title *Privacy on the Line*), visited Ann and me in Seattle. Susan is active in efforts to increase the numbers of women in computer science and mathematics. She commented that the misuse of student evaluations continues to be a big problem for her colleagues in academia, especially the women, who often ask her for advice in dealing with the university administration on this issue. On several occasions in recent years Susan has referred them to my article "Are Student Ratings Unfair to Women?"

During the academic year 1997-1998 I saw a level of irresponsibility on the part of the central administration at the University of Washington that left me deeply disillusioned; that was one reason why I was glad to get away for the following year on sabbatical.

In October my department voted unanimously to recommend our colleague Tatiana Toro for promotion to associate professor with tenure. The College Council, which advises the dean, supported the recommendation, and the dean approved it as well. Then it went to the provost, who normally rubber-stamps the dean's decision. This time, however, he refused to act on it for almost six months. Nothing like that had ever happened before to a tenure recommendation in our department.

The provost wasn't forthcoming about his reasons, but indirectly we learned what the issue was. The only thing negative in Toro's file was a couple of mediocre student ratings in calculus. At this time the central administration believed that the math department was insufficiently concerned about student rating numbers. This was the opinion, for example, of Debra Friedman, the Associate Dean of Undergraduate Education, who at the end of 1997 was being brought over to the provost's office as Associate Provost for Academic Planning. Apparently the provost felt that approving Toro's tenure would convey the message that it was okay to have low student ratings in calculus, and this was a message that he — and especially Friedman — didn't want to send.

In the 1990's calculus had the reputation of being one of the hardest — perhaps *the* hardest — freshman course at the university. In those years we used two workbooks that I had prepared that emphasized mathematical modeling exercises (word problems) that the students found challenging. Most of us did not get high student ratings.

Tatiana Toro's colleagues who had observed her teaching had written extremely positive evaluations. Students, in contrast, had used the ratings as a way to punish her for teaching a tough course. Her gender and Hispanic background had undoubtedly compounded the problem. Students don't like it when a white male is hard on them. But many of them are especially incensed when a dark-complexioned woman who speaks with a slight Latin American accent is equally tough. It should have been obvious to the provost that in such a case the raw student rating numbers should not be taken at face value.

Our administrators frequently pay lip service to principles of racial and gender equality: departments often receive memos reminding them of their obligation to try to increase the diversity of the faculty. But when the math department — for the first time in its history — proposed a minority woman for tenure, the response of the provost was to block her promotion for several months. The level of hypocrisy astounded me.

As the year wore on, our chairman repeatedly tried to get the provost to move on the promotion. Ann and I knew Tatiana fairly well — in fact, at our invitation in 2000 she wrote an autobiographical essay for the *Kovalevskaia Fund Newsletter* describing her education in Colombia and her early development as a mathematician. The unprecedented delay in approving her tenure was extremely stressful; she was pregnant with her first child, and she was worried about the effect of the stress on the pregnancy.

At one point I told our chairman that the administration should be informed that if they refused to approve tenure and if Tatiana decided to sue, then several people in our department — not only me — would be prepared to testify that the university had discriminated against her on the basis of gender, race, and national origin, all of which had clearly played a role in her student ratings.

Tatiana's closest colleagues in the department were especially upset because they knew that we could easily lose her since other universities would soon be making her offers. Finally, the chair and the department's executive committee wrote a strong letter to the president of the university. The following day the provost approved tenure for Tatiana Toro.

Sure enough, within a few months another university, knowing about the bad treatment she had gotten at UW, made her an offer. In order to keep her, UW had to come up with a big salary raise. So the story ended happily. However, none of that excuses the behavior of the provost —

who soon after became interim president when the incumbent was forced out because of sexual improprieties. I suppose that there are still some universities where the faculty can have confidence in the probity of top administrators, but UW isn't one of them.

Many people in academia get annoyed when women and minority faculty complain of discrimination or a "chilly climate." These skeptics have no trouble identifying and disapproving of the type of blatant prejudice that was common in the 19th century — for example, when the tsarist minister of education in 1880 wrote to Kovalevskaia that both Kovalevskaia and her daughter would grow old before the Russian government would allow a woman to receive a Master's degree and teach college — but it is not so easy for them to recognize the subtler forms of discrimination that occur in our day.

It wasn't until the late 1990's that I became aware that women faculty in many departments were disproportionately being assigned the most onerous jobs on campus. A few years later the *Chronicle of Higher Education* carried a long article about the problem of the scholarly careers of many women and minorities being held back because they were being pressured to take on extra service responsibilities. For example, minority faculty were often made to feel that it was their special obligation to advise large numbers of minority students and help them cope with all sorts of difficulties.

I thought that the article had missed a key point related to university service, and I wrote a letter to the *Chronicle* giving my analysis of the issue:

> Your article on the impact of excessive service obligations on the careers of women and minority scholars (Dec. 19, 2003) fails to make an important distinction. Service tasks are of two types, what I call "guy work" and "girl work." Let me explain.
>
> Several years ago an amusing illustration in *Newsweek* accompanied an article about theories that claimed that male/female divisions of labor are genetically determined. In the cartoon a helix-shaped Mr. DNA was manipulating two pictures, one showing the woman-of-the-house with a frying-pan in one hand and a baby in the other, the other picture showing the man-of-the-house hammering together a doghouse while Spot looked on in eager anticipation.
>
> Men's work — repairing the doghouse, taking out the garbage, watering the lawn, balancing the checkbook — takes a lot of time to say and little time to do. Women's work has the opposite characteristics: it is very time-consuming, but linguistically it can be conveyed in a few quick syllables: cook, clean, childcare.

I have observed the same contrast in the academic world. An example of "guy work" would be the Strategic Planning Committee for the University Honors Program. I had such a committee assignment. It sounds impressive, and it took all of three hours of my time, including a wine-and-cheese reception. At the opposite extreme is a job like T.A. training — a thankless, burdensome chore. Of course, it is not always women and minorities who are saddled with the onerous service tasks; nor is it always men who get to do the "guy work." But in many departments there is a gender difference in the types of service work that people do.

So the best advice for junior faculty — especially women and minority faculty — is to seek out service assignments that are interesting, take little time, and look good on your CV. Avoid like the plague ones that will take a lot of time, drain your energy, and bring you little professional recognition. In particular, avoid anything that involves advising large numbers of students or T.A.'s. Look for a committee where you get wine and cheese.

People who are slow to see the subtle (and not so subtle) forms of discrimination against girls and women are attracted to alternative theories that purport to explain why women are underrepresented in most areas of science and technology. One explanation that is popular even among certain feminists goes as follows. Science has historically developed as a highly competitive, individualistic enterprise. In contrast, women by their very nature value interpersonal interaction. They prefer to work with others, not alone, and have a more developed sense of community than men do. For this reason they are put off by the disciplinary culture of the sciences.

Like other essentialist theories that give psychological and biological reasons for inequalities in society, this one falls apart when one examines the true situation. In fact, the theory gets the difference between the two cultures — science and humanities — exactly backwards. As Ann explains in a chapter on "Gender and Science Theory" in her second book,

> A mathematician, chemist, biologist, or physicist probably needs and generally has more interaction with her/his colleagues than does, for example, a historian or philosopher. Concrete evidence of this can be found in the statistics on joint paper production. Although mathematics traditionally has been thought to have far less collaborative research than most of the other sciences, even in mathematics this image no longer reflects reality. From 1940 to 1994 the percentage

of single-authored papers abstracted in *Mathematical Reviews* (the mathematical world's largest and most-consulted abstracting service) has decreased from about 93 percent of the total to about 57 percent.

Ironically, the proportion of single-authored papers in most non-scientific fields — including the discipline of women's studies itself — is far higher than in mathematics. *Signs*, the premier journal of women's studies, has been in existence since 1975. Of the 489 regular articles published from volume 1 through the year 1997/1998... only 55 (11.25%) have had more than one author. In other words, mathematics appears to have roughly four times the proportion of multiply-authored papers of women's studies.

Thus, the essentialist hypothesis that women by nature shun fields where the tradition is to work alone and are attracted to fields where collaborative work is the norm would provide an explanation for the proportion of women in different departments only if mathematics departments were predominantly female and women's studies departments were predominantly male.

Indeed, Ann has been envious of the type of stimulating collaborations I have had over the years — with Benedict Gross and David Rohrlich in the late 1970's, with Mike Fellows in the early 1990's, with Balasubramanian and Joe Buhler in the late 1990's, with Edlyn Teske, Andreas Stein, and Michael Jacobson during my sabbatical at Waterloo in 1998-1999, and with Alfred Menezes during the past few years. (This will be described in the next chapter.) The only person Ann has coauthored articles with is me (and in one case Mike Fellows was also a coauthor). It seems that in order to experience the satisfaction of collaborative research, she has to step over to my side of C. P. Snow's two-culture divide.

Postscript

After reading this chapter, Ann commented that it's a bit one-sided because it suggests, rather arrogantly, that the accepted norms of behavior among scientists are always superior to those among non-scientists. For a more balanced picture I could have pointed out, for example, that living with a mathematician can be difficult.

Ann's favorite story to illustrate this is of our visit to the University of California at Irvine in the spring of 1981. I had given a talk at UCLA and, being too much of a cheapskate to rent a car or take a taxi, I insisted on going from UCLA to U.C. Irvine by public bus. It took us three hours and four buses to do this — actually, five buses because one of them broke down. During this odyssey we were lugging two heavy suitcases and several bags, since we were en route to a two-month stay in Princeton, where I would be working at the Institute for Advanced Study.

We finally arrived in Irvine and stayed with our friend Bill Messing (an algebraic geometer and protegé of Grothendieck; he is now at the University of Minnesota). Besides a common interest in arithmetic algebraic geometry, Bill and I shared a reputation for our ability to eat prodigious amounts of ice cream. But who was the best? It occurred to us that this was a good occasion to settle the matter once and for all — we would have a contest.

Bill's wife Rita, who had planned a nice dinner for that evening, objected, but to no avail. Ann went with Bill and me to a supermarket, where we purchased several containers of ice cream. Since Bill liked the rich, creamy kind and I much preferred ice milk, to level the playing field we agreed to use weight rather than volume. Bill, with his long frizzy beard, long frizzy hair, and pants held up by a rope rather than a belt, looked like somebody's guru. As the two of us were excitedly grabbing cartons of ice cream and running with them to the produce department to weigh them, Ann's embarrassment became unbearable and she went out to the car to wait for us.

As soon as we got home the contest started. Seeing us eat, Bill's son asked if he could have some. "Not until later," his father said. "That's contest ice cream." Rita was furious.

After I finished almost a half-gallon, I saw that Bill, still on his second pint, was unlikely to catch up. I paused from the contest and sat down to a big roast beef dinner with Ann, Rita, and the kid. After the meal I ate another pint or two of ice cream, just to be sure that Bill would have no chance. The whole spectacle — although I regarded it as a victory of sorts — made Ann and Rita wonder about what sort of profession would harbor grown men who indulged in such antics.

CHAPTER 14 **CRYPTOGRAPHY**

When I started at the University of Washington in 1979, my main area of research was p-adic analysis. I had just published a joint article with Benedict Gross about a p-adic formula for Gauss sums, and I was intrigued by the many p-adic analogies of classical formulas involving special functions and modular forms. I worked in this area for a while, but I found it difficult to prove results that had any depth. I could formulate interesting conjectures and maybe prove some special cases, but my results during the first few years in Seattle weren't as definitive or important as those in the Gauss sum paper. I started looking for a new area of research that would be more satisfying.

I initially decided to learn a little about cryptography just for teaching purposes. Introductory number theory was usually taught in a rather theoretical way with emphasis on the logical structure of theorems and proofs. I thought that students could be better motivated by discussion of real-world applications of elementary number theory.

Until the recent past number theory had been thought to have virtually no practical applications. In his 1940 book *A Mathematician's Apology* the great British mathematician G. H. Hardy wrote that "both Gauss and lesser mathematicians may be justified in rejoicing that there is one science [number theory] at any rate, and that their own, whose very remoteness from ordinary human activities should keep it gentle and clean." In Hardy's day most applications of mathematics were military, and as a pacifist he was pleased that number theory was studied not for its practical uses, but only for its intrinsic aesthetic appeal.

This image of number theory as "gentle and clean" took a big hit in 1977 when three MIT computer scientists — Ron Rivest, Adi Shamir, and Len Adleman — invented a radically new cryptographic system. An article in *Scientific American* by Martin Gardner described the RSA idea, explained its significance, and caused a sudden upsurge in popular interest in both cryptography and number theory.

In those years RSA was the most important way to achieve what came to be called "public key cryptography." Earlier systems for scrambling messages worked well in military or diplomatic applications, where there was a fixed hierarchy of people who were authorized to know the secret keys.

But by the 1970's, with major sections of the economy rapidly becoming computerized, the limitations of classical cryptography were coming to the fore. For example, suppose that a large network of banks wants to be able to exchange encrypted messages authorizing money transfers. In traditional cryptography any pair of banks must have its own secret set of keys that they agree on and exchange using a trusted courier. The number of possible pairs of banks could easily be in the hundreds of millions. So the earlier type of cryptography, called "private key" (or "symmetric key"), becomes extremely unwieldy.

In public key cryptography, the key needed to scramble a message is public information. Each user of the system (for example, each bank) has its own public key, which is listed in a directory much like someone's phone number. Anybody can encrypt a message using the public key. However, the unscrambling process requires knowledge of a totally different key, which the user keeps secret. The procedure for scrambling a message is called a "trapdoor one-way function." This means that once we look up the bank's public key it is computationally easy (with the help of a computer) for us to send it an encrypted message. If, however, we want to go the other way — unscramble the message — this is computationally infeasible unless we possess an additional bit of information, namely the secret key.

Rivest, Shamir, and Adleman devised a clever — but also simple — way to make a trapdoor one-way function using elementary number theory. Their construction is based on multiplication of two large prime numbers p and q to get a composite number $N = pq$. One has to assume that going from p and q to N is a one-way process in the sense that while anyone who's given p and q can quickly find N, factoring N to get p and q is very hard.

Thus, the security of RSA cryptography was entirely dependent on the presumed difficulty of factoring large integers. For this reason the invention of RSA gave a tremendous stimulus to the study of methods to factor integers. Whenever a new factorization technique was developed with a faster running time, users of RSA had to increase the size of the numbers N in order to keep them out of the range where anyone would be able to factor them. By the mid-1980's the so-called "index-calculus" factoring algorithms were being refined and improved to the extent that it looked like in the foreseeable future it would be possible to factor a 100-digit N.

Meanwhile, in 1984 Springer-Verlag published my third book, a graduate text called *Introduction to Elliptic Curves and Modular Forms*. A little later in the chapter I'll explain in intuitive terms what elliptic curves are. The number theory of elliptic curves had been a topic in my Ph.D. dissertation, and the subject continued to hold a lot of interest for me.

Late in 1984 Hendrik Lenstra sent me a one-page description of a new method he had developed for factoring large integers using elliptic curves. His clever and elegant algorithm was simple enough that I could understand it from the one-page outline, although a detailed analysis of its running time took many more pages. This was the first time that elliptic curves had been used in cryptography, and when I read the page that Lenstra had sent me I felt that at one stroke he had raised the mathematical sophistication in cryptography to a whole new level.

Shortly after that I left for the Soviet Union, where no one worked openly on cryptography. I continued to think about the subject, though, and soon it occurred to me that it should be possible to use elliptic curves in an entirely different way from what Lenstra had done, namely, to construct systems based on the hard problem of finding logarithms on the curve. Since I knew no one in the Soviet Union I could talk with about this, I wrote a letter to Andrew Odlyzko, then at Bell Labs, describing my idea for using the elliptic curve group to construct a cryptosystem. Odlyzko was one of the few mathematicians at that time who had done major work in both theoretical and practical areas. Nowadays it's not so unusual to bridge pure and applied mathematics, but in the mid-1980's Odlyzko was unique in this respect among the mathematicians whom I knew personally.

E-mail didn't yet exist, and letters between the U.S.S.R. and the U.S. took a couple of weeks in each direction. So it wasn't until a month later that I received a reply from Odlyzko. He said that my idea for a new type of cryptography was a good one, and in fact at the same time Victor Miller of IBM was proposing exactly the same thing. Victor and I had had similar mathematical training — he had written his Ph.D. dissertation on elliptic curves under Barry Mazur at Harvard at about the same time as I wrote mine at Princeton — and we both had come up with the same idea simultaneously and independently of one another.

The corporate world was very slow to embrace public key cryptography — even RSA — and in 1985 it appeared unlikely that the more esoteric-seeming mathematics of elliptic curves would be of commercial importance in the foreseeable future. At least that was my impression at the time. To me elliptic curve cryptography (ECC) was just a nice theoretical construction to study.

Later when I thought back on the invention of ECC, what I found surprising was not that I had had no notion of commercializing the idea — after all, despite my short-lived foray into helicopter repair when I was in the Army, I had never worked on real-world problems — but that Victor Miller also had not applied for a patent, although then as now IBM's reputation was that it strongly encouraged its employees to get patents for everything they possibly could, even on the flimsiest of grounds. I assumed that Victor, like me, had not been thinking in terms of a marketable product. However, recently Victor told me that even in 1985 he did view ECC as something practical, but that it was the bureaucracy at IBM (which then had no interest in promoting any type of cryptography other than DES) that discouraged him. In any case, the question of turning ECC into a commercial product would wait until other people became interested in it.

While I was in Moscow in February 1985, the mathematician and artist Tolya Fomenko invited me to give a talk at the Moscow Mathematical Society. He knew that I had become interested in cryptography and thought that it would be a lot of fun to have an American speak on that subject. There had never before been a Math Society lecture on cryptography, and Tolya loved the idea of the first presentation on secret codes being given by an American. There was also something ironical about the first lecture I ever gave on cryptography being delivered in Moscow.

I was very nervous about the talk, but that was not because of the strangeness of an American speaking about cryptography in the Soviet Union. Rather, it was because of all the famous mathematicians who would be there — luminaries such as Shafarevich, Manin, Kirillov, Gel'fand, and Arnol'd. My Russian was pretty fluent by then, but I had never given a talk in any language in front of such a large and distinguished group of mathematicians. Moreover, I was just a beginner in cryptography; my only justification for choosing this subject was that Soviet mathematicians, except for a few who worked in secret, knew even less about it than I did.

I did not mention ECC, since I had not yet received Odlyzko's reply to my letter. Most of my talk was devoted to index-calculus methods for factoring integers and for finding discrete logarithms in finite fields (which is much different from finding them on an elliptic curve). What I spoke on was for the most part well-known standard stuff — and of course none of it was secret.

I concluded my presentation by giving an illustration due to Gus Simmons of the possible use of public key cryptography for nuclear test ban verification. Suppose that two mutually hostile superpowers each need to place an instrument on the other's territory that will transmit seismographic readings about any earthquake or nuclear explosion. The two sides need to know that the device on their territory is not being used for espionage and that the device they've placed in the other country is transmitting information that has not been tampered with. All of this can be achieved using public key cryptography. At a time of great Cold War tension — this was the era when Reagan was referring to the Soviet Union as "the evil empire" — I thought that this was a fitting application to end my talk with.

Fifteen years later at an ECC conference in Essen, Germany, I met a Russian cryptographer named Igor Semaev who was now working in the West but in 1985 had been in the KGB (which was the Soviet equivalent of the FBI, CIA and NSA rolled into one) and in that capacity had gone to my lecture. I remarked to him that he must've already known everything I said and found the presentation boring. He replied that actually it had been at a higher level than he had expected, and the nuclear test ban example had been entirely new to him.

After I returned to the U.S., I started attending cryptography conferences. The most important were the annual Crypto meetings held each August in Santa Barbara, California. In the 1980's I found the atmosphere at Crypto to be refreshing and stimulating. It was a truly multidisciplinary meeting, with people from industry, government, and academia in fields ranging from math and computer science to engineering and business.

There was an element of "forbidden fruit" in the first decade of the Crypto conferences. At the beginning of the 1980's the National Security Agency had made a heavy-handed attempt to restrict open research in cryptography. In mathematics they asked for the right to review any journal submission in number theory and then block or delay publication if they thought that it might have cryptographic value. Among academic researchers the idea of restrictions on the freedom to publish one's results went over like a lead balloon. However, an outright refusal to cooperate would've provoked a confrontation with the U.S. government, which people wanted to avoid. So rather than saying "no," they did what amounted to the same thing — they set up a committee. Called the Public Cryptography Study Group, it included representatives of the American Math Society and other organizations. After lengthy deliberations, the committee decided that only a purely voluntary procedure would be acceptable. In practice this meant that the NSA's proposal died a quiet death.

Thus, the founding of the Crypto conferences in 1981 was itself an act of defiance. The free-spirited tone of the meetings in those years reflected the colorful and eccentric personalities of some of the founders of and early researchers in public key cryptography. One such person was Whit Diffie, a brilliant, offbeat and unpredictable libertarian who in 1976 had coauthored (with Martin Hellman) the most famous paper in the history of cryptography. Diffie used to run the "rump session," where informal, irreverent, and often humorous presentations were the norm. There was heckling, and at one point Whit had to impose some restrictions on what could be thrown at a speaker (empty beer cans were okay, but not full ones).

The corporate influence was much weaker then. There was a long lag between the invention of public key cryptography and its acceptance in the commercial world; until the late 1980's businesses generally had little interest in the issue of data security. Most researchers in cryptography had never signed a "nondisclosure agreement" limiting what they could say publicly — in fact, most of us had never heard of such a thing.

Even though most people at Crypto did not understand the mathematics of elliptic curves, they were open-minded and welcoming. Several of them complimented me on my recently published *A Course in Number Theory and Cryptography* (1987), which was the first book on cryptography that included an introduction to elliptic curves. A few people even told me of their interest in commercializing elliptic curve cryptography, but only one of them actually did it.

That person was Scott Vanstone, a mathematician at the University of Waterloo who had done extensive research in several branches of discrete mathematics, including the theory of finite fields. He led a multidisciplinary group that had implemented improved algorithms for arithmetic in finite fields. With that experience they were well equipped to work on ECC. After we met at Crypto, Scott invited me to visit Waterloo. I remember being very impressed with the group of people who were beginning to work on ECC, which included professors, graduate students, and even undergraduates in math, computer science, and engineering.

Scott's star graduate student was Alfred Menezes, who was writing a Ph.D. thesis on ECC. His *Elliptic Curve Public Key Cryptosystems*, published in 1993, was the first book devoted entirely to ECC. Alfred's family was from Goa, a former Portuguese colony that is now part of India. He was born in Tanzania, grew up in Kuwait except for a few years at a boarding school in India, and went to college and graduate school at Waterloo. Despite our very different backgrounds, one common element is that both Alfred and I spent a formative period of childhood attending Catholic schools in India. Could this be more than a coincidence? Well, not being a social scientist, I won't venture to posit a link between Indian Catholicism and cryptographic research.

Scott Vanstone, along with two other Waterloo professors, one in math and one in engineering, formed a company to develop and market ECC.

After several transformations and name changes, it became the Certicom Corporation. Based in Mississauga, near Toronto, it has been playing an increasingly important role in the data security industry.

In high school most people study conic sections — circles, ellipses, parabolas, and hyperbolas — but not more complicated curves. The equations of conic sections involve at most the square of x and y, never the third or higher powers. In contrast, the equation of an elliptic curve has an x^3-term; a typical example is $y^2 = x^3 - x$. Elliptic curves are the simplest type of curve that is more complicated than a conic section.

Although an elliptic curve is not an ellipse, there is a historical reason why the term "elliptic" is used. In the 19th century when people were looking for formulas for the arclength around an ellipse, they were led to a class of definite integrals in which the integrand came from a more complicated type of curve. The integrals were called "elliptic integrals," and the curves were called "elliptic curves."

An important feature of the most familiar conic section — the circle — is the *group of rotations*. Rotations of a circle can be added (in other words, combined one after the other) just as numbers can be added, and we call the set of all rotations a "group." It turns out that an elliptic curve, although it looks nothing like a circle, also forms a "group."

For purposes of describing the group, it's best to look not only at the real number solutions (x, y) to the elliptic curve equation, but also at the complex numbers (x, y) that satisfy this equation. The term "complex numbers," which arose for historical reasons, is a misnomer, because in many situations using complex rather than real numbers leads to a clearer picture.

If we look at complex number solutions (x, y) to the elliptic curve equation, we end up with a two-dimensional figure called a *torus*. This looks like the surface of a donut (or bagel or inner tube). In the same way as the xy-plane is the "product" of two axes (the horizontal and vertical ones), a torus can be thought of as a product of two circles. Thus, there are two types of rotations: rotations around the hole (the way a space station might rotate to create artificial gravity) and *into* the hole (meaning that each circular cross-section rotates through the same angle).

In cryptography we don't use either the real or the complex solutions (x, y) to the elliptic curve equation. Rather, we use solutions where x and y belong to a "finite field," a concept that was touched upon briefly in Chapter 6. The simplest example of a finite field is the integers less than a prime number. For example, let's take the prime 7. The field of 7 elements consists of the integers 0, 1, 2, 3, 4, 5, and 6. Any time we add or multiply, we remove multiples of 7 in order to get an answer in the range from 0 to 6 — in other words, we take the remainder after dividing by 7. In this number system, then, we have $4 + 5 = 2$ (because 2 is the remainder when 9 is divided by 7) and $3 \times 6 = 4$ (because 4 is the remainder when 18 is divided by 7). We call this "arithmetic modulo 7." Addition modulo 7 is used in everyday life when we want to know days of the week. For example, on a Wednesday in November you can say that exactly one month later will be a Friday, because adding 30 is the same modulo 7 as adding 2. Exactly one year later will be a Thursday, because adding 365 is the same modulo 7 as adding 1, since $365 = 52 \times 7 + 1$, in other words, a year equals fifty-two weeks plus one day. (Of course, one year later will be a Friday if it's a leap year.) However, there's one big difference between the finite fields used in cryptography and the field of 7 elements: in cryptographic applications the finite field is humongously large — for example, we might use the integers modulo a 50-digit prime.

Although my main area of interest in cryptography was ECC, for a few years I collaborated with Mike Fellows, a computer scientist then at the University of Victoria in British Columbia (about 80 miles/130 kilometers from Seattle by ferry), on a totally different approach to constructing cryptosystems. The idea was to use hard problems arising from combinatorics. The notion of constructing one-way functions in this way had been somewhat discredited by the experience of the so-called "knapsack" in the early 1980's. The knapsack problem (also called the *subset sum problem*) asks for a way to choose a subset of a given set of positive integers such that their sum has a desired value. This is similar to choosing items to fill up a knapsack with no space left over. In general this is a difficult computational problem, and in 1978 Merkle and Hellman used it to construct a cryptosystem that at first seemed promising. But just a few years later Adi Shamir, Ernie Brickell and others showed how to break the Merkle-Hellman system and its variants. For such a nice cryptosystem to be broken so dramatically was traumatic for the early researchers in public key cryptography, and the experience chastened them. After that they tried to be very careful about what problems could serve as a source for cryptographic one-way functions.

Mike Fellows and I thought that people had gone overboard, and some of these combinatorial problems should be revisited. We constructed a system based on a computational algebra question (called *ideal membership*) that involved polynomials, and we challenged people to try to crack it. The most attractive feature of our cryptosystem was the name that Mike thought up for it: *Polly Cracker*. It was very inefficient, and before long some papers were published that indeed cracked the code.

Mike and I also wrote a paper about a theoretical concept in the overlap between number theory and computer science called "self-witnessing problems" — this was the paper mentioned in Chapter 12 that I spoke about in Peru in March 1992. But probably the best part of our collaboration was developing ways of using cryptography to teach math and computer science to children. It turned out that a special case of Polly Cracker, called the *perfect code* cryptosystem, is ideally suited for children and others without mathematical training who want to understand what public key cryptography is all about. The perfect code cryptosystem can be broken by means of Gaussian elimination, but when used for recreational purposes by people who have never studied linear algebra it is reasonably secure. By the way, the term "perfect code" refers not to cryptographic codes, but rather to error-correcting codes of the sort used to remove noise and static in digital communication. The simplest example of a perfect code is the following: Send each transmitted symbol three times. Then if one of them is distorted on the way, the recipient can still tell what symbol was intended, because the other two occurrences are intact. Most perfect codes that are used in practice — for example, when images are transmitted from interplanetary spacecraft — are much more complicated than this one.

When I was on the program committee for Crypto '92, I lobbied to get Mike on the schedule as an invited speaker so that he could talk about what he called *kid crypto*. I was the session chair (moderator) during his presentation. He gave an entertaining and thought-provoking — if somewhat disorganized — lecture explaining his ideas on math education and the role that cryptography could play. At one point he observed that what security means is relative to the mathematical knowledge of the users. A system that can be cracked using college-level math (such as linear algebra) might be secure for use among high schoolers. Even adults shouldn't be too smug about our own knowledge, since to someone from a more advanced civilization it would seem that we're all doing kid crypto. To illustrate he projected onto the screen a front-page picture from an over-the-top tabloid called *Weekly World News* showing an extraterrestrial creature meeting with President Clinton. Mike said that this representative from another galaxy had told Clinton that they'd broken RSA three hundred years ago. I blurted out, "What about ECC? Did the alien tell him anything about that?" Everyone laughed at my question — and at my apparent inability to restrain myself from interrupting the speaker, which the moderator was not supposed to do. In reality, I had been with Mike when he bought the issue of *Weekly World News*, and we had discussed how he would use it in his talk and what would be a funny thing for me to interject. So my supposedly spontaneous outburst had actually been planned in advance with Mike.

The most important influence Mike Fellows had on me was in the area of elementary education; this will be described in the next chapter. He probably also had an effect on my style because he showed that even in a serious professional setting there is room for jokes and a camp sensibility. Mike, like me, is a product of the 1960's. But unlike me, he took part in the counter-culture; in fact, for a couple of years in the 1970's he lived in a hippie commune in California with his first wife. In his youth Mike was quite athletic and surfed a lot; in middle age he still has a wiry physique, although most of the time he settles for boogie-boarding.

In 2001 Mike married Fran Rosamond, a researcher in math education. Ann and I attended the wedding, which took place by the ocean near San Diego. Our role at the end of the ceremony was to blow soap bubbles over Mike and Fran as they put on wetsuits and surfed off into the sunset. Much as we like Mike, we have to admit that his behavior at times bears the telltale signs of his having grown up in southern California.

At Crypto '93 I gave a rump session talk on patents. For a long time I had had a dim view of the proliferation of patents in cryptography. Most of them were for contributions of questionable significance or originality, and many were for the type of things — mathematical concepts, for example — that should not be patentable. I had received information on the subject, such as examples of ridiculous patents, from Dan Bernstein, a graduate student of Hendrik Lenstra at Berkeley who a few years later became well known for suing the U.S. government over export controls and who now is one of the top researchers in cryptography. I put together a short presentation ridiculing the patent system and mocking people who said that they couldn't talk about their work because of company restrictions related to trade secrets. I argued that, like the "top secret" designation in government, patents and trade secrets allowed people who were doing mediocre research to avoid public scrutiny and create an inflated impression of the importance of what they were doing.

Many years later I heard an amusing story from Joan Feigenbaum, who's always a wonderful conversationalist. I have known Joan since the 1970's, when she was an undergraduate at Harvard working as a teaching assistant in my calculus course; now she is a professor of computer science at Yale. Joan told me that after my rump session talk in 1993 some people had asked her, "Is Koblitz really opposed to intellectual property?" She answered, "You have to understand about Neal. He's a communist, and he's opposed to *all* property." I complimented Joan on her clever answer, but pointed out that it's not strictly correct. No one would classify Ann and me as principled opponents of private property if they saw the 20-acre parcel we own on Copper Mountain in Arizona (about which I'll say more in Chapter 16) with its fences and locked gates festooned with No Trespassing signs.

By 1993 I was becoming disillusioned with the Crypto meetings. They had largely lost their quirkiness and sense of excitement, and were increasingly resembling the big computer science conferences known by acronyms such as STOC and FOCS.

Scott Vanstone and I had a discouraging experience that year. The movie "Sneakers" starring Robert Redford had just come out in late 1992. Public key cryptography was central to the plot — there was even a line mentioning the number field sieve factorization method — and the producers had hired Len Adleman (the A of RSA) as a consultant. While discussing the movie, Scott and I got the idea of inviting Robert Redford to the next Crypto to receive some sort of award — such as lifetime membership in the International Association for Cryptologic Research (IACR) — in recognition of his role in popularizing modern cryptography. Scott, who belonged to the IACR Board of Directors, presented the suggestion to the Board, which reacted negatively. They thought that such an action would be undignified and inappropriate for a serious academic conference. When Scott told me about the hostile reception of our proposal, I couldn't understand what their problem was. The IACR leadership must just be a bunch of humorless stuffed shirts, I concluded.

Each year when the Crypto participants arrived in Santa Barbara they were given a volume published by Springer-Verlag containing the conference papers, which had been written several months before. Almost all of the speakers essentially just repeated what was in the printed volume. Moreover, the contributed talks were highly specialized, and, unlike other big professional meetings, Crypto did not have parallel sessions — everyone was supposed to go to all the presentations. As a result between 80% and 98% of the people attending a given talk got little or nothing out of it.

I skipped Crypto '94 and '95. Just as Crypto was becoming less interesting to me, a large number of specialized conferences in the mathematics of public key cryptography were being organized, and some of them were becoming annual events. Crypto was no longer the only game in town, and there was no good reason for me to continue going to it.

But then I was invited to be the program chair of Crypto '96. That was considered an honor (although it was also a lot of work), and I was flattered to have been selected. So I accepted, even though it meant suspending my boycott of the meetings.

The program chair has primary responsibility for the scientific content of the conference. This means organizing the reviewing and selection of papers. In the mid-1990's only about 20% of the submissions could be accepted (the current figure is closer to 15%). Since Crypto is the most prestigious of the annual conferences in the field, a lot professionally is at stake.

To someone trained in mathematics the experience of being Crypto '96 program chair was unsettling. About two thirds of the submissions arrived by courier mail within 48 hours of the final deadline. Many had obviously been rushed and were full of typesetting errors. One author had sent me only the odd-numbered pages. A few had violated the requirement of anonymity (there was a policy of double-blind reviews). Several had disregarded the guidelines that had been sent to them. And in many cases the papers had little originality; they were tiny improvements over something the same authors had published the year before or a minor modification of someone else's work.

In some ways the situation has gotten even worse with electronic submissions. Alfred Menezes, the program chair for Crypto 2007, told me that of the 197 submissions, 103 arrived within eleven hours of the deadline and 35 arrived within the very last hour.

Mathematical publishing works differently. In the first place, most articles appear in journals, not conference proceedings — and journals don't have deadlines. In the second place, people in mathematics tend to have a low opinion of authors who rush into print a large number of small articles — the derogatory term is LPU (*least publishable unit*) — rather than waiting until they are ready to publish a complete treatment of the subject in a single article.

Cryptography has been heavily influenced by the disciplinary culture of computer science, which is quite different from that of mathematics. The gap might seem minor compared to the gulf that C. P. Snow observed between the sciences and humanities, but it is surprisingly large nonetheless. Some of the explanation for the divergence between the two fields might be a matter of time scale. Mathematicians, who are part of a rich tradition going back thousands of years, perceive the passing of time as an elephant does. In the grand scheme of things it is of little consequence whether their big paper appears this year or next. Computer science and cryptography, on the other hand, are influenced by the corporate world of high technology, with its frenetic rush to be the first to bring some new gadget to market. Cryptographers, thus, see time passing as a hummingbird does. Top researchers expect that practically every major conference should include one or more quickie papers by them or their students.

I don't want to give the impression that there are no scientific advantages to the hummingbird's viewpoint that characterizes cryptography and computer science. In general, cryptographers do not bear grudges as long as mathematicians do. I must've made a lot of enemies as a result of some of the decisions I made as Crypto '96 program chair. However, within a few years it seemed that hardly anyone still harbored ill feelings toward me — 1996 had long since faded from memory. Disputes in cryptography can be nasty, but most are quickly forgotten. In mathematics, on the other hand, there have been some well-known professional feuds that went on for decades. Hell hath no fury like a mathematician scorned!

Two of the decisions I made as Crypto '96 program chair were very unpopular among people who were submitting papers or had students who were. The first was that I required all authors to include with their submission a one-page statement describing what they would do in their talk. That was my way of driving home the point that presentations should not be just a rehash of what was written in the proceedings volume, and authors should think a little about what would benefit their audience.

Since I hadn't attended Crypto '95, it was Scott Vanstone who had had to make the public announcement at the IACR business meeting that the following year's Call for Papers would include the requirement that authors separately outline what they would do in their oral presentation. He described to me the overwhelmingly hostile reaction. People considered the one-page statement to be an unreasonable imposition on the authors, many of whom would be staying up all night to get their hastily put together manuscripts to FedEx before the deadline. Some at the meeting got very upset at Scott about it, and this surprised him. Why shoot the messenger? It was really me they should have been angry with — but I was half a world away, visiting Vietnam and India at the time.

My other controversial decision as program chair was to have five invited speakers at Crypto '96; in the past there had typically been only two. In order to make room in the schedule, I lengthened the conference day a bit and slightly reduced the number of accepted contributed papers. Some of the people who had organized earlier Cryptos thought that I had lowered the level by doing this, particularly since the invited speakers were asked to give talks that would be interesting and accessible to the four hundred attendees.

All five were good speakers and were well received by the audience. My favorite was Cliff Stoll, author of *The Cuckoo's Egg* and *Silicon Snake Oil*. The first was a best-seller that described how the author had caught a spy who was trying to hack into military computers. The second book was an anecdote-filled rumination about the down side of virtual reality and the computerization of society. A later edition of that book, titled *High-Tech Heretic*, has a chapter on education that is partly based on an article I wrote in *The Mathematical Intelligencer* opposing the use of calculators and computers in math instruction.

Cliff is an entertaining, provocative, and eccentric speaker. Watching him gesticulate and jump around the stage, one can't help thinking of a hyperactive adolescent. Ann and I enjoyed meeting him and found his talk delightful, even if he made liberal use of poetic license in some of what he said.

There was actually a third thing I did at Crypto '96 that most people thought was disastrous, although in that case I didn't get any grief about it because hardly anyone knew that it had been my idea. I suggested to the general chair, Richard Graveman, who was responsible for the other activities besides the talks, that the day before the conference begins we sponsor an excursion to the beautiful Channel Islands off the coast near Santa Barbara. We chartered a boat that included naturalist guides; on the trip we saw several types of whales, seals, and porpoises, and fish and invertebrates of every size and description. Unfortunately, the seas were rough, and most of the participants got badly seasick. Ann, who was one of the few hardy souls who didn't have to lean over the side to "feed the fishes," loved it. But most did not, and my nightmare during the voyage was that the people who were in agony, some of whom might also have been authors of submitted papers that I had rejected, would find out that this excursion had been my idea — and would toss me overboard. Fortunately, though, they never learned about my role in causing their misery.

By the mid-1990's the people who were selling RSA cryptography products were becoming worried about the threat posed by elliptic curve cryptography. The ECC company Certicom was tiny in comparison with their company and had not yet made any significant inroads into their market. However, they saw clouds on the horizon.

When ECC was introduced, some people were skeptical that the supposedly hard mathematical problem its security is based on — the elliptic curve discrete logarithm problem (ECDLP) — would stand the test of time. True, they would say, no one knew any way to apply an index-calculus style algorithm to this problem, and so the best one could do was to use a much slower method. However, the ECDLP had been studied less than the integer factorization problem. So perhaps the apparent security advantage of ECC was only illusory — it was just that people hadn't had long enough to think up attacks.

In contrast, integer factorization had been intensively studied. In the 1980's the best algorithms to factor an integer of n decimal digits would take roughly $\exp(n^{1/2+\epsilon})$ computer operations, where ϵ stands for an arbitrarily small number. Even when Hendrik Lenstra found a factoring algorithm using elliptic curves, it turned out that its asymptotic running time was essentially the same. This was a striking coincidence that suggested that $\exp(n^{1/2+\epsilon})$ might even be a lower bound on how much time is needed to factor an n-digit integer.

However, history sometimes plays jokes on us. Soon after — around 1990 — a major breakthrough, the "number field sieve," lowered the running time for factorization to roughly $\exp(n^{1/3+\epsilon})$. A casual reader might not think that changing 1/2 to 1/3 should be that big a deal — but to appreciate the scale of the improvement try plugging $n = 1000$ into both $\exp(n^{1/2})$ and $\exp(n^{1/3})$; you get $e^{\sqrt{1000}} \approx 54{,}149{,}865{,}250{,}000$ and $e^{10} \approx 22{,}000$, respectively.

Meanwhile, a great deal of work was also done on elliptic curve discrete logs. However, except for a relatively small set of elliptic curves that are easy to avoid, even at present — more than twenty years after the invention of ECC — no algorithm is known that finds discrete logs in fewer than $10^{n/2}$ operations, where n is the number of decimal digits in the size of the elliptic curve group. If, for example, we use a 50-digit group, then the number of computer operations required to break the ECDLP is greater than 10^{25}, that is, ten trillion trillion. That's well beyond the reach of the fastest computers (or networked systems of computers) available today.

By way of comparison let's look at the difficulty of breaking RSA by factoring N of various sizes. Recall that n denotes the number of decimal digits in N. I've tabulated the approximate number of computer operations both before and after the discovery of the number field sieve (NFS). Roughly speaking, each decade the number of digits needed to assure security doubled — 38 in the late 1970's, 75 in the 1980's, 150 in the 1990's, and so on. Table 14.1 gives the decade (first column), the number of digits in N (second column), the approximate number of computer operations to factor N with the best algorithms that were known before the number field sieve (third column), and the number of operations needed to factor N using the number field sieve (final column). It is the third column that gives the number of operations needed to factor an n-digit integer in the 1970's and 1980's, and the fourth column that is relevant starting in 1990. The remaining entries in those two columns are given

Table 14.1. Number of computer operations needed to factor an n-digit integer N

decade	n	pre-NFS	post-NFS
1970's	38	10^{12}	(10^{13})
1980's	75	10^{17}	(10^{16})
1990's	150	(10^{24})	10^{20}
2000's	300	(10^{35})	10^{25}

only for completeness, and I put them in parentheses since they can be ignored.

This table shows that in the pre-number field sieve days RSA would need to use N of about 150 digits to achieve the same security as a 50-digit ECC group. This means that ECC could use shorter keys than RSA, and most computations were also faster. But the advantage of ECC was still relatively small — a factor of three. However, in post-number field sieve cryptography, RSA needs to use N of about 300 digits for the same level of security — so now ECC's advantage is a factor of six. Moreover, as computing power increases the efficiency advantage of ECC will become even greater. For instance, if we want to choose our system so that it would take 10^{35} rather than 10^{25} computer operations to break it, that means increasing the size of the ECC group from 50 digits to 70 digits and increasing the size of the N in RSA from 300 digits to around 800 digits — more than ten times bigger than in ECC.

Thus, the main reason for the appeal of elliptic curves in cryptography is that one can work with much smaller numbers than in RSA while achieving the same level of security. Another attraction is that there's a tremendous variety of possible elliptic curves to use. On the one hand, for greater efficiency one can choose curves whose equations have a special form. In 1991 I proposed using certain curves whose equation is given over the field of two elements (consisting only of 0 and 1). This idea was developed and improved upon by Jerry Solinas of NSA. In fact, the very first paper ever presented publicly at a cryptography meeting by an NSA member was Jerry's analysis at Crypto '97 of an improved algorithm for computations on the curves I had proposed.

On the other hand, many people prefer to use elliptic curves whose equations have randomly generated coefficients. Even though working with random curves is not quite as efficient as using special ones as Jerry Solinas did, it is still a practical approach. In any case, the popularity of randomly generated elliptic curves is my justification for the title of this book.

Finally, elliptic curves are not the only kind of curves that can be used in cryptography. In a paper in the *Journal of Cryptology* in 1989, I proposed using certain groups coming from *hyperelliptic* curves. These are curves having even higher powers of x; for example, $y^2 = x^5 - x$ is such a curve. In recent years a lot of research, especially in Germany, has been devoted to hyperelliptic curve cryptosystems.

Elliptic curves have a long, rich history in several areas of mathematics. The so-called "chord and tangent" method for adding points on an elliptic curve actually goes back to Diophantus in the third century. Nevertheless, in the early years of ECC many cryptographers thought that elliptic curves were a difficult and abstruse topic in mathematics, viewed ECC with suspicion, and preferred a system based on more elementary mathematics. In the 1990's RSA tried to encourage this phobia of elliptic curves.

In June 1997 Alfred Menezes and I were teaching a two-week course for graduate students on cryptography at the Math Sciences Research Institute (MSRI) in Berkeley. I hadn't been following online debates — then, as now, I was only computer-literate enough for my essential needs — so Alfred showed me what RSA had put on its website. A special page called "ECC Central" was devoted to skeptical statements about elliptic curves by well-known experts in cryptography, several of whom had a financial stake in RSA. The most sophisticated and urbane comment was due to Ron Rivest (who's the R of RSA); he wrote,

> But the security of cryptosystems based on elliptic curves is not well understood, due in large part to the abstruse nature of elliptic curves. Few cryptographers understand elliptic curves, so... trying to get an evaluation of the security of an elliptic curve cryptosystem is a bit like trying to get an evaluation of some recently discovered Chaldean poetry.

I was impressed by Rivest's erudition — I didn't even know who the Chaldeans were. I asked Ann, who of course knew: they were a Biblical people who had lived in Babylonia. The Berkeley library had a volume called *The Chaldean Chronicles* that contained the deciphered fragments that scholars could read. Some of it sounded quite beautiful, even in translation.

At the MSRI short courses there was a tradition of having T-shirts with an appropriate logo made at the end. I proposed a shirt with a drawing of an elliptic curve and the words "I Love Chaldean Poetry." The shirt was popular among the graduate students — except for a couple of them who intended to get internships at RSA.

In March 1997 Scott Vanstone invited me to be a consultant to Certicom for $1000 a month. I was happy to agree, although I would have felt queasy about pocketing the money, since there was little that I did for Certicom during my five years as consultant that I wouldn't have been glad to do anyway free of charge. For the first 31 months I had Certicom donate the money directly to an internship program for minority science and engineering students at the University of Washington. When I returned from a sabbatical year at Waterloo in the summer of 1999, I learned that the university had not used the money for the intended purpose. In addition, at the end of the summer I received an irresponsible and insulting letter from University of Washington President Richard McCormick (which I'll discuss in the next chapter). So Ann and I decided never again to give any money to my university, and instead we had Certicom send all remaining payments to the Kovalevskaia Fund.

On a few occasions I did a sort of trouble-shooting for Certicom, writing a quick report on a mathematical development that might have implications for ECC. For example, in September 1997 three researchers Araki, Satoh, and Smart announced that they could quickly find discrete logarithms on trace-one elliptic curves. I studied the algorithm, which was fairly simple and very fast. The result could not be extended beyond an extremely rare subset of elliptic curves. It was important to make sure that people understood this as quickly as possible, since a researcher at RSA named Matt Robshaw had made a public statement greatly exaggerating the significance of the Araki-Satoh-Smart algorithm and suggesting that it provided a good reason to choose RSA cryptography over ECC.

In the competition between the two types of cryptography I was fascinated by the contrast between RSA's go-for-the-jugular approach and Certicom's laid-back style. The Certicom management felt that it was in their best interest to take the high road. If company people had started bad-mouthing RSA, then the main effect would have been to turn potential customers off to all forms of public key cryptography — "a plague on both your houses" would have been a common reaction.

Once a journalist for *Wired* magazine was doing a piece on the two forms of cryptography. Certicom gave him my name, but first they contacted me to be sure that I understood that it would not help matters if I criticized RSA. When I talked with the guy, he liked my explanation of elliptic curves and so asked me also to explain to him a little bit of the math of RSA. Having taught this many times to my number theory classes at UW, I could give him a basic mathematical description that apparently was much clearer than any explanation that the RSA people he talked to had given him. Most likely they had just plied him with buzzwords and hype, so that to get an understandable explanation of how RSA works he had to ask an ECC person.

Perhaps Certicom is less aggressive than RSA in part because of a national difference between Canada and the U.S. Canadians prefer to deal with the world in a less belligerent and militaristic way than Americans. Or maybe it's just that taking the high road makes good business sense.

In August 1997 a rumor was circulating at the Crypto meeting that the Russian mathematician Sergei Stepanov had told somebody that he had a very efficient algorithm to find discrete logarithms. If the claim were true, this would kill elliptic curve cryptography. Bob Silverman, who functioned at meetings as a sort of pit bull for RSA, was gleefully spreading the rumor to everyone he could. However, Odlyzko pointed out that Stepanov had made the claim in an informal way three months earlier and had not produced even an outline of his approach. After several more months with nothing from Stepanov, interest in the rumor died out.

The possibility that Stepanov might have an attack on ECC worried me. He had been a prominent mathematician in the Soviet Union. A protegé of the great analytic number theorist I. M. Vinogradov, he had been a permanent member of the Steklov Institute, the country's leading center of mathematical research. In 1972 he had given a striking new proof of the theorems of Hasse and Weil for elliptic and hyperelliptic curves, and as a result he was an invited speaker at the International Congress of Mathematicians in Vancouver in 1974. I had been at that Congress and had attended his talk. By the late 1990's Stepanov was working in Turkey, where he had moved after the collapse of the Soviet Union.

In December 1998, over a year-and-a-half after Stepanov made the original claim, I and several other cryptography researchers received an e-mail from him with the full text of his algorithm. I was in Chile at the time, and the e-mail came just as I was about to go give a lecture. Later in the day I studied Stepanov's paper, which was short, self-contained, and easy to read. The main body, where the validity of the algorithm was purportedly proved, consisted of less than five pages. Within an hour I found basic errors in both parts of his proof, and I e-mailed Stepanov. He replied with a patched up version, but that also contained similar errors. The mistakes were simple ones that a beginning graduate student should have been able to detect. After another month of e-mail correspondence, Stepanov finally conceded that his algorithm couldn't be patched.

When I returned to Waterloo in January, I gave a talk titled "The Russian Who Cried Wolf." A typical reaction to the episode was the statement by Alfred Menezes that "It's fair to say that this [Stepanov's preprint] is the most embarrassing paper ever written in the field of cryptography." I'm not sure that Alfred would still be of that opinion, since a lot of very embarrassing papers have been written since 1998. For me the farce was a sad reminder of the deterioration of Russian mathematics. Lo how the mighty have fallen!

In early September 1998, a few days before I was to leave for my year's sabbatical at the University of Waterloo, Ann got an urgent phone message for me from Scott Vanstone. Scott rarely called, and I hadn't even realized that he knew how to reach me when I was visiting Ann in Arizona. It had to be something serious. The next day I received an e-mail from Joe Silverman, a mathematician at Brown University who had written an excellent two-volume graduate textbook on elliptic curves (and no relation to Bob Silverman). I immediately knew what Scott had wanted to talk with me about. Joe's e-mail contained a brief outline of a new algorithm he was proposing to solve the elliptic curve discrete log problem (ECDLP) — in other words, to break elliptic curve cryptography.

Joe called his algorithm "xedni calculus." Later in the year I lectured on it in Chile, and when I arrived I learned that my host had tried without success to find "xedni" in books on mythology so that he'd know how to translate my title into Spanish. I apologized for not having sent him the correct Spanish translation, which was "secidni." In his paper Joe explained that he had called his algorithm "xedni" because that's "index" spelled backwards. His general idea was to perform steps that are similar to those in index-calculus algorithms, but in the reverse order.

The reason Joe thought that his algorithm might possibly be efficient was based on a deep and difficult relationship called the Birch and Swinnerton-Dyer Conjecture. Ironically, in a book titled *Algebraic Aspects of Cryptography* that I had published just a few months before, I had included a discussion of this conjecture in a section that I called "Cultural Background." My tone was apologetic to my readers for taking their time with mathematics which, while of great interest to theoreticians, was unlikely, I said, ever to be applied to cryptography. Then within a year I was intensively studying Silverman's attack on ECC that was based precisely on the idea behind that conjecture. This shows that it is unwise to predict that a certain type of mathematics will *never* be used in cryptography.

Scott Vanstone and the others at Certicom were extremely worried about Joe Silverman's algorithm. They badly wanted a convincing argument that it wouldn't be practical. Joe himself was agnostic on the question — he made no claim about practicality. He also was discreet about his discrete logarithms. He wasn't going to spill the beans to RSA that he had an attack on elliptic curve cryptography. However, Scott knew that word would get out soon, and there'd be little time to subject the xedni algorithm to a careful analysis before the RSA people started proclaiming from the rooftops that ECC had been broken.

I saw that analyzing Silverman's algorithm would be no simple matter. I had no idea how to measure the practical consequences of his use of the Birch and Swinnerton-Dyer Conjecture. Even among mathematicians who worked with elliptic curves there were few who had any experience with the type of computational questions that arose in Joe's xedni calculus, and I certainly wasn't one of them. I was about to leave for Waterloo, and I hated the thought of disappointing my friends there.

Very early the morning of my flight I couldn't sleep, and returned to thinking about the xedni algorithm. There was no way I could quickly rule out the possibility that it could be made into a successful attack on ECC.

Then suddenly I saw that I could do something almost as good. A minor modification of Silverman's algorithm could be used to factor large integers. In other words, if xedni turned out to be efficient in breaking ECC, then it would break RSA as well! I breathed a sigh of relief — I would have something useful to bring to Canada after all. I had bought us time to study and analyze xedni calculus, since the RSA people would now be in no hurry to trumpet the power of xedni.

The first few months of my sabbatical year were devoted to a thorough analysis of the Silverman algorithm. In October I found a theoretical argument, using the concept of the "height" of points, that showed that for very, very large elliptic curve groups the xedni approach would be extremely inefficient. However, with this general line of reasoning I couldn't be specific about the sizes for which the algorithm would be impractical. It was conceivable, although I thought it unlikely, that the algorithm would not be totally infeasible for curves in the size range that's used in cryptography.

It is important to understand that an asymptotic result — such as my argument that established the inefficiency of xedni in the limit as the size of the group increases — cannot be relied upon as any kind of guarantee of security. Rather, one must analyze the algorithm for elliptic curves of the size employed in cryptography. The asymptotic argument might be helpful as a guide — and certainly it made us hopeful that we would be able to show that xedni is impractical for the curves used in the real world — but it cannot serve as a substitute for a concrete security analysis. It turned out to be much harder and more time-consuming to carry out this analysis than it had been to come up with the theoretical argument for the asymptotic result.

In order to answer the crucial question of efficiency of xedni for elliptic curves in the practical range, I worked with a multidisciplinary group of young mathematicians and computer scientists at the Centre for Applied Cryptographic Research at Waterloo, especially Edlyn Teske, Andreas Stein, and Michael Jacobson. We were in constant communication with Joe Silverman, who gave us suggestions on how best to test his algorithm. We held a seminar in which the members of our team spoke on different aspects of the algorithm, both theoretical and computational. Finally, by mid-December enough computations were in, and Joe agreed that his algorithm was impractical. In fact, that's an understatement — it turned out that his algorithm was probably the *slowest* one that had ever been thought up to find elliptic curve discrete logarithms. Xedni was extremely clever, though, and none of us had any regrets about having worked on it.

If xedni calculus had turned out to be efficient, it would have destroyed both ECC and RSA, which were the only two types of public key cryptography that were at the stage of commercial use and acceptance. Thus, when I visited different places speaking on our work on xedni I sometimes used the title "How Pure Math Almost Killed e-Commerce." That was a bit of an exaggeration, but it did help boost attendance at my talks.

My sabbatical year was enjoyable as well as productive. Waterloo has the best center for mathematical cryptography anywhere; a constant stream of visitors as well as permanent people in a wide variety of fields make a superb environment for collaborative work.

I was also impressed with the undergraduates at Waterloo. The university has more majors in the mathematical sciences than any other institution in the world (with the possible exception of UNAM in Mexico) — roughly 4,500 (including computer science) — and many of them are very good. I gave a talk to the math club that I was able to pitch at the level of American graduate students.

After Alfred Menezes received his Ph.D. at Waterloo, he spent five years at Auburn University in Alabama before returning to his *alma mater* as a faculty member. He commented to me that it took some adjustment to get used to teaching at Waterloo. Even in his large freshman classes the students were constantly asking thoughtful and provocative questions that, if he wasn't careful, would lead him into long digressions on matters that went beyond the scope of the course. In contrast, during five years teaching in Alabama he never got a single intelligent question from an undergraduate student. ("Will this be on the exam?" doesn't count as an intelligent question.) That is not to say that his years at Auburn were a dead loss; Alfred became (and still is) an ardent fan of the football team. Alfred would be the first to admit that Waterloo is a much better university than Auburn only in academics; in virtually any intercollegiate sport the Auburn Tigers would wipe the floor with Waterloo.

Just as I've been impressed with the students at Waterloo during my many visits there, I have also had good experiences with Mexican students on several occasions. In 1999 my colleague Horacio Tapia-Recillas at the university in Iztapalapa asked me if I would give a short course on elliptic curve cryptography for a group of selected graduate and undergraduate students. Horacio's own area of research is error-correcting codes, but he has also done a lot to increase interest in and support for research on cryptography in Mexico.

I told him that I would do that, but it had to be somewhere other than Mexico City, which is not one of the pleasanter cities in Mexico. We settled on the Pacific Coast towns Ixtapa/Zihuatanejo, where a former student of his named Juan Manuel taught at the Polytechnical Institute. Ann and I arranged to spend two weeks in a timeshare condo in May 2000. The first week I gave a five-day intensive course. I lectured for four hours each morning, and the afternoons were devoted to studying and problem-solving. I held "office hours" on the beach. I suppose it was risky to do that, but the students were highly motivated and worked hard all day long despite the temptations.

The second week Ann and I relaxed. We snorkeled, boogie-boarded, and socialized with Juan Manuel, his fiancée, and a student named Hector who stayed a few days after the course. Juan Manuel took us to meet his family and see a little of Guerrero, which is one of the poorest states in Mexico.

It has been gratifying to see the growing influence of elliptic curve cryptography. In 2005 Certicom organized a meeting in Toronto, mainly for people in industry and government, that included a celebration of twenty years of ECC. They gave out espresso cups with the slogan "small but strong." At the annual RSA conferences the tradition is to give out large coffee mugs to everyone; Certicom's response is to distribute little espresso mugs and point out that elliptic curve cryptography is equally strong even though it uses smaller keys than RSA.

Although Certicom is a Canadian company, its biggest customer has been the U.S. government's National Security Agency. In 2003 NSA paid Certicom a $25 million licensing fee for 26 patents related to ECC. Behind the scenes NSA had been supporting the use of elliptic curve cryptography for about a decade. For me one of the ironies of my work in cryptography is that the super-secret U.S. government agency has been ECC's biggest friend. The leading NSA researcher on ECC, Jerry Solinas, has been great fun to talk with. Jerry doesn't even mind when I tease him about coming to Hanoi or Havana with me to lecture on cryptography. Poor Jerry had to jump through bureaucratic hoops at NSA even to get permission for an extended visit to Canada.

At the Certicom conference in 2005 Vic Miller and I were given awards on the occasion of the twentieth anniversary of the invention of ECC. In my remarks I recalled the early years of ECC and noted how surprised people were to see the number theory of elliptic curves acquire a practical use. I cited the famous comment of G. H. Hardy about number theory remaining "gentle and clean," and asked whether he would be upset about today's applications to cryptography. I speculated that he probably wouldn't be, because it was really military rather than commercial uses of mathematics that bothered him. I then said, "If G. H. Hardy were to come back from the grave, what I'm sure would really upset him today much more than the use of number theory in cryptography would be to see that the world's most powerful nation had unleashed a cruel, foolish, and unnecessary war in a faraway land." I had thought for a while before deciding to comment on the war in Iraq. Knowing that NSA, which is part of the U.S. Department of Defense, was Certicom's best customer, should I be biting the hand that feeds me, or at least feeds Certicom? My hunch, however, was that not even the NSA employees in the audience would take offense at my remarks. And anyway, this was Canada. Any inhibition I might have had about speaking my mind was left behind at the border.

Later that day the President and CEO of Certicom at that time, Ian McKinnon, came by my table and said that he had liked my remarks about the Iraq War. He mentioned that his wife was refusing to take vacations in the U.S. as long as Bush was President. Military contractor or not, Certicom is still a Canadian company that was founded mainly by mathematicians. One should not be surprised to find more enlightened attitudes than in a similar U.S. company.

In recent years most of my papers have been joint with Alfred Menezes. The ones that have had the most impact are three that critique the subfield of cryptography known as *provable security*. Although the papers have been widely downloaded and most of the reaction has been favorable, our work in this area has not been welcomed by everyone. Many specialists in theoretical cryptography have resented our intrusion into their field.

In the 1980's it seemed that all cryptographers were glad to see the influx of mathematicians. Twenty years later, however, I have the impression that some of them wish that we would just go away.

The idea of "provable security" is to give a mathematically rigorous proof of a type of conditional guarantee of the security of a cryptographic protocol. It is *conditional* in that it typically has the form "our protocol is immune from an attack of type X provided that the mathematical problem Y is computationally hard."

Here the word "protocol" means a specific sequence of steps that people carry out in a particular application of cryptography. From the early years of public key cryptography it has been traditional to call two users A and B of the system by the names "Alice" and "Bob." So a description of a protocol might go as follows: "Alice sends Bob..., then Bob responds with..., then Alice responds with...," and so on. When I wrote my first book on cryptography I tried to change this anglocentric choice of names to names like Alicia and Beatriz or Aniuta and Busiso. However, the Anglo-American domination of cryptography is firmly entrenched — virtually all books and journals are in English, for example. My valiant attempt to introduce a dollop of multiculturalism into the writing conventions of cryptography went nowhere. Everyone still says "Alice" and "Bob."

The form that proofs of security take is what is known as a *reduction*. Reductions are the main tool used in computer science to compare and classify problems according to their difficulty. Reductions occur implicitly throughout mathematics. For example, in high school algebra we learn that solving quadratic equations reduces to taking square roots. In other words, if you know how to take square roots (or have a calculator with a square root button), then you can solve such equations thanks to the quadratic formula. Conversely, taking square roots reduces to solving quadratic equations, because if you can solve such equations then, in particular, you can solve the equation $x^2 - N = 0$ for any N, and that means finding $\sqrt{N}$. Problems that reduce to one another like that are considered to be of equivalent difficulty.

In provable security papers the authors try to prove that a mathematical problem that is widely believed to be computationally hard, such as factoring large integers or finding elliptic curve discrete logs, *reduces* to a successful attack of a prescribed type on their cryptographic protocol. This means that anyone who could break their cryptosystem could also, with only a little extra effort, solve the supposedly hard math problem. Since that is assumed not to be possible, the conclusion is that the protocol is *provably* secure.

For mathematicians who study the provable security literature, as Alfred and I did, there are several reasons to be uneasy. Most obviously, a provable security theorem applies only to attacks of a specified sort and says nothing about clever attacks that might not be included in the theorem. Moreover, the result is conditional in a strong sense. Unlike in mathematics, where conditional theorems usually mean something like "assuming that the Riemann Hypothesis is true" (which it almost certainly is), in cryptography the condition is of the sort "assuming that no one finds an improved algorithm for a certain math problem" — and that's anyone's guess. As we saw, history has not been kind to the latter type of assumption. For example, in the late 1980's and early 1990's the development of the number field sieve for factoring an n-digit RSA modulus resulted in a dramatic decrease of the running time from $\exp(n^{1/2+\epsilon})$ to $\exp(n^{1/3+\epsilon})$.

There's also a difficulty that comes from the disciplinary culture of cryptography that I commented on before. People usually write papers under deadline pressure — more the way a journalist writes than the way a mathematician does. And they rarely read other authors' papers carefully. As a result even the best researchers sometimes publish papers with serious errors that go undetected for years.

In 1994 two of the leading specialists in the new area of provable security, Mihir Bellare and Philip Rogaway, proposed an RSA-based encryption method that they called OAEP (the O stands for "optimal," a much overused word in the over-hyped high tech world). They held the view that security proofs should be sufficiently detailed so that one can get concrete guarantees for specified key sizes and choices of parameters. Partly because of the security proof that accompanied OAEP, it was adopted for use in a new standard of Visa and MasterCard. It turned out, however, that the proof was fallacious, as Victor Shoup discovered seven years later. This was a bit of a scandal and caused many people to wonder about quality control in provable security papers.

If a careful and astute reader is watching closely — and Alfred Menezes is such a reader — then errors in proofs are discovered much more quickly. A case that in many ways is even more striking than that of OAEP is the recent flap over an "improved" set of key agreement protocols designed by Hugo Krawczyk. (A *key agreement protocol* is a sequence of steps that Alice and Bob follow in order to generate a secret key that they have in common and can use for various cryptographic purposes.) In February 2005 Krawczyk, who works for IBM and is a top researcher in provable security, submitted a paper to Crypto 2005 in which he claimed to have found flaws in the Menezes-Qu-Vanstone (MQV) key agreement system. He replaced it with a modified version (HMQV) that he maintained was both more efficient and *provably secure*. If his claims had been valid, this would have been a major embarrassment not only to Alfred and his coauthors, but also to NSA, which had licensed MQV from Certicom and whose experts had studied it carefully.

Krawczyk did not send his paper to Alfred or the other designers of MQV before submitting it, although to do so would be considered a standard courtesy in the scientific world. But what to me seemed more scandalous was that neither did anyone on the Crypto 2005 program committee. They apparently rushed to accept the paper after only a superficial reading. When Alfred finally got a copy of the paper — after it had been accepted by the program committee — he immediately saw that the so-called flaws in MQV that Krawczyk listed either were based on misunderstandings or else were picayune theoretical points that had no practical significance.

More importantly, Alfred found that the paper's main argument was fallacious. Krawczyk claimed that in his modified key agreement system he could increase efficiency by discarding a certain security check (called a "public key validation") that had been put into MQV so as to prevent known attacks. It was his security "proof" that gave him the confidence to do this. But Alfred quickly found that certain of the HMQV protocols succumb to the same attacks that MQV would have if those security checks had not been put in. After seeing that some of the conclusions of Krawczyk's theorems were false, Alfred started reading the "proof" carefully until he came upon a blatant gap in the argument.

Both Krawczyk and the referees on the program committee had been so mesmerized by the "proof" that they failed to use common sense. Anyone working in cryptography should think very carefully before dropping a validation step that had been put in to prevent security problems. Certainly someone with Krawczyk's experience and expertise would never have made such a blunder if he hadn't been over-confident because of his "proof" of security. As with many other over-hyped ideas — fallout shelters in the 1950's, missile shields in the 1980's — "proofs" of the security of a cryptographic protocol often give a false confidence that blinds people to the true dangers.

In our first paper on provable security, Alfred and I objected to the terminology:

> There are two unfortunate connotations of "proof" that come from mathematics and make the word inappropriate in discussions of the security of cryptographic systems. The first is the notion of 100%

certainty. Most people not working in a given specialty regard a "theorem" that is "proved" as something that they should accept without question. The second connotation is of an intricate, highly technical sequence of steps. From a psychological and sociological point of view, a "proof of a theorem" is an intimidating notion: it is something that no one outside an elite of narrow specialists is likely to understand in detail or raise doubts about. That is, a "proof" is something that a non-specialist does not expect to really have to read and think about.

The word "argument," which we prefer here, has very different connotations. An "argument" is something that should be broadly accessible. And even a reasonably convincing argument is not assumed to be 100% definitive. In contrast to a "proof of a theorem," an "argument supporting a claim" suggests something that any well-educated person can try to understand and perhaps question.

Alfred and I also investigated some subtler problems of interpretation of provable security results. Even when the proofs are correct, they often mask a big "tightness" gap. This means that in the reduction argument the attack on the protocol must be repeated millions of times in order to solve the hard computational problem. In this case the practical guarantee that one gets is very weak. Alfred found some extreme examples of this "nontightness" problem in a few well-known papers on random number generators. In one paper it turned out that, if you carefully follow the authors' argument with recommended parameter values, all they've really proven is that an attacker would need at least 10^{-40} nanoseconds to break the system. That's much less time than it takes light to travel a micron. (Alfred also found a case where the formula gives a *negative* lower bound on the adversary's time — but you don't need a mathematical theorem to tell you that the amount of time required for a successful attack on your cryptosystem is greater than a negative number!)

What had happened was that some influential people (such as the Internet Engineering Task Force) had made recommendations for parameter values that were based on an asymptotic theorem. That theorem says that in the limit as N approaches infinity, you can securely generate $O(\log \log N)$ pseudorandom bits each time you perform a squaring modulo the composite number N. (Here "securely" means, roughly speaking, that no one can distinguish between the sequence and a truly random one by an algorithm that runs in reasonable time.) However, as I mentioned when discussing Joe Silverman's xedni calculus, it is fallacious to use an asymptotic result as a practical guarantee of security. Rather, one needs to perform a detailed analysis using realistic ranges for the parameters. It is often a lot harder (as it was in the case of xedni) to carry out this concrete analysis than it was to prove the asymptotic theorem, and sometimes the conclusions are not what one would hope for. In the case of the pseudorandom bit generator the analysis (if one assumes that $\log_2(\log_2 N)$ bits are taken in each iteration, as recommended) leads to an absurd lower bound on the amount of time that an adversary would need in order to successfully attack the generator.

When we started our collaboration, Alfred and I approached the subject in some sense from opposite directions. Without having read any of the papers, I was immediately skeptical — the whole idea of "proving" security struck me as bogus. Alfred, on the other hand, had a lot of respect for the researchers who were filling their papers with security proofs, and he thought that at the very least there must be a baby in the bath water somewhere. He explained to me what was in some of the most important papers in the field, and in a few cases convinced me that what they achieved was useful. In other cases I deconstructed them and showed that they had little of substance. Initially I had wanted to title our paper "Pitfalls of 'Provable Security'," but Alfred correctly insisted on a neutral title, "Another Look at 'Provable Security'." Eventually the use of the words "another look" to start a title became our trademark (although hardly a legally enforceable one).

The more Alfred explained to me about the provable security papers, the more I was ready to admit that my initial dismissal of the whole field was a little simplistic. But Alfred also changed his stance and became more skeptical as time went on. By the end of the second "another look" paper, he allowed me to put the following passage in our conclusion:

> Embarking on a study of the field of "provable security," before long one begins to feel that one has entered a realm that could only have been imagined by Lewis Carroll, and that the Alice of cryptographic fame has merged with the heroine of Carroll's books: "Alice felt dreadfully puzzled. The Hatter's remark seemed to her to have no sort of meaning in it, and yet it was certainly English." The Dormouse proclaims that his random bit generator is provably secure against an adversary whose computational power is bounded by a negative number. The Mad Hatter responds that he has a generator that is provably secure against an adversary whose computational resources are bounded by 10^{-40} clock cycles. The White Knight is heralded for blazing new trails, but upon further examination one notices that he's riding backwards. The program committee is made up of Red Queens

screaming "Off with their heads!" whenever authors submit a paper with no provable security theorem.

Lewis Carroll's Alice wakes up at the end of the book and realizes that it has all been just a dream. For the cryptographic Alice, however, the return to the real world might not be so easy.

To me the fundamental issue was a familiar one — mathematics as propaganda — that I had first written about a quarter century before. Just as in the debate with Herbert Simon in the late 1980's, Lewis Carroll provided some excellent material for an article about the misuses of mathematics.

I was pleasantly surprised that Alfred allowed the above passage to stay in the final version of our paper. Ann, who often understands people better than I do and who reads the non-technical parts of our papers, had thought that he would. Alfred's demeanor — cautious, respectful, modest, even deferential — leads people to think of him as a moderate in all things, as the type of person who strives to avoid confrontation at all costs. But there's another side to Alfred that people ignore at their peril — a hint of which is revealed when he gleefully sends his friends links to the latest scandals in intercollegiate football or the Bush administration.

At one point Alfred and I had the idea of structuring the conclusion of our second provable security paper as a parody of the rather pretentious concluding section of a well-known paper in theoretical cryptography in which the different coauthors wrote different conclusions in subsections titled "Ran's Conclusions," "Oded's Conclusions," and "Shai's Conclusions." Our subsections would be "Alfred's Conclusions," "Neal's Conclusions," "Ann's Conclusions," and "Joshua's Conclusions" (Joshua is Alfred's little nephew). I tried to capture both Alfred's style and his true feelings about provable security in a passage (never published, of course) that I proposed for the "Alfred's Conclusions" subsection:

> The field of provable security has had a profound and lasting impact on cryptographic research, and the enhanced rigor that has been introduced into practice is nothing short of miraculous. The leaders of the field are brilliant researchers, and, because of their humility and generosity of spirit, are wonderful people to have as colleagues.
>
> Indeed, they write papers so prodigiously and announce breakthroughs at such a breathtaking pace that it has become difficult for us ordinary mortals to keep up with the field. My only objection — and it's so minor that I hesitate even to mention it — is that when

I read their work with careful attention to detail, I find major errors and questionable interpretations in almost every paper.

The "another look" papers have been a wonderful learning experience for me, thanks mainly to Alfred's patience as a teacher in explaining technical points to me. In the world of cryptographic research these papers have had more impact than anything else that either of us has written in many years.

But not everyone is thrilled with what we're doing. Just before our first provable security paper was due to appear in print — and almost two years after it was accepted for publication — the editor-in-chief of the *Journal of Cryptology* informed us that a member of the editorial board had strenuously objected to its being published. To placate this guy he wanted to insert the subtitle "Position Paper," which would suggest that it's merely an opinion piece and not a serious article. He asked us if this was okay, and we emphatically said that it was not. So he published the paper without the subtitle and at the beginning of the issue wrote an unprecedented "Editor's Note" explaining and justifying his decision to publish such a controversial paper.

When we sent our second and third "another look" papers to the IACR online preprint server — which by policy immediately posts absolutely any submission from reputable authors with no vetting or review — there was a week's delay while the manager of the server consulted with four top specialists. He was thinking about barring one or both papers, but fortunately none of the four thought that he should censor our work from the eprint service.

The posting and the publication of articles that critique earlier work are fundamental for the healthy development of any branch of science. It is regrettable that in cryptography there are some prominent researchers who see nothing wrong with using their position as eprint service manager or editorial board member to try to block the dissemination of work that they regard as threatening to their interests.

The story of our first "another look" paper has an amusing postscript. The editorial board member who objected to its publication was Professor Oded Goldreich of the Weizmann Institute, who is one of Israel's leading computer scientists and a top name (some would say *the* top name) in theoretical cryptography. When he realized that it was too late to prevent our article from appearing in the *Journal of Cryptology*, he posted on the IACR eprint server a 12-page essay titled "On Post-Modern Cryptography" that lashed out at us on philosophical grounds. He accused

Alfred and me of being "post-modern" and "reactionary" because our criticisms of provable security "play to the hands of the opponents of progress."

The part of our paper that seems to have incensed Goldreich the most was our explanation of why we were not persuaded by certain arguments that he and others had made in order to undermine the so-called "random oracle" assumption. The random oracle assumption relates to what are called "hash functions" (short strings of symbols that act as a sort of "fingerprint" of a message). This assumption essentially says that the fingerprint that a well-constructed hash function gives is in practice indistinguishable from a random string of symbols. This is an intuitively reasonable assumption, and in our paper we argued that all attempts to undermine it — even ones that the authors claimed to be of practical relevance — in fact use constructions that violate basic cryptographic principles and so have no relation to real-world cryptography. We concluded our discussion by saying that "our confidence in the random oracle assumption is unshaken."

Goldreich responded to this by bringing down the wrath of the Old Testament upon us. Accusing us of turning the random oracle into a "fetish," he recounted a story from the Bible that our paper reminded him of (in what follows I've preserved the emphasis, capitalization, and spelling of the original):

> Indeed, what happened with the Random Oracle Model reminds us of the biblical story of the Bronze Serpent, reproduced next. (See *Numbers* (21:4-8) and 2 *Kings* (18:4).) During the journey of the People of Israel in the dessert, the prophet-leader Moses was instructed by the Lord to make a "fiery serpent" as a symbolic mean for curing people that have been bitten by snakes (which were previously sent by the Lord as a punishment for some prior sin). Several hundred years later, the bronze serpent made by Moses has become an object of idol worship. This led the righteous King Hezekiah (son of Ahaz) to issue an order for breaking this bronze serpent to pieces. Let us stress that the king's order was to *destroy an object that was constructed by direct instruction of the Lord,* because this object has become a fetish. Furthermore, this object no longer served the purpose for which it was constructed. This story illustrates the process by which a good thing may become a fetish, and what to do in such a case.... [G]iven the sour state of affairs, it seems good to us to abolish the Random Oracle Model.

Goldreich sees himself as a 21st-century righteous King Hezekiah defending the provable security researchers against infidels and post-modern fetishists such as Alfred and me. It is clear from his essay that he had not read our paper carefully before writing his response; nor does he seem to have been aware of our other two posted papers criticizing provable security. But of course it was not necessary to actually read the technical details in our three articles in order to denounce us on religious and philosophical grounds.

The angry reactions of a few researchers who seem to perceive our work as inimical to their interests are not the type of thing one normally encounters in theoretical mathematics, where usually the only issues that could cause someone to object to a paper would be an error or omitted acknowledgment of earlier work (neither of which has been found in any of our three papers on "provable security"). But far from being bothered by the accusations made by Goldreich and others, I am encouraged by them, because they at least show that people are paying attention.

Cryptography has the excitement of being more than just an academic field. Once I heard a speaker from NSA complain about university researchers who are cavalier about proposing untested cryptosystems. He pointed out that in the real world if your cryptography fails, you lose a million dollars or your secret agent gets killed. In academia, if you write about a cryptosystem and then a few months later find a way to break it, you've got two new papers to add to your résumé!

Drama and conflict are inherent in cryptography, which, in fact, can be defined as the science of transmitting and managing information in the presence of an adversary. The "spy vs. spy" mentality of constant competition and rivalry extends to the disciplinary culture of the field. This can get to be excessive — and even childish at times — but it also explains in part why it can be so much fun to do research in cryptography.

CHAPTER 15 **EDUCATION**

Before 1990 I'd had some interest in undergraduate education, but no more than what was typical for mathematicians at research universities. At both Harvard in the 1970's and the University of Washington in the 1980's I had advocated a more applications-oriented approach to teaching first-year calculus and had developed handouts and workbooks for this purpose. And, as mentioned in the last chapter, I'd changed the number theory course at UW to emphasize applications — especially RSA cryptography — rather than theory. But I had no involvement in K-12 education until I met Mike Fellows in 1990.

Mike first contacted Ann and me in 1989 after hearing about the Kovalevskaia Fund. Ann stayed with him and his family, who were then in Moscow, Idaho, when she was driving across the country that year to begin her job at Hartwick College. Mike was interested in Kovalevskaia Fund activities and said he'd like to join us on one of our trips to Latin America. When I met him the following year, I discovered that we also had mathematical interests in common, and I learned that he had become passionate about improving math teaching in elementary schools.

Mike believed that math should be taught as an exciting branch of science where even beginners can get a taste of what it's like to work on the frontiers of knowledge. He pointed out that in some other sciences — such as astronomy and natural history — children enjoy thinking about the unsolved problems. How old is the universe? What caused the dinosaurs to die out? But math is presented in the schools as if it were a dead subject, as if anyone with the proper training knows all the answers (or can get a computer to cough them up).

Using his own areas of research, graph theory and algorithmic complexity, Mike devised problems for children that had nice stories and also connections to important questions in research mathematics and computer science. For example, a puzzle called Tourist Town consisted of a schematic map of a village with lines for the streets and circles or big dots for the street corners. Suppose that in preparation for the tourist season the town decided to build enough ice cream stands at the street corners so that someone who was hot and hungry and needed an ice cream cone would find one either at the street corner where they were standing or, if not, at one of the neighboring street corners. What is the minimum number of ice cream stands you'd need, and where would you put them? In graph theory this is known as the problem of finding a minimum dominating subset of vertices. A solution to this problem is called a "perfect code" (this concept was used for a type of "kid crypto" mentioned in the previous chapter) if every street corner is in the neighborhood of only one ice cream stand; in that case you never have a choice of two different ice cream stands if you're only willing to walk at most one block. No efficient algorithm is known — nor is one likely to exist — for finding a minimum solution for a large Tourist Town. But it's an open question to design really good algorithms for solving small and medium-size Tourist Towns.

Mike started visiting his own children's school to present math enrichment lessons and then branched out to other elementary and middle schools. He wanted to make similar visits when he came to Latin America with us. To prepare for this, I learned more about what he was doing and went along with him on a couple of visits to schools near Victoria, British Colombia. (Mike had just moved from the University of Idaho to the University of Victoria.)

In June 1992 Mike came to Peru and El Salvador with us. In Lima the mathematician Michel Helfgott set up a visit to a third-grade math class, which turned out to be fascinating and a lot of fun. I had to help Mike with the explanations, because his Spanish was atrocious. Ann was sitting at a table with four kids, who were soon relating to her in a very informal way. At one point a boy asked her whether Mike could do long division. She didn't understand him at first, since she didn't know the Spanish term, but he showed her by an example what he meant. That was clearly the hardest task he knew of. Ann said yes, she was sure that Mike could do long division. "Then why can't he speak Spanish?!" the boy asked indignantly.

Although most of Mike's examples and stories carried over well, the Tourist Town didn't translate completely into the Peruvian setting. Ice cream was sold either in permanent cafés or else from carts on wheels. The latter were a source of income to people who would otherwise be unemployed, and there was no social reason to try to have as few as possible in a resort area. After a while, I replaced *Ciudad Turística* (Tourist Town) by *Aldea con Pozos* (Village with Wells). Now the question was how many wells would have to be dug at street intersections in the village so that someone would have to go at most one block to fetch water. In many Third World countries this story, which is mathematically equivalent to Tourist Town, would make more sense to children. In 1994 Mike, Ann and I published an article "Cultural Aspects of Math Education Reform" in the

Notices of the American Mathematical Society that was based in part on our experiences presenting math enrichment topics to kids in Peru and elsewhere.

Mike got Ann and me hooked on school visits. We have found that this is a wonderful way to deepen our relations with people when we travel. Foreigners who are interested in the education of their children are put in a different category from other tourists and academic visitors. The classroom visits are good for morale, since teachers and administrators see them as a symbolic recognition of the importance of what they are doing. Our experiences with school children in other countries have also contributed to our understanding of pedagogical issues. Unfortunately, most Americans who write about math education have observed classrooms only in the U.S., and so lack a broader perspective.

Starting in the 1990's Ann and I have visited schools and given math enrichment lessons in twelve countries besides the U.S. and Canada: in Asia (Vietnam and India), Africa (Zimbabwe, Malawi, and South Africa), and Latin America (Peru, El Salvador, Belize, Chile, Mexico, Cuba, and Puerto Rico). There are few generalizations about people's attitudes that apply universally across cultures, but here is one: wherever we've been, without exception, people are interested in issues of math education and are not satisfied with the job their schools are doing. In particular, parents and teachers everywhere are frustrated that so many kids dislike math.

One of our most memorable visits was to an isolated village called Pacaycasa in the Peruvian Andes, a half hour or hour drive (depending on the condition of the road) from the provincial city of Ayacucho. We spent an afternoon with a large class of middle-school-age children, working with them on graph theory story problems, arithmetic games, and a proof-without-words of the Pythagorean Theorem. Spanish was as much a foreign language for them as for us, since their mother tongue was Quechua. The kids liked the geometry most of all, and we speculated that this might be connected with the rich geometrical tradition of the ancient Inca and Wari. As the afternoon wore on, we noticed that, although the children showed no sign of tiring or losing interest, they were having difficulty seeing the blackboard. The sun was going down, and the school did not have electricity.

On another occasion Ann and I visited a school in Cape Town for Xhosa children (the largest black African ethnic group in the Cape Province). A Xhosa mathematician from the University of the Western Cape named Loyiso Nongxa translated for us and expanded upon our explanations. At one point he couldn't help laughing — he had heard one of the kids comment that "this American sure speaks good Xhosa!" As he explained to us, the kids found it easier to imagine that a black American visitor spoke unaccented Xhosa than that a member of their tribal group had become a professional mathematician.

In the mid-1990's Ann and I visited the resort town of San Pedro, Belize on several occasions to go snorkeling. From there we went out on dive boats to some of the most beautiful coral gardens in the Americas, including the Great Blue Hole, which had been popularized by Jacques Cousteau.

Once as we walked with our luggage from our hotel to the town's airstrip three blocks away, Ann pointed out that a school was in session right there, and perhaps the next time we came we could give some math lessons. I was skeptical — we were, after all, on vacation — but the next year when I phoned from the U.S. to make our hotel reservation I asked the guy at the desk whether he knew anyone at the school who could arrange for us to visit some math classes. He said sure, his wife taught there and would be happy to do that.

We spent a full morning giving two-hour sessions to two different classes. They went so well that the teacher prevailed upon us to do it again before we left. When the second class was about to start, a girl went up to Ann and explained that she would not be able to stay for the whole time, but would leave promptly at 11:00 for a family matter. But 11:00 came and went, and the girl didn't leave. The children were wrapped up in a prime number game with five dice. They had to combine the numbers on the dice according to prescribed rules so as to form the largest prime they could. Finally someone stopped in from the main office to find out why the girl hadn't come. She said that she couldn't leave just yet, because the first class had gotten a prime greater than 1000, and it was important for her class to get above 1000 as well.

The language of instruction in the school was English, which is the main national language of Belize. Most of the kids knew both English and Spanish, but two girls who had recently emigrated from Guatemala knew only Spanish. So I presented the math enrichment topics in both languages. The kids found it funny that visiting Americans would be speaking Spanish with them. In San Pedro, Spanish was viewed as a language for use at home and in informal settings among the immigrants. The people who worked in the stores, even if Spanish was their first language, preferred that tourists use English with them. At first I spoke Spanish in the shops, but people seemed almost insulted, as if I was being overly familiar with them. So I switched to English.

While we were waiting for our flight out of San Pedro, we noticed that the father-in-law of the teacher who had set up our visits was in the airport seeing off a group of scuba divers. He had been one of the first dive masters in Belize and had been well known in scuba circles since the time of Cousteau. When he saw us, he broke off his conversation with the divers and came over to shake our hands and thank us profusely for having spent some of our vacation with the children of the town.

There was no need to have thanked us, since we had benefited from the experience as much as the kids had. The deep respect toward education and the teaching profession of this famous dive master was typical of the attitude that we have encountered in many parts of the Third World among people in all walks of life. It is a feature of their cultures that is sadly lacking in the United States.

Sometimes the adults' responses are almost as interesting as the kids'. On occasion — especially in the more elite schools — the teachers are uneasy about our presenting material that they themselves don't understand. Usually those who feel insecure just sit quietly in the back, but a couple of times they have gone around the class pretending to be experts and giving wrong instructions. In contrast, one of the best responses from a teacher was at a school I visited in 1999 in downtown Havana not far from the Plaza de Armas. In material terms it was a poor school, but the children were hard-working and full of energy. The teacher, a middle-aged Afro-Cuban man, sat down at a table just like one of the kids and learned the material side by side with them. He didn't have any problem with the children knowing that the subject was as new to him as it was to them. Clearly his pupils' respect for him was on a deep enough level that he didn't feel insecure about putting himself on an equal footing with them during the lesson. I was impressed.

In 1993 Ann and I gave a lesson to seventh-graders at a crowded school in what was euphemistically called a "high-density suburb" of Harare, Zimbabwe. Although the kids did well on the worksheets and everyone participated, they seemed subdued, and we thought that they hadn't particularly liked the material. To our surprise, afterwards their teacher commented that she'd never seen the children so rambunctious, even to the extent of speaking when they had not been called upon. We then realized that Zimbabwean schools were still modeled on an old-fashioned — almost Dickensian — version of the British system. Behavior that to us seemed quiet and reserved to them seemed boisterous and hyperactive.

Some of the most impressive students Ann and I have encountered were in Vietnam. We visited three schools in Hanoi, two of which were special schools that attracted the best students. But it was in a fifth-grade class at the third school, which was described to us as an average one, where we had one of the most stimulating math enrichment lessons ever.

Like their counterparts in Belize and many other countries, the children especially took to the prime number dice game, and they got very competitive about it. We had them working in groups, and a few of my colleagues from the Hanoi Math Institute had come along to observe and also translate. One group of kids formed a large odd number by multiplying and squaring the numbers on the dice to get 72, then squaring 72 and subtracting 1; the result was 5183. We let them use a calculator for the verification of primality by trial division and also to compute the square root of their number. (They understood that you have to divide the number by all primes up to its square root before you can be sure that it's prime.) The group split up among themselves the task of trial division by all the primes up to 71. As they worked, I told them that their number was not going to be a prime. They saw that I hadn't done any trial division, so they didn't see how I could know that — and 5183 certainly wasn't divisible by any of the small primes. As they neared 71, they thought that they were headed for success — that I was wrong and they had the class's largest prime. But at the very last step they found that 71 goes evenly into 5183. Then they were curious about how I could have known that without doing all the work.

I got the attention of the whole class and showed everyone the number that the group had come up with: $5183 = 72^2 - 1$. I asked if they had seen equations with the letter X, and I wrote $X^2 - 1 = (X + 1)(X - 1)$ on the blackboard. The teacher, who was observing in the back, seemed worried and explained to me that the children had not yet done any algebra. I asked the class if parents or older brothers or sisters had ever mentioned X, and many had heard about it. I told them to replace X by 72 and watch what happens. That's how I knew that 5183 was not a prime. If they remembered that algebraic identity, I said, they could avoid wasting time on such a number in the future when they play this game. The children eagerly jotted down the identity.

This had not been planned in advance. But what I had done was solve a difficult pedagogical conundrum that has plagued algebra teachers for centuries — how to make algebraic identities, which are intrinsically among the most boring things in mathematics, seem interesting to children. Despite the teacher's apprehensions, her pupils didn't at all mind an algebraic explanation that went beyond the scope of fifth-grade math, because it was presented as a tip that would help them play the prime number game more efficiently.

Math education in Vietnam is quite traditionalist, with heavy emphasis on computation and drill. During the entire lesson Ann and I didn't see a single student make an arithmetic error. Their hand calculations were organized in a meticulous format with lines always drawn with a straight-edge — we thought they were being almost anal in their neatness. But contrary to a stereotype of Asian students that one often hears in the West, these kids were as imaginative and receptive to new ideas as any we've seen anywhere.

The one topic that they had trouble with was a statistical example I'd designed to illustrate the difference between the median and mean. Given a list of the workers' wages and the much higher pay of a small number of managers and owners, what are the mean and median salaries, and which of the two would be used by the workers who want to argue that they need a raise and by the managers who want to claim that the average pay is already pretty good? The children easily computed the two types of averages, but they didn't get the point of the final question, and in fact got the answer backwards. The teacher later explained to us that the children were not well educated about social problems in the adult world, such as labor and salary disputes. We told her that was fine — we hadn't meant to imply that they needed to know about such things.

In Vietnam youngsters are not expected to be aware of adult concerns. Ann and I said that we thought that it was nice that Vietnam allowed children to remain children. In the U.S. we go to the opposite extreme. Most American youngsters are poorly prepared in basic skills, but they have knowledge of social problems that in earlier times would never have been talked about in front of children. Once in a Seattle seventh-grade classroom I was reading one of the book reports that the teacher had proudly posted on the bulletin board. The spelling and word usage, while typical of U.S. middle schoolers today, were what in earlier times one might have expected of third or fourth graders, but the content certainly was not: "This book is about a girl and her friend. The priest came to live with the girl's Mom. He raped the girl. When they grew up her friend killed the priest." (I've cleaned up the spelling.) This book was not what the Vietnamese would have considered appropriate children's literature.

When Ann and I visited a class of older kids in Hanoi, I gave them a simple but challenging problem: estimate to the nearest power of 10 how many liters of water are in the Pacific Ocean. This is a very nontraditional math problem for several reasons: no information is given for the students to compute with, they have to draw on their knowledge of other subjects in order to get started, an order of magnitude approximation rather than an exact answer is asked for, and numbers of astronomical size are involved. The kids first guessed the average depth of the Pacific Ocean and then came up with a clever way to figure out its surface area. They knew the approximate area of Vietnam, and they knew that Vietnam occupies 0.2% of the earth's land, which, in turn, is about half as great as the area covered by water. They also knew that the Pacific Ocean takes up roughly a third of the earth's surface. They put all this together — and, needless to say, had no trouble accurately converting the metric units — and gave the correct answer, which is 10^{21} (a billion trillion).

Through our university contacts in Peru we learned of a school called Colegio Gauss (the Gauss School) that had a special orientation toward mathematics, as one might guess since it was named after the great mathematician Carl Friedrich Gauss. Located in a working-class suburb of Lima called Carabayllo, the Gauss School is private but low cost. Although it draws pupils from the lower-middle class, the quality of its teachers and academic program is as high as in Peru's two pricey elite institutions (the American School and the Humboldt School). It was founded in 1984 by a couple named Felix Maldonado and Ruth Naveda, and they have directed it ever since.

Maldonado, who was trained as an engineer, has a tremendous love for mathematics. He collects compendia of challenging problems as well as recreational math books, and he encourages innovation among his teachers. Ann and I have visited the Gauss School during each of our biennial trips to Peru starting in 1994, when we initiated a scholarship for girls at the school. The Argentinian-American mathematician Cora Sadosky, who was then the president of the Association for Women in Mathematics, came with us and spoke at the inauguration of the scholarship at the Gauss School.

The scholarship is named after the mathematician Ruth Ramler Struik (1894-1993), and the Struik family supports it through donations to the Kovalevskaia Fund as a memorial. We had known the daughter Rebekka, who's a mathematician at the University of Colorado, and the husband Dirk (1894-2000) for many years. Dirk Struik was a well-known historian of mathematics, as well as a life-long Marxist and a founding member of the Dutch Communist Party (in 1919).

The Gauss School, like many in Peru, has a quasi-military style. When Felix Maldonado brings Ann and me from our hotel (it used to be in a Volkswagen Beetle, but he now has a different car that's less fun to ride in), he stops about a mile away to phone and let people know that we are about to arrive. As we drive up to the school, the marching band starts playing, and the littlest children are lined up in front of the gates waving homemade American flags. We could do without the flags, but at least they are only rough approximations — the number of stars seems to vary from ten to thirty — and the toddlers have a good time waving them. Typically we watch a musical or dance performance by the students, hear a couple of short speeches, and then address the large group — the enrollment at Gauss grew from about 300 to over 500 during the decade after our first visit. Next we go to a classroom to present a math enrichment lesson to twenty or thirty youngsters with teachers observing in the back.

The most ambitious and sophisticated topic we ever did was perfect code cryptography. We started outside so we could sketch a giant chalk illustration on the basketball court, and when the sun got too hot we moved to the recently built school library, which had been named after Ruth Struik. The kids worked on the topic for three hours without a break.

The resources that Peruvian schools have to work with are meager compared with those of a typical American school. But in some cases what they accomplish is impressive. Ann and I are frequently struck by the informal but respectful way the teachers and pupils relate to each other. There is a spirit of excitement about their studies and a realization of the importance of hard work.

During our travels I have occasionally lectured about the problems we have with math education in the United States. I like to illustrate one of the difficulties by comparing two coffee mugs. One of them was a souvenir from the Gauss School on the occasion of its tenth anniversary. The mug shows the school's symbol — an icosahedron — surrounded by the words *estudio, trabajo, disciplina* (study, work, discipline). The other mug I received from an American educational supply company after ordering a bunch of geoboards for teaching geometry. Around the side of the cup are listed "101 Ways to Praise a Child" — way to go!, super!, you're special!, well done!, you're unique!, etcetera, etcetera. The self-esteem movement in American education has reached a level that would seem absurd to people in most parts of the world. When I mentioned to a Cuban audience that a preschool in Seattle had the name "The Hurray for Me School," they couldn't believe it. In Spanish a school named ¡*Bravo Por Mí!* sounded particularly goofy. But I wasn't making it up.

Through Felix Maldonado, we learned of a school in the provincial city of Ayacucho that shares the educational philosophy of the Gauss School. It was named after the 19th century German educational reformer (and creator of the word "kindergarten") Friedrich Froebel. Ann and I knew that

Peru was a much different country in the provinces than in the capital, and we liked the idea of developing ties with people outside Lima. In 1996 we made our first visit to Ayacucho and initiated a Kovalevskaia Scholarship there that is similar to the Struik Scholarship at Gauss. We visited again in 1998, 2000, 2002, 2004, and 2007; each time the Lagos sisters who run the school put us up in the Hotel Santa Rosa, a block away from the city's main plaza. A beautiful example of Spanish provincial architecture, the hotel was built in 1630 and has been carefully restored and maintained. The Santa Rosa is by far the oldest building we have ever stayed in.

When we started going to Peru in 1992, we were cautioned about travel to the provinces, especially areas such as Ayacucho where the Shining Path guerrillas were strong. In fact, the Shining Path had been founded in Ayacucho and for a long time the city had been a center of the rebellion. However, by 1996 the Shining Path had been largely defeated, and it was safe for us to travel there.

When I visited Lima for the first time, I was surprised to find that even leftist academics referred to the Shining Path as "terrorists" and had no sympathy for them. This contrasted with what I was used to in Central America in the 1980's, where most intellectuals had sympathies with the FSLN of Nicaragua, the FMLN of El Salvador, and similar groups. While I was in the apartment of Nicolás Lynch, a well-known progressive political scientist at San Marcos National University in Lima, he received a phone call informing him that Shining Path had murdered a prominent community activist in one of the slums. By the 1990's Shining Path had truly degenerated into a terrorist organization and was being compared to the Khmer Rouge under Pol Pot — fanatical and violent against anyone not under their control.

When we visited Ayacucho, however, we noticed that people would talk about the guerrillas in various oblique ways, but never once did we hear them referred to disparagingly as "terrorists." It seemed that many Ayacuchans nurtured fond memories of the early days of Shining Path, when it was regarded as the only mass organization defending the rights of the poor and the Quechua-speaking population.

We also found it curious that the Froebel School had been allowed to function as an educational institution with high academic standards (that is, it was not a guerrilla training school) during all the years when Shining Path effectively controlled the region.

The school was started and directed by three sisters of the Lagos family. We learned that they had had a younger sister named Edith. At the age of 16 she had become a *comandante* of the guerrillas, and three years later, in September 1982, she was captured, tortured and killed by the military.

She became the most famous martyr in the history of the Peruvian revolutionary movement. Even though her funeral was declared illegal by the government, 30,000 people attended. (Her grave to this day is decorated with fresh flowers, many of which are put there by market vendors who believe she brings good luck. In 2007, Ann and I went with Becky Lagos and, on behalf of the Kovalevskaia Fund, lay a wreath at Edith's grave in commemoration of the 25th anniversary of her martyrdom.)

In 1996 we traveled to Ayacucho with another couple, Bruce Schneier (the author of a best-selling book on cryptography) and his fiancée Karen. While we were at the Froebel School, they walked around town. Americans were still a novelty in Ayacucho, and they attracted a lot of attention from the children, who practiced their English with them. Bruce and Karen noticed that several of the girls in their early teens were named Edith. They asked us about it, and then we realized how deep the veneration was for Comandante Edith in the region. And what the surviving sisters do — running the best school in the city — is regarded as in some sense a continuation of Edith Lagos' credo of sacrifice for the common people.

In monetary terms the amount the Kovalevskaia Fund gives in scholarships is small — just $700 per year (recently increased to $1000) at Froebel and the same at Gauss. But the value of a scholarship funded from abroad is also symbolic, since it's a form of recognition for what the teachers and directors are working hard to accomplish. In 2004 the parents' organization that supports the Froebel School gave us a "Parchment of Appreciation." The beautiful suede wall hanging has a painting of the school with a message of appreciation written ornately in an old-fashioned style of Spanish. The members of the parents' committee signed their names underneath. This tapestry, which Ann has in her office at Arizona State, has been much admired by the Spanish-speaking campus workers who come in to clean. One of them specially brought over her friend who worked in another building so that she could look at it.

Because the U.S. has the reputation of being a leader in science and technology — and a Ph.D. from an American university is a ticket to success in many parts of the world — people in other countries often assume that the U.S. must have a first-class educational system. Even those who are strongly opposed to U.S. foreign policy and have a critical point of view on American popular culture nevertheless figure that American schools must be pretty good, or the country wouldn't have produced so many Nobel prizewinners.

People with experience teaching in the U.S. know differently. The U.S. is a society of extremes, and nowhere more than in education. A relatively small number of people get a first-rate education and become successful scientists. In addition, the U.S. profits from the "brain drain" from other countries. Most of the best doctoral dissertations in the sciences are now written by people who received their education through college in another country, and many of them decide to stay in the U.S.

The level of an *average* American school would come as a complete shock to most foreigners. When I give talks on education in America, as I have in China, Peru, South Africa, Vietnam, and Cuba, I try to give people ammunition that they can use to argue against the mindless imitation of American models that has become popular among education bureaucrats in many countries. This was definitely my objective when I spoke at the Beijing Normal University (teachers college), where mathematicians had told me of their concern about the Americanization of education in China.

When Ann and I visited Cuba in 2003, I gave a talk on "the decline of learning in America, and the danger for other countries of letting their educational system be unduly influenced by the U.S. model." This was a provocative and heavy-handed title, and I thought that the Cubans, who dislike the U.S. government but admire American achievements in science and technology, might think that I was exaggerating our problems. On the principle that "a picture is worth a thousand words," I brought a video of eighth-grade math lessons that had been produced by the Third International Math and Science Study (TIMSS). The TIMSS had been the most carefully conducted and the most extensively publicized of the comparative studies of math and science performance of children in different countries. The results, which put the U.S. near the bottom, were widely covered in the American press. But much more interesting to me than the statistics on test scores were the videotapes that TIMMS produced of typical eighth-grade geometry and algebra lessons in the U.S., Germany, and Japan. The most dramatic contrast was between the geometry lessons in the U.S. and Japan, and this was what I showed to the Cubans.

The director of TIMSS was UCLA's James Stigler, one of the few academics in the education field who has systematically observed math classes in other countries. When introducing the video, Stigler explained that they had made every effort to choose typical geometry and algebra classes in each of the three countries. A particularly helpful feature of the video was that English subtitles were provided for the English as well as the Japanese and German dialogues. The Cubans could follow them well because their reading knowledge of English tends to be much better than their oral comprehension.

What shocked the Cubans was the same as what shocks American mathematicians who see the film. The U.S. class is bored and apathetic. The teacher expects the students to answer only very simple questions based on memorized formulas, and he compliments them on the most trivial correct responses, almost as if he's afraid of the kids. The material is below grade level; moreover, there is no actual *thinking* going on.

In the Japanese geometry class, on the other hand, the students are involved in solving an imaginative problem that requires that they apply basic geometric concepts in a novel setting. The teacher has an easy rapport with them, and jokes with them while some are discussing their solution with their classmates. The contrast was day and night, and no one who saw the video of the two geometry classes would think that I was exaggerating in my complaints about math education in the U.S.

A great deal of space in newspapers and magazines has been devoted to the fierce debates between "reformers" and "traditionalists." In California journalists dubbed this controversy the "math wars." (This was the title of a two-part article by Allyn Jackson in the *Notices of the American Mathematical Society* that gave an excellent, balanced treatment of the subject.) But to many of us who spend time in math classrooms in the U.S. these debates seem almost to be irrelevant, to be orthogonal to the reality of the public schools.

When I began visiting math classes in the early 1990's, what I found most unexpected and disorienting in the American schools was the frequent interruptions that teachers have to put up with during a 50-minute math class: announcements over the public address system, telephone calls to the teacher during class, messengers with notes from the main office. Someone enters the room to pick up attendance sheets. Someone else interrupts to bring in a late student or pull a student out of class for some reason. It is no wonder that "attention deficit" has reached epidemic proportions among American middle-schoolers. When the administration shows little respect for the integrity of the class hour, the children inevitably get the message that the academic content of the lesson is not very important.

One incident stands out in my memory as a sad and poignant illustration of the lack of common sense in our schools. One day a boy named Jamaal was wanted in the main office. Jamaal looked alarmed, and his classmates all stared at him. Had he done something terrible? What was in store for him? Jamaal was escorted out, and fifteen minutes later was brought back to the class. His return caused even more commotion among the sixth-graders than had his departure. His friends, burning with curiosity, had lost all interest in the math lesson. Jamaal had tears in his eyes and was trying hard not to cry in front of the others. I later learned from the teacher that Jamaal had been taken away to be told that his grandmother had just died. What could possibly have been the point of yanking the child out of class rather than waiting until later, when he could have been discreetly called aside and sent home? Why the needless cruelty to Jamaal and the senseless disruption of the math class?

Writing fifteen years ago in *The American Educator*, educational researchers Stigler and Stevenson compared math classes they had observed in Asia and in the U.S.:

> American lessons are often disrupted by irrelevant interruptions. These serve to break the continuity of the lesson and add to children's difficulty in perceiving the lesson as a coherent whole. In our American observations, the teacher interrupted the flow of the lesson with an interlude of irrelevant comments or the class was interrupted by someone else in...47% of all fifth-grade lessons... In fact, no interruptions of either type were recorded during the eighty hours of observation in Beijing fifth-grade classrooms. The mathematics lesson in one of the American classrooms we visited was interrupted every morning by a woman from the cafeteria who polled the children about their lunch plans and collected money from those who planned to eat the hot lunch. Interruptions, as well as inefficient transitions from one activity to another, make it difficult to sustain a coherent lesson throughout the class period.

In my naiveté I would have thought that this problem would be easy to solve. It wouldn't require any money, just a new policy district by district that no interruptions or non-academic digressions are to be allowed during a class in a basic subject. But in America there seem to be major cultural and sociological obstacles to implementing such a simple reform.

In November 1992 I started making biweekly visits to sixth-grade classes at Washington Middle School, which is located in an inner-city neighborhood of Seattle. Although the teacher was happy with the visits, I felt that they would be much more effective if I could bring some University of Washington students to help me by working with the kids in small groups.

Meanwhile, in February 1993 I was informed that I had been chosen by the College of Arts and Sciences as a "Liberal Arts Professor" for one year. I had been nominated for this teaching award by the department chair, Ramesh Gangolli. During my first decade or two at the University of Washington, several of the older professors, including Gangolli, Bob Phelps (who was chair when I was hired and when I was tenured), and the late Bob

Warfield, put great effort into being supportive of their younger colleagues and creating a congenial atmosphere within the department. Gangolli is also a talented musician; after retirement from the math department he became a professor of musicology, and he has played a central role in promoting Indian music both locally and nationally.

The Liberal Arts Professor designation meant that I was encouraged to devote some thought to possibly developing a new course, in which case I would receive a small grant to do it. I designed a junior-level two-quarter sequence for math majors who intended to teach middle or high school. It had several features which marked a radical departure from anything the math department had offered in the past. In the first place, the five-credit course included a "lab" consisting of visits to an inner-city school (at first Washington and a few years later Meany Middle School). Each week we presented math enrichment material to several sixth- or seventh-grade classes. The topics ranged from cryptography to graph theory to arithmetic games to geometry. After one of us briefly explained the topic, we passed out worksheets and circulated among the kids, helping them individually or in small groups.

In the second place, each week the UW students were required to turn in a short paper (two pages or less) discussing some aspect of the class readings or the school visits. I corrected the essays, and the students were expected to hand in a revised version of each paper. Guest speakers, assigned readings and videos, and class discussions covered a variety of controversies: gender equity in the classroom, single-sex classes, ethnic and racial stereotyping, labeling of children and ability grouping, accountability and assessment, academic standards and grade inflation, the role of the media and of computers, and conflicting approaches to math education reform.

The writing component of the course was difficult and time-consuming not only for the students, but for me as well. Careful correction of the first and revised versions took me roughly a half hour per student per week. Despite all this effort, the results were often disappointing. It is hard to change entrenched bad habits; a student is less malleable in the third or fourth year at the university than during the first. Several students said that in their fifteen years of schooling never before had their written work been handed back covered in red for bad grammar and style. And never before had they been required to hand in revisions that were also corrected.

Despite the mixed results, I decided not to give up on the writing part of the course. As I explained to the students, when they become teachers they will write memos to the principal, notes to parents, and letters of recommendation. We wouldn't want them to appear semi-literate.

In the mid-1990's a new administration came in at the University of Washington at a time of great pressure from the state legislature to increase the proportion of students who finish in four years. University President Richard McCormick created an Office of Undergraduate Education, headed by a former psychology professor named Fred Campbell, one of whose key tasks was to improve four-year graduation rates. The main causes of the decline in these rates had been the poor high school preparation of incoming students and their tendency to work part time for more than twenty hours per week while in school, and there was little the new Dean of Undergraduate Education could do about that. What he could do, however, was to get faculty to water down requirements, inflate grades, and make their courses easier.

After a few years academic standards had taken a hit. For example, UW is one of very few colleges in the country where a student can meet an upper-level distribution requirement with a grade of D-minus — at other institutions the minimum grade is a C. At a meeting of the committee that was revising the list of courses for the science requirement, the math department representative objected to the business math course being on the list. He pointed out that that's a course in what used to be considered high school math, and moreover, all science applications had been removed to make the course easier for business students. His point of view was inconsistent with the goal of increasing four-year graduation rates, so he was overruled. Thus, at UW a student can meet the upper-level science requirement by getting a D-minus in business math and similar courses.

What bothered me most was the watering down of the writing requirements. At the freshman level, the English department's writing course suffered from extreme grade inflation and a very uneven quality of adjunct instructors (as at many large universities, regular faculty do not deign to teach basic writing). The upper-level requirement was becoming a joke: the faculty committee that used to approve the "W" designation for courses was abolished, and any department that was under pressure to boost enrollments could list whatever courses it wanted as W-courses, even if writing was only a minor component. In addition, instructors did not have the authority to decide who deserved the W-credit; any student who received a C or above was guaranteed W-credit.

In the 1990's I met with several deans and faculty committees in a desperate attempt to get someone to pay attention to the deterioration of writing standards, but I eventually came to realize that I was just hitting my head against a brick wall. The only encouraging comment I heard was a remark by the divisional dean of humanities that my complaint about student writing was one that was not often raised within the university, but was frequently mentioned by people on the outside who hired our graduates. One would think that the administration would be able to put two and two together.

I've never fully understood why UW can't get better people in administration. Even though faculty salaries are lower than those at comparable universities, top administrators at UW are well paid. For example, our current president is the fourth highest-paid public university president in the country. So one would expect the quality of the administrators to be higher than that of the faculty — but the reverse is true.

Once I was at the same table as the Dean of Undergraduate Education at a reception for students in the Honors Program. I was astounded at his clumsiness in interacting with young people — in fact, I had never seen anyone as nervous and awkward around students as he was. Thus, in his case it must have been not only the higher salary, but also the minimal contact with students that provided an incentive for him to leave the classroom for a deanship.

In August 1999 I wrote to the university's president, Richard McCormick, about my concerns. I started the letter by saying that Ann and I had donated about $30,000 (the Certicom consulting fees) to an internship program for minority students, but that we were not sure that we would continue because of "concerns [that] relate to certain tendencies in undergraduate education at UW." My letter was polite and carefully written — I showed it to Ann and several colleagues before sending it — but it included a sharply critical assessment of the Office of Undergraduate Education. On the subject of writing, I recalled my repeated attempts to "call the attention of the appropriate faculty committees and administrators to an alarming decline in writing standards at the university... The trend is for more and more of our students to graduate without the basic writing skills that they need to be competitive in the global economy." I thought that since I was a senior member of the second largest department on campus and also a donor to the university, McCormick would take my carefully argued letter seriously. Perhaps he would delegate someone to meet with me for a more detailed discussion. Even if he did not have any sincere interest in what I was saying, I thought that at the very least a university president was supposed to be a savvy fundraiser, and I'd heard that most colleges keep tabs on aging childless professors with a history of donations and try to maintain good relations with them.

I couldn't have been more wrong. At the end of August I received a reply from President McCormick in which he rejected my complaints; in fact, most of the letter seemed to have been ghost-written by the very people I had implicitly criticized. That's the only way I could explain the insulting words directed against me and my departmental colleagues. Concerning our math courses McCormick wrote:

> More telling are years of comments from students complaining of poor course materials, untrained teaching assistants, indifferent faculty and even a hostile classroom environment.

The phrase "poor course materials" refers to the applications-oriented calculus workbooks that I had prepared over a period of years (and that had received an enthusiastic evaluation from the College of Engineering's curriculum committee). Our department had been one of the first to institute a T.A. training program, and McCormick (and his ghost-writer) had no basis at all to accuse us of "indifference" and "hostility" toward students. Perhaps this was Dean Campbell's projection onto us of his own attitude toward students. In any case, even if the allegations had been true, I couldn't understand what McCormick thought he was accomplishing with such a nasty letter.

On the subject of writing, the letter did not show the President as someone whose own literacy was exactly at the highest level. He mentioned "a statewide effort to evaluate the level of student writing," and continued:

> I hope that this work will lead to the development of measurable standards for assessing the writing ability of our undergraduates and will serve as a basic learning outcome for the University of Washington.

Huh? How can an "effort to evaluate" something "serve as an outcome"? A good high school English teacher could have explained to President McCormick that there were several word usage errors in his letter.

Coming just a year after the administration's cruel and irresponsible treatment of my colleague Tatiana Toro (discussed in Chapter 13), McCormick's letter was the last straw. Immediately after receiving it, I contacted Certicom about redirecting my consulting fees to the Kovalevskaia Fund rather than to UW. A few months later when the math department at Queen's University in Kingston, Ontario sounded me out on a special research professorship there, I was much more receptive than I would have been just a couple of years before. It was not only the high reputation of Queen's University, but also my disenchantment with the UW administration that led to my receptivity to their offer. In May 2000 my university responded with a 25% salary raise in order to keep me from leaving. This sudden generosity came from the same administration whose top dog had just nine months before accused me of being a lousy teacher!

A final note concerning our esteemed university president: on October 25, 2002 Richard McCormick unexpectedly announced that he was going to leave the University of Washington a month later in order to become President of Rutgers University in New Jersey. The full story didn't come out until a year after that, when an article in the November 2, 2003 issue of *The Seattle Times* revealed that McCormick had resigned under pressure from the Board of Regents because of ethical questions related to an extramarital affair with a subordinate.

Because they believe that private industry operates much more efficiently than a traditional educational institution — an assumption that could easily be disputed, but rarely is — many administrators like to talk about following a "business model" in running a university. In their version of this model, professors are the employees and students are the consumers. It follows that a central aim is to attract more students and keep them happy. One result of this philosophy is the reliance on student rating numbers as the most important measure of the quality of teaching. Another consequence is what some have called the "amenities arms race" and the "countryclubization" of the college environment — the construction of ever more luxurious dormitories, eating places, and recreational facilities. This is one reason why college costs in the U.S. have outpaced inflation. (Another reason is that university budgets have to pay for a mushrooming bureaucracy that includes an army of deans, associate deans, assistant deans, associate provosts, assistant provosts, and so on, all of whom demand high salaries, nicely furnished offices, and secretarial support.)

The fallacy in the administrators' business model is that students are *not* the consumers. If one insists on drawing an analogy between a university and a factory, then the students (when they graduate) are the *product*. The *consumers* are the students' future employers and society at large. Thus, rather than "Are the students happy?" the central questions should be, "Are people in the outside world happy with the quality of what we are producing? When prospective employers examine the writing skills, technical competence, and work ethic of our graduates, do they want to hire them? Or will they outsource the jobs to a part of the world where a healthy work ethic is alive and well?"

One feature of the modern American university bureaucracy is a love affair with buzzwords and insipid slogans. For example, at the University of Washington the new motto is "Creating Futures." I sometimes wonder whose futures are being created. When we dumb down courses, inflate grades, and treat our students like spoiled little children, we're not helping to build *their* futures. Rather, we're giving a competitive advantage to the graduates of universities in India, China, Mexico, and elsewhere who are eager to take up challenging work in the high-tech global economy. So to me the words "Creating Futures" have an unintended irony.

In 2005 I had been giving my math education course for twelve years, but times had changed and my department had to respond to new pressures. Teaching the course was a lot of work, and it counted for almost two-thirds of my teaching load. The education students who took the course were harsh in their ratings. In part, as future teachers they considered themselves to be good judges of other people's teaching ability or lack thereof. Moreover, most of the students intensely disliked the weekly essays. They objected to my "picky" corrections of their many errors, they dreaded having to revise and rewrite them, and they thought that I graded much too low. Not only were my student ratings low, but enrollments were also not particularly high (between 15 and 20 in the first quarter and between 10 and 15 in the continuation).

In autumn of 2005 I was told that the department was considering reducing the teaching credit that I would get for the course. Such a move would have amounted to a negative judgment on what I was doing, since in practice it would have meant an increase in my load as a kind of penalty for low enrollment and student evaluation numbers. I wrote a long note to my department chair explaining in detail the rationale for the course and the reasons why it took so much time and effort to teach.

The department never made a final decision about reducing teaching credit for that course, but it did something that seemed to me just as ominous. That year I was denied a supplemental salary increase that was apportioned on the basis of merit. I wasn't privy to the committee discussions or the chairman's thinking about my case, but what stood out was that I was given a below-average merit evaluation for teaching. It appeared that I was being punished for the way I was teaching the math education students. At that point, not being a masochist, I decided to stop giving the class. The chairman and I agreed that a change in my teaching assignment would be mutually beneficial. In a memo I analyzed the demise of my course:

> In 1993 high student satisfaction levels were not perceived to be nearly so important as they are now. From the beginning [this course] was not designed to provide students with a high comfort level — quite the contrary.

For example, for our practice teaching I chose schools at the lower end of the socio-economic spectrum. Meany Middle School, where we currently work with five seventh-grade math classes, is top in the District in the percent of students in the subsidized-lunch program, and it is bottom in the District in WASL [standardized test] scores. Over 75% of the students are African American and other minorities. There is a lot of stress involved in working at the school. The UW students I take there almost always work well with the kids, but I know from their papers and comments that the experience is often frustrating and troubling to them. Most of them had never been in such a school before. If I had wanted to provide the students with a more comfortable experience, I would have taken them to teach math to wealthy white kids.

In the second place, I chose to make student writing an important component of the course. Here again I did not do this in a way that was easy on the students. I could have simply collected their essays and graded them on content alone — an A if I liked the content, a B if I didn't. Instead, I marked all the grammatical mistakes, word usage errors, and awkward, unclear, or illogical places, and I required them to revise each paper. If many of the errors were not properly corrected in the revised version, then they would receive a low grade. This was also stressful to the students, most of whom had never had their writing corrected and graded in this way.

In the third place, I expected the students to show the type of professional conduct that will be expected of them when they become teachers. In particular, I deducted points from their grade for repeated absence or tardiness. I was not sympathetic to the argument that it's hard to come on time to an 8:30 class. A teacher's workday starts early. At Meany the first period bell rings at 7:45.

Despite the effect on student satisfaction levels, I do not believe that I was wrong in these policies.... But times have changed.... I fully understand the need to reassess whether the department wants to put resources into such a course. The components that take the most faculty time... are the types of activities that are hardly central to the mission of a university math department — correcting student writing, outreach to inner-city schools.

In the mid-1990's, a few years after my teacher prep course started, my department conducted interviews of all graduating math majors. Concerning my course the most common sentiment was that they disliked me but thought that the class was the most valuable one they had taken in the department.

The course was also a rewarding one for me; I often felt that I was learning at least as much as my students. When I started visiting Washington Middle School in 1992, the teacher I worked with was an African American named Ethel Jackson. She ran a tight ship, was deeply respected by her sixth-graders, and, despite her traditionalist approach to math, was remarkably successful. She would tell me about some of the children's problems at home; she was as much a parental figure as a teacher for many of them.

I soon learned that even though she had no detectable accent in English, she was actually Afro-Cuban and had immigrated to the U.S. as a young adult about thirty years earlier. Before classes started we would talk in Spanish. She would share with me her memories of Cuba, which she enjoyed recounting in Spanish much more than she would have in English, and it was good language practice for me as well.

One day she pointed out to me two girls who were leaving the room chatting in Spanish. She told me that they were cousins, and they never spoke Spanish in front of the other children because they thought they would be looked down upon. We discussed how sad it was for children not to have pride in the language of their ancestors and decided that we'd continue speaking Spanish to each other after the students came in and the bell rang. It was important for the English-speaking children — and also the two cousins — to see their teacher and her colleague conversing in Spanish.

In addition, when I passed out the Muddy City problem (one of Mike Fellows' favorite ways to teach kids about algorithms), some worksheets were labeled Muddy City and some *Ciudad Lodosa*, because, as I explained to the kids, I also used these sheets in schools in Latin America. When we came to Caesar ciphers (encryption by an alphabet shift), some of the secret messages were in English and some were in Spanish. One of the geometry worksheets involved finding the area of a map of Peru. There is no reason why a math class can't also teach a little about linguistic diversity and pride in one's Latino heritage.

The impact of school visits on the kids is hard to gauge. What is clear, however, is that the time spent both in the inner-city middle schools in Seattle and in the various classrooms in other countries has greatly influenced my way of thinking about education. I have drawn upon those experiences when writing and talking with people about controversies in math pedagogy, and they form some of the nicest memories of our travels that Ann and I have.

CHAPTER 16 **ARIZONA**

During the years when Ann taught at Hartwick College, I rarely went there to see her; rather, she would regularly come to Seattle. A few times when I was visiting Waterloo, we'd rendezvous for the weekend in Brighton, Ontario, where there was a secluded inn on the lake where we liked to stay. The Canadian side of Lake Ontario was prettier than the American side, although it was a long drive for Ann to get there. I used to joke that the proprietor must've thought it suspicious that we arrived in separate cars from different directions — telltale signs of an illicit assignation. Ann scoffed and said that of course he figured that we were married. "Yes," I said, "but not to each other."

In autumn of 1997, after Scott Vanstone invited me to spend at least a semester at the University of Waterloo, I arranged a sabbatical for the following academic year. I figured that, with me in the Northeast, Ann and I would see each other more frequently, meeting for the weekend somewhere in Ontario. I never imagined that a few months later she would land a job in the West. Her first year at Arizona State University coincided with my sabbatical, and so there was one final year when we were separated by the Great Divide. However, despite the distance between Waterloo and Phoenix it turned out that I was able to spend quite a bit of time with her in Arizona that year.

The Phoenix area is in many ways the antithesis of Puget Sound. Seattle is rainy and cool, while the Valley of the Sun is hot and dry. Seattle is sandwiched between a lake and the sea; Phoenix has only the Salt River, which most of the year is bone dry. The hills and mountains of the Pacific Northwest are covered with green, whereas in central Arizona the skyline is dominated by browns and burnt reds, perhaps with a solitary saguaro silhouetted against the horizon.

From the beginning Ann and I were entranced by the physical beauty and fascinating history of the Southwest. We traveled extensively and visited some of the wonderful local museums and historical sites in Arizona and New Mexico. We particularly enjoyed off-road trips to ghost towns, abandoned mines, and remote mountain overlooks.

In March 2005 we rented a houseboat on Lake Powell for a few days. During the summer the lake is mobbed, but off season it is deserted and the scenery is just spectacular. Most of the time we were the only people in sight as we explored the vast wilderness, sailing in the side channels surrounded by cliffs. Because of drought and patterns of water use in Arizona and California, the lake was at its lowest level since the original flooding of Glen Canyon in the 1970's. No maps had yet been made showing the new contours and the beachheads where the houseboat could be anchored for the night. So each day was an adventure. When we finally located a cove with a sandy shore just before sunset and I jumped out in my bare feet to plant the two anchors, I joked to Ann that I felt like Robinson Crusoe. The photos that I got with our new digital camera were the best pictures that I have ever taken.

Soon after buying a house in Phoenix, we started thinking about also getting some rural property in Arizona. Originally we had had the vague idea of purchasing a plot of land near Seattle on which to eventually build a retirement home. But land prices anywhere around Puget Sound had climbed ridiculously, and in any case the house where we already lived in Seattle would be perfectly fine for retirement. So Ann suggested that maybe we should buy land in Arizona, where real estate prices were much lower.

In 2000 we decided to look for some remote property in the Sonora desert having varied terrain and beautiful views. It would be nice to get a fair amount of acreage and not have to pay too much. So where do people who know nothing about real estate go to find a bargain on land in the desert? Why Safeway, of course. I picked up one of the many "homes and land" magazines on the rack, and Ann found an ad for twenty acres (in the metric system that's a little over eight hectares) of hilly desert for $40,000. We contacted the owner, visited the property, which was located in the Picacho Mountains half-way between Phoenix and Tucson, and fell in love with it. We paid full price — the owner must've thought that we were the most gullible city slickers he had ever seen.

Our motives in buying the land were not totally rational. It was waterless and many miles from power and paved roads. It had gorgeous rock outcroppings, a tremendous variety of cactus, including several dozen saguaros, and hillsides and washes that were great fun to climb around. Of course, we could have hiked in the desert without owning any of it. But somehow having a proprietary interest in this piece of the Sonora desert made a psychological difference. The closest analogy might be with people who like to buy original artwork and feel that the experience of ownership is fundamentally different from just viewing art in museums.

There are many mysteries, big and small, in the history of Arizona, and one of the small ones is how it came about that half of a square-mile section in the middle of the Picachos — one sixteenth of which was our property — was privately owned. All of the surrounding land belonged either to the state or to the federal Bureau of Land Management. Although our acreage was pristine, there was plenty of evidence of human activity within a short distance — crumbling sheds, troughs, windmills, rusted mining equipment and pipes.

In hopes of answering some of our questions, we visited the museum in Florence, which is the county seat. The librarian was helpful, and gave Ann a folder of old newspaper clippings. She insisted that Ann wear gloves when handling the documents. Ann loved it — the idea of this sort of detective work (wearing gloves, no less) appealed to the historian in her. That was the start of Ann's interest in the history of Arizona.

The best local museum in the state — and perhaps in the country — is located in Prescott, about 90 miles (145 kilometers) northwest of Phoenix. It was founded by Sharlot Hall, who was Arizona's first state historian and one of the key people who lobbied for Arizona to be admitted to the Union as a separate state rather than as a part of New Mexico. In the Sharlot Hall Museum archives Ann found material for the book she was writing, *Sex and Herbs and Birth Control*, which was going to deal with fertility control in different cultures and time periods. (I had thought up the title, which is intended as an allusion to the phrase "sex and drugs and rock 'n roll.")

Much of what Ann found fascinating in the old newspapers and documents was not relevant to her book, however. She used that material to write articles on other subjects, which were published as part of a weekly column called "Days Past" in the newspaper *The Yavapai Courier*.

When my visits to Arizona coincided with Ann's work in the Sharlot Hall archives, I would look for something to keep myself busy in Prescott. By 2002 we were having doubts about our property in the Picacho Mountains. We still enjoyed hiking around it, but a quarry operation less than a kilometer away had grown in size and become unsightly. And the Pinal County zoning board had recently given permission to a landfill company to build one of the largest dumps in the Southwest — 800 acres (325 hectares) of it — just three miles (five kilometers) from our property. In addition, we had spent enough time in Arizona to realize that the high country of the Colorado Plateau has a pleasanter climate than the lower desert.

I contacted a real estate agent named Dawn who downloaded listings of reasonably priced acreage between Prescott and Phoenix. I found four that looked promising, and Ann and I ended up visiting three of them. One parcel especially intrigued me because it was priced much lower than the others — twenty acres (eight hectares) for only $10,000 — and because it was described as "steep rock" on the flank of Copper Mountain with "no insurable access," as if even the real estate agent who had listed it thought that it was hopeless.

I had studied the USGS topographical map of the area and didn't believe that the property was nothing but a steep rock. To anyone who either had experience with topo maps or else had taught calculus for over twenty years — and I was in the second category — it was apparent that there must be a saddle point somewhere in the middle of the parcel. The map showed level lines at twenty-feet (six-meter) intervals, and, to put it in mathematical terms, there was clearly a point where the partial derivatives vanished. I also conjectured that there should be an unobstructed view to the east from that point.

As it turned out, in fact the whole property resembled a giant saddle. There was a poorly maintained dirt road that crossed the parcel far down the western flank of the saddle, where a large mine shaft had been dug, and from that perspective the property indeed seemed to extend indefinitely up a steep hillside. But from the map I could see that it would have to flatten out as it reached the mountain's ridge line, which came down from the main summit and then went up again to reach a secondary summit at the north end of the property. On the other side of the ridge was the eastern flank of the mountain.

I returned to Copper Mountain and bushwacked my way up to the center of the parcel. There was no trail, and I had to watch out for prickly pear spines, but juniper branches are sturdy, and I could haul myself up over steep and crumbly terrain with their help. As expected, there was a saddle point there, although it wasn't the nice smooth place I had imagined. Rock outcroppings and juniper trees created a complicated landscape, and to my complete surprise another large shaft had been dug, and there was evidence of extensive blasting, digging, and camping out by miners. I brought a few rusted pieces of miners' equipment back for Ann to examine. And the views — in all directions, not just to the east — were much more stunning than I had anticipated. I took photos to show Ann, and she was hooked. At Dawn's suggestion Ann offered $8000 for the property (we had thought that the listing price was already a bargain), and the owners accepted.

In Chapter 14 I mentioned that Joan Feigenbaum was not quite correct in saying that I'm a communist who doesn't believe in private property. In fact, between 2002 and 2005 Ann and I owned all or part of six properties and were paying real estate taxes in six counties: Maricopa, Pinal, and Yavapai Counties in Arizona (where our Phoenix house and the two twenty-acre parcels are located), King County in Washington (which includes Seattle), and Kauai and the Big Island in Hawaii. The Hawaiian properties are really just timeshare condos; we purchased one week in 1996 and another week in 1997, and since then we have been spending two weeks a year in what we believe to be the most beautiful vacation spots in the U.S.

Of course, the price we have to pay for such unabashed hedonism is that many people would consider us to have sold out to bourgeois temptations. How can principled opponents of capitalism permit themselves to indulge in the types of activities for which Hawaii is famous? When we snorkel in the waters near Captain Cook's monument, watch the surfers from our balcony on the Poipu coast, or stargaze from the summit of Mauna Kea, we try not to think about such questions.

When Ann examined one of the rusted cans I had brought down from our Copper Mountain property, she found faded letters on the bottom spelling out "blasting caps." I hadn't realized what was in the can when I put it in my jacket pocket and set out down the mountain with the contents shaking up and down. At that time even after reading the words we didn't know exactly what blasting caps were. I thought of the cap-guns we had loved when I was a kid and figured that they had something to do with starting a fuse. But by May 2006 when we were getting ready to pack up Ann's office for her sabbatical, we had learned much more about mining and knew that blasting caps had more explosive power than the caps used in children's toy guns. The other rusted artifacts in Ann's office we just boxed up or threw out, but we thought that something potentially hazardous should be discarded at a special facility for such things. So Ann called the ASU police asking for instructions on where to dispose of the blasting caps.

As soon as they heard the words "blasting caps," they told her to immediately vacate her office and wait for a police officer. The police came, met us outside Ann's office, and asked everyone on her floor to leave. A few minutes later they decided to evacuate the entire building and call in the city's bomb squad.

We were shocked — we'd had no idea that they were so dangerous. Ann stood outside the building profusely apologizing to all of her colleagues who couldn't get to their offices and in some cases had classes they had to move to another building. Ann's "Women As Healers" course was also meeting that afternoon to hear student presentations. The police wouldn't let her leave for another building, though, because she had to be available when the bomb squad arrived in case they had questions for her. So she held her class on the sidewalk by the building — in the afternoon heat of Phoenix in May.

At one point a student presenter was having a hard time being heard because of a helicopter flying overhead. Ann asked me what it was doing there, and I pointed out to her that it was a news helicopter from a local TV station. Ann said "Oh no! That's all I need!" But fortunately that Tuesday was not a slow news day in Phoenix. There had been a couple of car crashes and murders that made better TV footage, so Ann was spared an embarrassing interview on the local news.

After about an hour the bomb squad came, and an officer in full hazmat gear removed the can and brought it to an explosion-proof vehicle. They later detonated the caps in a special facility they have for that purpose.

The episode was the top story of the day on the university's website, and soon after someone from the administration investigated the details. The news reports about the dangerous explosive material removed from a professor's office had indicated the building — which was part of the engineering complex — but not the name of the professor or department involved. When the administrator learned that this had happened not in Chemical Engineering or Physics but in Women's Studies, she said, "Well, this gives a whole new meaning to 'Take back the night'!"

Two days later Ann got an e-mail message from bomb squad Detective Chuck Corning. "I would like to speak with you [in] reference [to] the blasting cap container we disposed of on May 2nd," he wrote. "You should buy a lottery ticket." When Ann phoned him, he was cordial and curious about where the caps had been found and how old Ann thought they were. She said that mining activity on our claim probably stopped in the 1950's, and the caps presumably dated to shortly before that time. Det. Corning said that the blasting caps that were made then were very dangerous because they were unstable — and that's why they stopped being produced in the mid-1950's. Someone in Oregon had been killed when a can he was carrying exploded. It had contained just three caps, whereas Ann's had had fifteen. Ann was lucky — that's why Det. Corning had suggested that she buy a lottery ticket.

The police were remarkably nice about the whole thing. Det. Corning was obviously pleased that they had used their knowledge and equipment to take care of a potentially dangerous situation. What people like him hate is false alarms that just waste everyone's time. This time the danger was real, and they were happy to be able to do their job.

There was an interesting side story to all this. Ann's students gave their presentations near a doorway where two campus police officers were stationed to prevent anyone from entering the building. We noticed that they were paying careful attention to what the presenters were saying. A student named Dawn who worked as a registered nurse gave a talk about the benefits of breast-feeding. She herself was pregnant, and for her it was particularly unpleasant to stand and speak in the heat. After she finished, one of the police officers came up to her and said that his wife was pregnant and wanted to breast-feed after the baby was born. He added that he had been against it but now had changed his mind thanks to Dawn's presentation and would support his wife's decision. In Dawn's e-mail to Ann about this, she said that this made all the discomfort she had endured seem worthwhile after all. She wrote that this confirmed her belief that even things that seem pointless — like Ann's blasting caps and the evacuation of Engineering A — happened for a reason.

Ann is usually more observant than I am. In spring of 2002 when she saw the photographs I had taken of the huge rocky ledge at the north end of the Copper Mountain property, she commented that the stones seemed to form walls. I had assumed that the piles of rocks were rubble from the blasting, but when we got a closer look it became clear that she was right. At first we thought that maybe the miners had put them up, perhaps for shade, but they were not in the right position to provide any significant protection from the sun. Then Ann conjectured that before the miners got there perhaps the Yavapai Indians, who had lived in the area in the 19th century, had built them.

One day in December 2002, a few months after Ann purchased the property, I was in a bookstore in Prescott browsing through a history of the region that is now called Yavapai County. The first page, which was devoted to prehistory, showed a picture of a hilltop site that was located twenty or twenty-five miles (thirty to forty kilometers) to the south of our land, and it had walls that looked very much like ours. The description said that they had been constructed in the 11th century and used for three or four centuries after that. A shiver went down my spine — this meant that our property had Native American ruins from the time of Marco Polo and the Norman Invasion! I hurried to show this to Ann, and we immediately started looking for all the information we could find about the archaeology of the hilltop sites in central Arizona. We later found our ruins in a list of several dozen archaeological sites in the region.

A historian Ann knew at the Sharlot Hall Museum told her that among living archaeologists the person who had worked the most on hilltop sites was David Wilcox of the Museum of Northern Arizona. While Ann was teaching, I went to the Arizona Room of the ASU library and found an article by Wilcox and two coauthors in a journal called *Plateau*. In it they depicted a dramatic scenario of epic battles between the prehistoric inhabitants of the Phoenix basin and those living in the mountain ranges to the north. According to Wilcox and his collaborators, the purpose of the hilltop sites was military.

Although I knew nothing about archaeology and had no idea how solid their evidence was, I was bothered by the tone of the article. For example, they described how they had come up with some of their theories while talking with one another late at night around a campfire. This was an odd thing to put in a scholarly paper. Once while jogging with another mathematician in my department, I told him about the article and the authors' campfire references. He said that it was as if a mathematician had written in a paper that "the proof of the following lemma came to me while I was on the toilet."

Moreover, Wilcox's passages about one of his collaborators, an amateur archaeologist and decorated Vietnam War veteran named Gerald Robertson, were very troubling to me. The description of what Robertson had done during the War made it clear that he had been complicitous in torture and other war crimes. Yet Wilcox gushed with sycophantic admiration for Robertson's war record, which Wilcox thought gave him special expertise in judging the extent of warfare in prehistoric Arizona.

I showed the article to Ann and asked her if I was overreacting. She agreed that there was something fishy in all this and proceeded to do further reading on the archaeology of north-central Arizona. This turned into a project of several months that resulted in a paper analyzing the methodology of the "Dr. Warfare" school of archaeology. Published in the journal *Men and Masculinities*, Ann's article was unusual in exposing the *machista* biases in a currently fashionable trend in academic research. It was Ann's first paper that could be categorized not as "women's studies" but as "gender studies." (By coincidence, the same year as Ann's article appeared her department changed its name from Women's Studies to Women and Gender Studies.) The second postscript to this chapter is an abridged version of her article "Male Bonding Around the Campfire: Constructing Myths of Hohokam Militarism."

By late 2004 we were thinking of selling the Picacho Mountain property, since we were much more likely to want to do something with the Copper Mountain mining claim. We thought that we had probably overpaid for it and so should expect to take a loss. I commented to Ann that we should be prepared to accept $30,000. Even so, we had no regrets about having bought it four years before. We had spent many enjoyable mornings exploring the property and nearby hillsides, and our ownership of that remote parcel had stimulated our interest in Arizona history.

In early 2005 we were contacted by a real estate agent who initially told us that she could get us $60,000, but then backed off from that figure and tried a bait-and-switch with us. Soon after this discouraging episode we met another agent named Valerie who turned out to have the Midas touch. This was the peak of the Arizona real estate bubble (which burst a few months later), and Valerie found a buyer who offered $230,000 for our twenty acres in the middle of nowhere. In August 2005 we sold the property for almost six times what we had paid for it.

With all this money burning a hole in our pockets, we thought that the time was right to think seriously of having a dream house built on Copper Mountain. Three months before we had found someone to put in a dirt road leading to the saddle point at the top of the property. Before we could have a new road cut across state land, we had to go through a procedure that lasted almost a year and cost several thousand dollars, mainly for a survey and a perpetual lease for the state land the road was on. We were very relieved when the Arizona State Land Office finally approved the road.

The previous summer I had wandered around Prescott and Phoenix stopping in at various architectural firms. Most seemed either too mundane and conventional or else too expensive (or both). But one place I visited, called CCBG Architects, gave me a brochure that showed some striking residential designs that appealed to us. In one case they boasted of having built on a hillside that "everyone else passed on as impossible to build upon."

Soon after the sale closed on the Picacho Mountain property, I called CCBG and spoke with one of the principal architects, Joe Groff (the G of CCBG), who agreed to look at our land. When we took him to Copper Mountain, he was enchanted. He had grown up in Prescott and when he was young had done a lot of off-road driving in areas not far from where we were, so he thought he knew what to expect. But when we arrived he was amazed: he hadn't imagined that the terrain would be so dramatic and the views so magnificent. Within minutes he sited the house in a portion of the property we hadn't thought of. He took out a pad and quickly sketched a concept for it — a type of butterfly roof configuration over a mostly glass structure in the shape of a T. The central part of the T would end in a great room and large balcony facing the Native American ruins that run along the ledge that marks the north end of the hilltop. The mountain has a natural gap in its ridge line, and the roof would appear as if it were filling it in. The sketch was a perfect example of a famous remark by Frank Lloyd Wright that a house should not be "on a hill," but rather "of the hill."

Ann and I had read several books about architecture. We had especially enjoyed *How Buildings Learn* by Stewart Brand, which contains some highly critical comments about the profession. Indeed, working with an architect was entirely different from anything we had done before. An architect is a creative thinker — like a mathematician, one could say — and, like mathematicians and other creative people, sometimes has brilliant ideas and sometimes has dumb ones. Even the great Frank Lloyd Wright had his foolish moments — kitchens and bathrooms that were tiny and impractical and roofs that leaked, for example.

In our case, the design of the fireplace was one of the areas where we had to intervene and put the kibosh on Joe's idea. The fireplace separating the great room from the dining room would be the focal point of the public areas, and Joe wanted it to open in both directions. He said that that would even give a view of the ruins from the dining room. Ann and I were skeptical, and after reading two books on fireplaces and examining over a hundred pictures we confirmed our impression that two-way fireplaces almost never draw air properly unless they have glass inserts. But we felt that a true wood-burning fireplace should have nothing between us and the flames (except for an optional screen that can be pulled across when no one's watching the fire); we did not want inserts. In the second place, the opening in the fireplace would be too low to serve as a window to the ruins on the other side — unless one crouched far down. So we decided that a two-way fireplace was a dumb idea and asked Joe to design the dining room side of the fireplace to be solid stone with niches for the books that we'd sometimes want to spread out on the dining room table — flora and fauna identification guides, sky charts, maps, and photo albums.

There were more important things than fireplace design that our architect was wrong about. For a long time Joe insisted that our main energy need would be heating in the winter, not cooling in the summer. Situated at an altitude of 5000 feet (1500 meters), the house wouldn't even need air-conditioning — an evaporative cooler would suffice, Joe thought. But from the beginning Ann said that because of all the glass we'd need major air-conditioning. The first time she hiked up to the site (before we had the road put in), the temperature reached almost 100°F (38°C). If we had been inside our glass-and-steel house on that June day with nothing more than an evaporative cooler, we would have boiled like lobsters, Ann said. But Joe maintained that with the northern exposure and "passive cooling" (by which he meant partial window shades and roof overhangs), we wouldn't have a problem. We put our foot down, and Joe designed the house with air-conditioning.

We got confirmation that Ann was correct when we were in Barcelona in January 2006. We were attending a conference in honor of the 60th birthday of the number theorist Pilar Bayer, at which the three invited speakers were Ann, me, and Gerhard Frey.

For many years Frey has had a research group in cryptography at the University of Essen; the main focus of much of their work has been cryptographic applications of hyperelliptic curves. In the broader mathematical world Frey is best known for his crucial role in the proof of Fermat's Last Theorem. It was he who conjectured that the elliptic curve $y^2 = x(x - A^n)(x + B^n)$ is not "modular" if $A^n + B^n$ is a perfect n-th power. The proof of this conjecture by Ken Ribet in 1986 paved the way for Andrew Wiles' proof of Fermat's Last Theorem.

When we mentioned our planned house to Gerhard, he told us that he lived in an architect-designed home on a hill that had large windows. He commented that it heated up quickly in the sun, and even in the winter (and Essen is quite a bit colder than central Arizona at that time of year) he rarely turned on the heat. When we told him how much glass would be in our house on Copper Mountain, he cautioned us that keeping cool in the summer would be a major concern.

In the history of mathematics the most elegant proofs usually have a certain inevitability and timelessness, so that in retrospect they look natural and almost obvious, even though it might have taken a stroke of genius to discover them. In a similar way the best that one can hope for in architecture is a creation that fits in so well with its environment that it seems as if no one could have considered constructing anything else there. Ann and I believe that the plan for our house on Copper Mountain will achieve this objective.

The main section of the T is all glass on the sides, and its roof slopes upward to give the great room 19-foot (6-meter) windows at the front. The supports are steel, creating a sleek, modernist overall appearance. The final design was similar to the rough sketch that Joe made the first time he saw the site. The most important change was suggested by a colleague of his named Craig Stoffel. In the original drawing the main floor of the house formed a bridge from the back of the property to a large rock outcropping that continues the mountain ridge. What Craig did was to raise the floor of the house several feet and cantilever the north side out over the outcropping. This not only affords a view of the mountain range in the distance beyond the ruins, but also considerably reduces the damage caused by the construction: the house will "have a gentler footprint," as they say.

The design of the house evokes three time periods — the ruins of Copper Mountain's prehistory, the copper mines of the historical period, and 21st-century technology. Someone driving to the house or walking up to it along the stone walkway will see a massive column in the center that is made up of local stones and cultured stone and extends upward to form the fireplace and then a large chimney; those elements are intended to echo the style of the Native American walls. The roof of the passageway under the main level and the part of the sides of the house that are not glass will be corrugated metal, which suggests the construction style of late 19th- and early 20th-century mining outposts. The roof will contain sixty large solar panels to support a 10-kilowatt photovoltaic system. With no power, cable, or telephone lines anywhere near, anyone who wants to live a comfortable and "connected" life on Copper Mountain will have to depend on the latest in modern technology: a satellite dish will capture TV and internet, and cell phone reception is excellent. Thus, the base of the house, the middle area, and the roof will evoke the three time periods that are suggested by the site.

From the beginning Joe had a free hand; we said that his main client was the site and our role was secondary. We knew that many architect-designed homes appear weird and quirky because the owners imposed their own bizarre ideas and tastes. Often such houses end up selling at a loss because most people who are wealthy enough to purchase them would rather design something according to their own preferences than buy a monument to someone else's idiosyncracies. In contrast, we wanted our house to have a timelessness and inevitability — like an elegant proof in mathematics — that would seem to transcend anyone's personal tastes, and we trusted Joe to do this for us.

We're obviously limited in what we can spend on a house on Copper Mountain; that is, the project is contingent on finding someone to do it competently and reliably at reasonable cost. We told Joe that if it can't be built for less than 50% over his original estimate, then it won't get built. As of this writing, Joe is batting 0 for 2 in the search for a general contractor.

One of the main purposes of constructing the house would be as an investment. After living there for about a decade we'd sell it, probably when Ann retires from ASU. Located between one and two hours drive from three of the main population centers of Arizona (Phoenix, Prescott, and Flagstaff), it is an excellent location for a wealthy person's weekend home. At the end of a four mile ($6\frac{1}{2}$ km) dirt road, it has as much seclusion as anyone could want. Ann's fantasy was that maybe a retired NFL quarterback would buy it (she's a football fanatic), and we insisted on having an elevator partly for this reason — and also because her knees have almost as much trouble with stairs as an old quarterback's. The house is in fact well designed for the elderly or moderately disabled.

Ultimately, the profit from selling the house is what will endow the Kovalevskaia Fund in perpetuity. When we die, the Fund as such will no longer be an independent entity. Rather, it will have been incorporated into a professional organization — most likely the American Math Society — which will coordinate a grant program for women in Third World countries. The size and number of these grants will depend on how much our house on Copper Mountain will sell for.

The prototype for such a program is a $5000 per year system of Kovalevskaia Grants for junior researchers that started in 2005 and is administered through the Mexican Math Society. It is financed through the donations of a former student of mine and his wife, who is an attorney at Microsoft; they give $2500, which is matched by Microsoft. Similar programs could be established in many countries if we're able to leave a multimillion-dollar endowment.

Thus, in order to eventually bring the most benefit to the Kovalevskaia Fund, the house has to have the type of high-end features that will appeal to a wealthy buyer such as Ann's quarterback. We have stoically resigned ourselves to accepting amenities that we would normally consider too luxurious to be in a house of ours but are necessary for maximum resale potential. When Joe suggested a gourmet kitchen, a whirlpool tub in the bathroom, and a walk-in closet bigger than our bedroom in Seattle, we sighed and reluctantly agreed — that's the cross we have to bear, the altruistic sacrifice we have to make for the good of our charity!

Many people would undoubtedly find it annoying that I make light of serious subjects, such as charitable work. I'm sure that Serena (the Peace Brigades representative we met in El Salvador in 1989) would say that it would be ridiculous to compare our lives to those of people who live among the poor and help them fight against injustice. She'd be right about that. And she'd be horrified to see that our place on Copper Mountain will be even more luxurious — far more — than the hotels where we stayed in Escalón. It is also clear what the reaction would be of the *New York Times* reporter who ignored the debate on South African investments and described the Tenth Reunion of the Harvard class of 1969 as proof that radicals lose their edge and become more conservative as they grow older.

To him, Ann and I would be just another example of ex-activists seeing the light and adopting a bourgeois lifestyle.

Whatever other people might think, I prefer to view the house we're building, which is our investment in the future of the Kovalevskaia Fund, as my answer to a question that has been on my mind ever since I was a little kid spending a year in Baroda, India. Is it possible to lead a privileged life and not feel guilty about it? The home on Copper Mountain is admittedly not a perfect solution to this conundrum, but it is probably better than that of the child who picked the jacket pockets of his parents' guests and handed out the money on the streets of Delhi.

Postscripts

The Sharlot Hall Museum in Prescott contributes a weekly column called "Days Past" to the regional newspaper The Yavapai Courier. *The article that follows, titled "They Kept Digging 'Til Their Hope Ran Out," is the latest of six that Ann has written for the Museum.*

When people think of mining, typically they conjure up images of the large enterprises of Virginia City in Nevada or Jerome and Globe in Arizona, where prospectors and miners could become rich almost overnight, and millions of tons of high grade ores were extracted and processed during the course of decades-long operations. The elaborate 19th-century mansions and the mountains of tailings in and around Jerome bear mute testimony to large scale mining activity in Yavapai County. But there was another kind of enterprise — much more common and much less successful economically — that was equally important to the social history of mining in our state.

A few years ago I purchased a twenty-acre mining claim on Copper Mountain not far from the town of Mayer. The land bears the remnants of at least four shafts, a 600-foot tunnel, and other relics such as antique nails and barrel staves, blasting caps, and rusted-out miners' lunch pails and food cans. The nearby hills, both private and government lands, are similarly dotted with exploratory shafts, tailings, claim markers, and other artifacts. But except for one medium-size mine a short distance to the north which produced copper in the first half of the 20th century, most miners apparently had as little success with their claims as did those who sank shafts on my property.

As a historian, I was naturally interested in learning more about the abandoned shafts that add such a ghostly presence to my land and that of my neighbors. As a result of discussions with surveyors and research in Sharlot Hall Museum archives and Bureau of Land Management records, I have pieced together a story of hope, desperation and dreams.

My property was registered in 1936 as the "Defiance Claim." The patenting documents declared that it was worth $36,000, divided into twelve shares of $3,000 apiece. Now, 1936 was the height of the Great Depression. At first I was astounded that the mine owners thought they could get anyone to invest so much in a small new mine at such a time — $3,000 was a year's salary for a Harvard professor. But one of the surveyors I spoke with said that probably it wasn't so much money being invested as time and labor. In other words, a group of unemployed men got together, pooled scarce monetary resources, and contributed most of their investment in the form of "sweat equity." According to the surveyor, "They kept digging 'til their hope ran out."

The hills northeast of Mayer feature some beautiful quartz and onyx formations, and many of the tailings contain enough "peacock" (bright teal-colored rock faces that indicate traces of copper) to make it clear why the miners might have had their hopes raised. But the fact is, there is barely enough metal there to justify extraction even with today's sophisticated mining techniques. In the 1930's, the situation would have been even more bleak.

This does not, however, mean that the Defiance enterprise was unsuccessful from a social point of view. At a time of enormous unemployment, the men would at least have kept themselves busy. They had the companionship of their fellow miners, they were getting fresh air and exercise, and the wild hope that they might, after all, strike it rich probably kept them sanguine for quite a while. If they had families in the vicinity of Mayer, their steady occupation at least made it unlikely that they were abusing their wives and children. And they might have brought back the occasional rabbit or quail or rattler to sweeten the stew pot at home.

In fact, late 19th- and early 20th-century reports of mining throughout Yavapai County tell a similar story: mining was socially useful and at times profitable, but most of the profit did not come from the ore itself. From its beginnings in 1865, the *Arizona Miner*, as befitted its name, devoted a lot of attention to mining enterprises, and its tone was almost always optimistic. But most of the time the *Miner* spoke of mines in the early stages of development and the glorious results the editor was certain could be obtained if only a bit more money were invested and more equipment were brought in. And it was usually the money being poured into mining operations in Yavapai County, rather than any windfall profits being brought out, that was the theme of enthusiastic reports in the *Miner*.

This is not to say that nobody got rich. A clever entrepreneur like "Diamond Joe" Reynolds or Professor Poland (for whom the settlement of Poland was named) could milk a weak strike or even a salted claim for far more than it was worth; there were plenty of gullible people (mostly from the East) who were willing to invest in Arizona mines so long as the front man sounded plausible and the stock certificates looked impressive.

Most of the money, however, was in the spin-off enterprises of a mining boom. Miners and prospectors needed supplies, and merchants were willing to furnish them, usually at a premium. Miners had to eat and sleep, get their laundry done, be entertained — and so rooming houses, restaurants, laundries, bathhouses, saloons and bordellos all sprang up quickly in the vicinity of mining activities, whether or not the earth ever gave up much valuable ore. Freight companies and railroad branch lines grew if the minerals lasted long enough, and a real town with enough momentum to withstand the playing out of the ore could emerge.

Mayer itself is an example. Joe Mayer founded the settlement in 1882 to service the prospectors who trekked hopefully into the hills around Copper Mountain as well as the better-known mining operations that stretched into the Bradshaw Mountains toward Tip Top and Crown King. At one point Mayer was a railroad junction, and thousands of tons of ore were shipped through there. But the ore ran out, the mines closed, and the miners on Defiance and neighboring claims found other things to do. Mayer, however, survives to this day.

In 2006 Ann published "Male Bonding Around the Campfire: Constructing Myths of Hohokam Militarism" in the journal Men and Masculinities, *Vol. 9, No. 1. The following is an abridged and slightly edited version of that piece.*

Life Imitates Art

Once while waiting in the gate area for a flight from Miami to Lima, Peru, I was whiling away the time idly watching my fellow passengers gather. One particular group caught my attention. They were obviously colleagues of some sort meeting for a joint venture; some of them claimed previous acquaintance with one another, while others were introduced for the first time. Eventually, the group numbered some fifteen persons, all male, and I could contain my curiosity no longer. I walked over and asked one of them whether by any chance they were archaeologists. With some astonishment, he said yes, and asked me how I could possibly have known. Not wishing to offend him, I laughed, told him it was a lucky guess, and returned to my seat. But in reality I had not been in much doubt about the men's profession. Almost all of them were dressed in battered hiking boots, cargo pants or jeans, stained leather jackets, and slouchy brown fedoras. In fact, all that was missing was a coiled leather whip to turn them into credible fascimiles of Indiana Jones (the intrepid adventurer-archaeologist played in the film *Raiders of the Lost Ark* and two equally popular sequels by the ruggedly handsome Harrison Ford)! Here was an excellent example of life imitating art and of members of a profession (sub)consciously shaping their image to fit a heroic mold.

This essay presents a case study of the effect that a culture of aggressive masculinity can have on the content of a scholarly discipline. I will argue that testosterone-driven fantasies appear to have influenced the theory formation of a significant group of archaeologists studying the American Southwest. On the basis of scant evidence, these archaeologists have created a story of prehistoric militarism that harmonizes well with early 21st-century U.S. political culture. Whether this warlike image has much bearing on the actual lives and pursuits of indigenous Southwest populations of the 11th through 15th centuries is, however, open to doubt.

The Hohokam

The most commonly told story of the Hohokam is that they flourished in central Arizona from north of present-day Phoenix to the south past present-day Tucson in the period from the early centuries of the common era to around 1450. They had a sophisticated culture with an intricate irrigation system, a series of "great houses" possibly constructed for astronomical observation, and complex trading networks. Remains of this civilization include extensive petroglyph fields, occasional traces of irrigation ditches, and the ruins of great houses, platform mounds, and ballcourts.

For most of the 20th century, archaeologists almost always portrayed the Hohokam as basically peaceful. Cultural evidence from petroglyphs and pottery supports that view. The Hohokam had numerous decorative motifs. Humans, animals, birthing scenes, religious and clan symbols, celestial objects, geometric figures, even possible relief maps are featured in their art. But depictions of war are conspicuous by their absence — of the 2500 petroglyphs in the North Pass and Picacho Point sites, for example, there is not one that shows a human fighting a human. Ceramic evidence of militarism is equally lacking. The Hohokam were accomplished potters and painted much of their clayware with human and animal motifs. But there are no recognizable portrayals of warfare on the countless pottery vessels and shards that have been recovered from Hohokam archaeological sites.

Dr. Warfare

In recent years, the consensus view of a basically peaceful Hohokam civilization has been challenged. At the nationel level, Harvard University's Steven LeBlanc has decreed that warfare is ubiquitous in human history, and that researchers "should assume warfare occurred among the people they study, just as they assume religion and art were a normal part of human culture." LeBlanc is proud of the fact that some of his colleagues, resistant to his sweeping claims, have taken to calling him "Dr. Warfare." He positions himself as an objective, disinterested scientist who has "raised the hackles of National Park Service personnel unwilling to accept anything but the peaceful... message peddled by their superiors," and he portrays his opponents as sentimental advocates of political correctness over hard facts. Yet LeBlanc himself seems particularly prone to dramatic and sensational generalizations, such as his claim that warfare was "endemic throughout the entire Southwest" or his remark, quoted above, that warfare is as much a universal characteristic of human societies as is spiritual or aesthetic expression.

The War Hero

An avid regional proponent of the view that warfare was pervasive in the Southwest is David R. Wilcox of the Museum of Northern Arizona. Wilcox has conjured up a late Hohokam world filled with epic battles involving large armies, military alliances, punitive expeditions, and castle-like fortresses. To help (re)create this past, Wilcox has enlisted the aid of amateur archaeologist and former military man Gerald Robertson, Jr. It is worth quoting at some length Wilcox's explanation of the need for a "military science approach" and his enthusiastic description of Robertson's qualifications:

> Military science should provide many insights [into Hohokam warfare in the 14th century]. But the knowledge of this field by most southwestern archaeologists is secondhand, derived mainly from reading. Wilcox had long thought that collaboration between archaeologists and people with real military experience, both formal training in military science and firsthand knowledge of war, would be a fruitful

methodological course to follow. So, beginning in 1992, when he first met the avocational archaeologist Jerry Robertson and learned he had been the commander of an infantry rifle company in Vietnam, a captain in the 101st Airborne, Wilcox began to ask Robertson to apply his military knowledge to the interpretation of archaeological sites. Not only had Robertson been awarded a Silver Star, three Bronze Stars, two Purple Hearts, and — most unusually — an air medal for valor, he had organized, developed, trained, and led a group of twenty-eight former Viet Cong, operating for five months in the same area where they had previously been guerrillas. Though he has tried to forget those days and thus found it a little difficult to do as Wilcox asked, he became interested and has managed to keep his bad memories at bay.

One could make many observations about the implications of the above quotation. At the very least one could question the assumption by Wilcox and his coauthors that decorations given by the Pentagon in a 20th-century war automatically indicate exceptional qualifications for judging the extent of military activity seven hundred years ago. Yet Wilcox sets great store by Robertson's military experience. In fact, he gives Robertson credit for two "seminal" ideas connected to Wilcox's proposal that the network of hilltop sites on Perry Mesa (a large canyon and mesa complex north of Phoenix) constituted a military "confederacy" at war with the Phoenix basin Hohokam.

Campfires and a Big Stick

Wilcox and his collaborators attach great significance to the exciting conclusions they see as emerging from informal socializing on long field trips to remote archaeological sites. On at least three occasions they write about the insights that were derived from late night male bonding sessions around the campfire. One such narrative starts out: "Over the campfire that night, we stayed up late talking about the behavioral processes of attack and defense."

Wilcox is also enthusiastic as he describes Robertson's stick drawings during one of their field trips: "When Robertson was first telling us about his ideas, we had stopped on a dirt road, and it was natural to find a stick and sketch out a map, the questions being, Who were their neighbors, and who were they afraid of?" The idea of archaeologists having to rely on a stick in the dirt to draw a map is peculiar, to say the least. Even Indiana Jones had his field notebooks and pencils, while most modern-day archaeologists go into the desert equipped with palm pilots and Internet-ready laptops as well as paper and pencil! The stick in the dirt is an affectation that one would sooner expect to find in a description of how to earn a Boy Scout merit badge than in a scholarly article.

Basking in the pleasant afterglow of the campfire sessions, the shared treks through desert terrain, and the maps drawn in the dirt with a stick, Wilcox and his coauthors seem unable to distinguish between cascading speculations and concrete archaeological evidence. There is nothing in ceramic, petroglyph, or ethnographic records to support the theory that the Phoenix basin Hohokam ever fielded an army of a thousand warriors under any circumstances, let alone that, as Wilcox claims, they went into battle with such an army against the supposedly warlike inhabitants of Perry Mesa. But Wilcox and his coauthors seem reluctant to abandon their militarist fantasy, at least in part because it harmonizes with heroic masculine images of the Indiana Jones variety.

Looking-Glass Logic

Proponents of LeBlanc's theory of ubiquitous warfare often fall into the trap of assuming that just because one sometimes finds A in the presence of B, that means that one can assume B whenever one finds A. Two examples of this logical fallacy concern "no-man's lands" and the link between trade and war.

Throughout his career, LeBlanc has insisted that areas of prehistoric depopulation ("no-man's lands") are a primary indication of warfare. Certainly it is true that depopulated zones often emerge in times of conflict, and that sometimes the areas remain uninhabited long after the war has ceased. But it is illogical to assume, as LeBlanc and his followers do, that the existence of an unpopulated or sparsely populated area between two population centers necessarily proves that those two populations were ever at war.

We can see the logical fallacy here if we look at modern Arizona. Between the sprawling population centers of Phoenix and Tucson lies Pinal County, an area of largely desert scrubland whose innumerable dirt roads crisscross a ghostly landscape of abandoned mines, farms, ranches, orchards, and homesteads of European settlers, as well as remnants of ancient Hohokam settlements. A large swath of the county — the area around Route 79 and the Picacho Mountains — could be characterized as a "no-man's land" between the two cities. But Phoenix and Tucson are not now nor have they ever been at war. Rather, the 20th-century depopulation of much of Pinal County had a variety of nonmilitary causes: unequal distribution and depletion of water resources, falling prices of agricultural products, the depression of the 1930's and the recessions of the 1970's and 1990's, and the general malaise that has affected many rural agricultural regions in the United States and around the world. Warfare, in other words, had nothing to do with the abandonment of large areas of Pinal County.

Equally obvious should be the illogicality of making the existence of trade relations evidence for warfare. In the book *Deadly Landscapes*, LeBlanc and Glen Rice insist that trade connections are a key element that proves the ubiquity of prehistoric warfare. They note: "Trade is also frequently an area for competition, and at times such contests can be only slightly less extreme than war." In his chapter "Giving War a Chance" another contributor to *Deadly Landscapes*, University of Illinois anthropologist Lawrence Keeley, says: "Universally, exchange has been a productive source of war — for all kinds of societies, trade partners have been the most common enemies." Keeley further boosts his position with a quote from the famous anthropologist Claude Levi-Strauss: "Warfare... is closely connected with barter." Yet sweeping pronouncements of this sort cannot be relied upon to tell us what happened in specific instances. It should be obvious that just because trade disputes can sometimes be a cause of war, it does not follow that evidence of complicated trade patterns means that war must have been a common occurrence.

Archaeology in the Service of the State

It might not be a coincidence that the theme of warfare in precolonial America has become popular among certain archaeologists at the same time as militaristic policies have found increasing favor among U.S. political leaders. Indeed, some of the proponents of pervasive prehistoric warfare seem to go out of their way to link their theories to current political concerns. One archaeologist who has allowed his own political viewpoints to show through in his archaeological writing is Dr. Warfare himself, Steven LeBlanc. An article he published in 2003 in the popular journal *Archaeology* ended with a peculiar comparison between Hitler and Saddam Hussein, the relevance of which to archaeology was tenuous at best. This juxtaposition of Hitler and Saddam Hussein mirrored the phraseology being used at the time by the U.S. government in its justification of its war with Iraq. It is curious that LeBlanc went out of his way to make this belabored reference to Saddam. The homage to U.S. policy in LeBlanc's article in *Archaeology* suggests that he might be pushing his theories of ubiquitous prehistoric militarism in part because they harmonize nicely with the current agenda in certain American political circles.

Conclusion

In several scientific fields that are closely related to human physiology and behavior — such as primatology, sociobiology, and endocrinology — the content of the subject has been shown to be deeply influenced by the ideological and gender biases of the (usually male) scientists. These biases have often caused researchers to confidently put forward theories which are at variance with the scientific evidence but which fit in nicely with the popular conceptions and stereotypes of their time.

The main line of argument of *Deadly Landscapes* is an example of such a breakdown in scientific methodology, this time in archaeology. Despite their reliance on the supposed expertise of a much-decorated Vietnam War veteran and their self-image as intrepid scientists battling against politically correct park service personnel, what Dr. Warfare and his followers have constructed is a myth about Hohokam militarism that seems more suitable for retelling around the campfire than for scholarly publication. The story of warfare of epic proportions between Perry Mesa and the Phoenix basin appears to be more a modern American masculinist narrative than an accurate description of the interactions among peoples in 14th-century Arizona.

Index

A Mathematician's Apology, 297
ABCs of Divestiture, 133, 134
abortion, 249, 262–267
Absent Without Leave (AWOL), 56
Abu Ghraib, 62
academic freedom, 67, 82, 83, 256, 281, 283
Ace Ventura: When Nature Calls, 272
Acta Mathematica Vietnamica, 181
Adleman, Len, 297, 307
Aeroflot, 21, 172, 178, 184, 198, 199
Aeronica, 233
affirmative action, 126, 127, 277
Afghanistan, 107
　war in, 61, 241
AFL-CIO, 25
African National Congress (ANC), 129
Agent Orange, 189, 201
Ahaz, 328
AIDS, 42, 202, 203
Air Vietnam, 210
Albania, 45
Alfaro, Lilia, 232
algebra, 57, 273, 281, 284, 285, 321, 336, 342
Algebraic Aspects of Cryptography, 316
algebraic geometry, 163, 168, 175
Alice in
　cryptography, 321, 325
　Wonderland, 287, 325
all but dissertation (ABD), 279
Allen, Bruce, 36
Alva, Jossy, 274
Alvarez, Lilliam, 248
American Airlines, 275
American Association for the Advancement of Science (AAAS), 202
American Association of University Professors (AAUP), 227
American Civil Liberties Union (ACLU), 57, 60
American Communist Party, 23, 31, 229

American Educator, 344
American Embassy in
　Bangkok, 210
　Karachi, 179
　London, 109
　Moscow, 103, 108, 170, 184
　New Delhi, 7
　Saigon, 170
　San José, 255
　San Salvador, 235, 256, 257, 262
American Mathematical Society (AMS), 17, 49, 117, 301, 365
　Bulletin, 92
　Notices, 333, 343
American School in Lima, 271, 338
American Southwest, 121, 231, 353, 369, 370
Aníbal, Vice-Technical Adviser, 244, 247
Ancud (Chile), 94
Andersen, Hans Christian, 34
Andrei Rublyov, 172
Angkor Wat, 211, 213, 222
Angola, 210, 212
Annals of Chinese Mathematics, 181
Annals of Mathematics, 87
Another Look at Provable Security, 325
Ansara, Michael, 42
Anti-Apartheid Act, 141
anti-semitism, 10, 77, 98, 100, 111, 114, 127
apartheid, 129, 130, 132–134, 137–139, 141, 242, 268
Apocalypse Now, 76
Apollo–Soyuz, 108
Aquinas, Saint Thomas, 250
Aquino, Corazon, 204
Araki, K., 314
Aranow, Phil, 138
archaeology, 359, 360, 368, 369, 371–374
Archaeology (journal), 373
architecture, 340, 361–364
ARENA (El Salvador), 255, 257

Argueta Antillón, José, 269
Aristotle, 249
arithmetic algebraic geometry, 85, 123, 296
Arizona, 232, 273, 307, 316, 353–355, 357, 359–361, 364, 366, 369, 372, 374
Arizona Miner, 367
Arizona State Land Office, 361
Arizona State University, 80, 93, 122, 231, 271, 281, 353, 358
Army, 4, 41, 48, 50, 52, 54, 55, 57, 59, 60, 75, 77, 80, 90, 163
   Advanced Individual Training, 51, 53–55
   Basic Training, 50–53, 55, 64
   discharge from, 57–60, 63
   fragging, 52
   Intelligence, 41, 53, 61
   Military Occupation Specialty (MOS), 50
   National Personnel Records Center, 60
   Officer Candidate School, 51
   Uniform Code of Military Justice (UCMJ), 58
Arnol'd, V. I., 300
Arteaga, Mélida, 253
Ash, Arlene, 201, 204, 205, 245
Association for Women in Mathematics (AWM), 289, 291, 338
Atlantic Monthly, 66–68, 82
attention deficit, 343
Auburn University, 318
   Tigers, 319
Augustine, Saint, 249
Averianova, Natalia, 204
Axler, Sheldon, 286
Ayacucho (Peru), 333, 339–341
Aymara, 273
Azize, Yamila, 264, 266, 268

baby-boom generation, 242
Babylonia, 313
Bạch Mai Hospital, 177
Bachop, Marty, 79
Balasubramanian, 208, 295
Bangkok, 107, 182, 185, 200, 202, 210
   Nation, 202
Barcelona, 363
Bard College, 3, 5
Baroda, India, 4, 6, 366
Bayer, Pilar, 363
Bearded Lesbian Circus, 234
beauty contests, 208, 275

Beijing Normal University, 342
Belize, 333, 334, 336
Bell Curve, 70
Bell Labs, 299
Bell, Daniel, 25
Bellare, Mihir, 322
Bellow, Saul, 178
Berg, John, 27, 33
Berganza, Blanca Elba, 268
Berkovich, Volodya, 96, 114, 169, 172
Berlitz, 171
Bernstein, Dan, 306
Bethune, Norman, 221
Bibi Khanum, 106
Bible, 57, 265, 266, 278, 313, 328
Big Island, 357
Biko, Steven, 129, 132, 133
Bilazarian, Pete, 22, 25, 27, 59
Bình Trị Thiên province, 224
biotechnology in Cuba, 245
Birch and Swinnerton-Dyer Conjecture, 316, 317
Birkhäuser, 119
Black Consciousness, 129
Black Panther Party, 44
Black Power, 12
Black Sea, 120
black studies, 33, 39
blasting caps, 357–359, 366
Bok, Derek, 127, 130, 131, 133, 135–137
Boles, Alan, 19
bomb squad, 357, 358
boogie-boarding, 306, 319
Borevich, Z. I., 14, 17
Borge, Tomás, 235, 236, 240, 249, 256
Boss, Valentine, 29, 33, 35
Boston Globe, 33
Boston University, 89, 91, 93, 109, 134, 284, 288
Botswana, 268
Bott, Raoul, 14, 26, 35, 134
Boy Scouts, 372
Boy Who Cried Wolf, 111, 316
Boyer, Richard, 26
Boyle
   Bill, 235, 256, 258
   Lastenia, 258
   Mayra, 259
Bradshaw Mountains, 368
brain drain, 342

Brand, Stewart, 362
breast-feeding, 359
Breslow, Doris, 4, 57
Brickell, Ernie, 304
Brighton, Ontario, 353
Bromfield, Louis, 6
Brothers Karamazov, 14
Brown University, 316
Brutus
   Dennis, 132
   Tony, 132, 139
Buhler, Joe, 123, 295
Bulgakov, Mikhail, 116
Bulgaria, 120, 121, 178
Bulletin of the Atomic Scientists, 9, 26, 163
Bundy, McGeorge, 22
Bureau of Land Management (BLM), 355, 367
Burma, 107
Burt, Cyril, 69, 81
Bush, George W., 229, 241, 320, 326
business model, 349
busing, 124, 228

Cabral, Esperanza, 204
Caceres, Ana, 255
Cal State Fullerton, 181, 190
calculus, 3, 223, 290, 291, 306, 331, 348, 356
Calderón, Neyssa, 244
Calkins, Hugh, 137, 138
Callejas, Emilisa, 237
Cambodia, 49, 186, 209–214
Cambridge Rent-Control Campaign, 25
camp sensibility, 40, 306
Campbell, Fred, 346–348
Canada, 103, 315, 320, 353
   flights to Cuba, 248
   freer speech in, 320
   sanctuary in, 47, 48
Canadian Football League, 229
Canadian Woman Studies, 122, 219
Cantor set, 63
Cape Cod, 4
Cape Town, 141, 333
Carabayllo (Peru), 338
Caravelle Hotel, 218
Carrey, Jim, 272
Carroll, Lewis, 287, 325, 372
Castillo, Fabio, 269
Castillo, Salvador, 256, 269, 275

Castro, Fidel, 243
   sexism of, 244
Catholic
   schools in India, 1, 233, 302
   stand on abortion, 264
Caucasus, 105
CCBG Architects, 361
Central America solidarity campaign, 230, 240–242, 258, 270
Central American University (UCA), 235, 253, 258
Central Asia, 105, 107
Central Intelligence Agency (CIA), 92, 190, 234, 301
Certicom, 303, 310, 314–316, 319, 320, 323, 347, 348
Chaldean poetry, 313
Chamorro, Violeta, 239
Channel Islands, 310
characteristic of a field, 86
Chechen-Ingush, 50
Chekhov, A. P., 15
Chern, Shiing-Shen, 180
Chile, 94, 315, 316, 333
Chiloe (Chile), 94
China, 45, 176, 177, 220, 342, 344, 350
China Airlines, 275
Chinese Revolution, 229
Christie, Renfrew, 141
Christmas bombings, 70, 177
Chronicle of Higher Education, 293
Chùa Hương Pagoda, 197
civil rights movement, 10, 12, 13, 242, 268
class prejudice, 13, 25, 41, 56, 80, 91
Clifford algebras, 26
Clinton, Bill, 305
Cold War, 14, 15, 102, 103, 108, 136, 301
College of New Rochelle, 257
Colorado Plateau, 355
Columbia University, 21, 24, 28, 29, 101, 183
Committee Against Racism (CAR), 124
Committee for Health Rights in Central America (CHRICA), 266
Committee for Scientific Cooperation with Vietnam, 181, 189–192, 200, 258
communism, 24, 119, 306, 356
   on one floor, 120
Communist Party of India Marxist-Leninist (CPI M-L), 209
complex numbers, 303

Complexities: Women in Mathematics, 291
computer science, 305, 307, 308, 327
CONAPRO (Nicaragua), 232, 236, 239, 240, 243
conditional theorems, 322
Continental Congress, 74
Contra war, 213, 230, 231, 235, 240, 252, 253
Cook, Captain, 357
Cook, Jul, 44
cooking, 96, 97, 115, 124
Cooper, John Sherman, 7
Cooperman, Ed, 181–185, 188–191, 199, 258
Copper Mountain, 307, 356, 357, 359–361, 363–366, 368
Coptic Church, 178, 280
Corning, Det. Chuck, 358
corporate withdrawal from South Africa, 129–131, 133, 134, 138, 139
Corrie, Rachel, 243
corruption, 272, 273
Cosgrove, Serena, 270, 365
Costa Rica, 234, 239, 247, 255, 269
Council of Ministers (Vietnam), 225
counter-culture, 39, 306
Country Joe and the Fish, 52, 54
Course in Mathematical Logic, 99
Course in Number Theory and Cryptography, 302
court-martial, 55, 58–61
Cousteau, Jacques, 334, 335
cover your ass (CYA), 54
Creating Futures, 349
Crimean Tatars, 50
Crown King (Arizona), 368
Crusoe, Robinson, 354
Cruz, Fredy, 232
Crypto, 301, 307, 308
   Crypto 2005, 322
   Crypto 2007, 308
   Crypto '92, 305
   Crypto '93, 306
   Crypto '95, 309
   Crypto '96, 307, 309, 310
   Crypto '97, 312, 315
cryptography, 117, 121, 208, 248, 291, 297, 299–304, 306, 312, 313, 315–319, 322, 323, 326, 328, 329, 331, 341, 345, 363
   elliptic curve (ECC), 245, 299, 302, 304, 310–314, 316–319
   private key, 298
   public key, 297–300, 302, 305, 307
   RSA, 298, 299, 305, 310, 312–315, 317, 318, 322
   symmetric, 298
Cuba, 44, 45, 213, 243–245, 247–249, 262, 333, 335, 342, 352
Cuban
   Academy of Sciences, 244, 245, 248, 249
   Communist Party, 249
   Embassy in Managua, 243, 244
   Embassy in Mexico, 246
   exiles in Miami, 53, 245
   Interest Section in Washington, 246
   missile crisis, 14
   revolution, 243
Cubana airlines, 190, 244
Cuckoo's Egg, 309
Culotta, Charles, 91
Cultural Revolution, 180
Cunningham, Ellen, 238
Cuscachapa Lagoon, 261, 262

Đà Nẵng, 224
Daily Princetonian, 71, 75, 77, 81
Daley, Mayor Richard, 23, 40
Dali, Salvador, 116
Damon, Nancy and Fred, 72
Dance, Bob, 261, 262
Dangerfield, Rodney, 212
Darwin, Charles, 89, 91, 93
   racism of, 89, 91, 93, 94
   social darwinism, 88, 89, 118
Data Encryption Standard (DES), 300
Davenport, Charles, 88
Davis, Chandler, 169
De Gaulle, Charles, 21
Deadly Landscapes, 373, 374
Defiance mining claim, 367, 368
Deligne, Pierre, 98, 99
democratic centralism, 50
Deutsch, Karl, 135, 277
Diamond, Malcolm, 76
Điện Biên Phủ, 188
Diffie, Whit, 291, 301
dinosaurs, 331
Diophantus, 313
Discharge Review Board, 60
Discover magazine, 288
discrete logarithms
   in finite fields, 300
   on elliptic curves, 299, 310, 316, 321

Disneyland of the left, 234
divestiture campaign, 129–131, 133, 135, 137, 140
Đoàn Quỳnh, 165
Dobrushin, R. L., 16
Dohrn, Bernadine, 43, 44
Đống Đa Hill, 176
Dormouse, 325
Dostoevsky, F. M., 13, 14
Dow Chemical, 22, 36, 66
Dr. Warfare, 360, 370, 373, 374
draft, 48, 241, 242
    in Nicaragua, 252
    lottery, 48, 49
    resistance, 22, 47, 48
    student deferment, 48
drugs, 28, 39, 56, 64, 272
Duchamp
    L. Timmel, 231
    Tom, 230, 231, 244
dumbing down, 346, 349
Dương Thị Cương, 189, 201, 202
Dương Thị Duyên, 197
Dutch Communist Party, 338
Duvall, Robert, 76
Dylan, Bob, 18, 40
d'Arbuisson, Roberto, 255

earthquake of 1986 in El Salvador, 253
Education and Politics at Harvard, 283
education in
    China, 344
    Cuba, 249
    El Salvador, 276
    India, 1
    Nicaragua, 233, 234
    Peru, 272
    the Soviet Union, 17, 18, 105
    the U.S., 18, 66, 67, 186, 310, 318, 331, 333, 335, 339, 341–344
    Vietnam, 166, 175, 187, 219, 336, 337
Eisenhower, Dwight, 10
Ejsberg, Denmark, 7
Eklof, Ben, 102
El Al, 178–180
El Nuevo Diario (Nicaragua), 244
El Salvador, 124, 229, 234, 237, 241, 247, 253–260, 262, 264–268, 270, 275, 332, 333, 340, 365
electronic battlefield, 70, 75, 76

Electronics and Telecommunications Research Institute (South Korea), 275
elephant (time perception), 308
Ellacuría, Ignacio, 235, 253, 258
ellipse, 303
elliptic curve, 117, 186, 298, 299, 302, 303, 313, 315–317, 320, 363
    cryptography, 245, 299, 302, 304, 310–319
Elliptic Curve Public Key Cryptosystems, 302
elliptic integrals, 303
Ellis, Richard, 14
embargo against Cuba, 244, 246
Emerson, Ralph Waldo, 58
endocrinology, 374
Engelhard, Charles W., 130
Engels, Friedrich, 250
English only, 231
equivalence relation, 233
error-correcting codes, 305, 319
Escalón (El Salvador), 270, 365
Escher, M. C., 116
Esenin, Sergei, 102
Esenin-Volpin, A. S., 102
Esquipulas peace accords, 236
Esquivel, Olga, 235
essentialism, 294
Estrada, Jorge, 248
Ethnic Diversity Requirement, 277, 278
eugenics, 88
evangelical Christianity, 264, 265
Evergreen State College, 243
evil empire, 301
Ewha University, 276
Exodus, 265
export controls, 306

Face the People (Nicaragua), 237, 256, 263
factorization of integers, 274, 298, 310, 311, 317, 321
    using elliptic curves, 299, 311
Faculty Senate, 230
Fair Housing Act, 128
Falk, Richard, 169
fallout shelters, 323
Fanshen, 26
Farabundo Martí National Liberation Front (FMLN), 235, 253, 270, 340
    November 1989 offensive, 259, 260
Farquhar, Alex, 49, 63, 65, 78

Federal Bureau of Investigation (FBI), 47, 74, 182, 183, 190, 301
Feigenbaum, Joan, 306, 356
Fellows, Mike, 274, 295, 304, 331, 332, 352
feminism, 90, 96, 99, 236–238, 250, 252, 254, 263, 267, 280, 294
Fermat, Pierre de, 92
Fermat's Last Theorem, 123, 363
fertility control, 355
fetishist
   post-modern, 329
   underwear, 86
Fields Medal, 98
finite fields, 86, 302, 303
Finnair, 184
fireplaces, 362
First Amendment, 61
Flagstaff, 364
Florence (Arizona), 355
Fomenko
   Tanya, 116
   Tolya, 116, 117, 300
Fonda, Jane, 178, 207
Ford (in South Africa), 138, 139
Ford Foundation, 196
Ford, Harrison, 369
Foreign Affairs, 287
forensic medicine, 249
Fort Belvoir, 56, 57, 59
Fort Dix, 41, 50, 53, 59, 64
Fort Eustis, 41, 51, 53, 59
Fort Leavenworth, 59
fragging, 52
France, uprising of 1968, 21
free speech, 26, 61, 67, 71, 115, 127, 256
Freud, Sigmund, 89, 250
Frey, Gerhard, 363
Friedan
   Betty, 79
   Dan, 79
Friedman, Debra, 291
Friendship Medal, 218
friendship, American notions of, 113
Froebel School, 339–341
   Parchment of Appreciation, 341
Froebel, Friedrich, 339
Frontline (India), 209
Fujimori, Alberto, 273
Fulbright fellowships, 1, 95

Gaborone, 268
Gaetsewe, John, 134
Gaita, Cecilia, 273
Gallagher, Pat, 183, 189
Gandhi
   Indira, 107, 204
   Mahatma, 13, 28
Gangolli, Ramesh, 344
Gardner, Martin, 297
Gates, Bill, 221
Gauss School, 273, 338, 339, 341
Gauss, C. F., 297, 338
Geison, Gerald, 91
Gel'fand, I. M., 97, 98, 99, 108, 300
gender studies, 360
generation gap, 25, 38, 208, 220, 221, 242
Geneva Accords of 1954, 9, 222
genocide, 22, 23, 26
gentrification in
   Cambridge, 25, 27, 140
   Princeton, 73
geometry, 333, 339, 342, 345
Gerschenkron, Alexander, 34, 77
Gilbert, Alan, 23, 27, 38, 50
Gillispie, Charles C., 89, 91, 94
girl work, 293
Gissen, Mel, 12
Glazman, Mary, 237
Glen Canyon, 354
Glick, Thomas, 93
Globe (Arizona), 366
Goa, 302
Gogol, N. V., 15
Goheen, Robert, 67, 75
Goldreich, Oded, 327, 329
Goldwater, Barry, 9
Gonzaga, Emilio, 273
Gonzalez, Mariano, 273
grade inflation, 346, 350
Granma (Cuba), 249
graph theory, 331, 333, 345
Graveman, Richard, 310
Great Blue Hole, 334
Great Depression, 4, 242, 367, 373
Great Society, 9
Great Wall, 220, 271
Grenada (Mississippi), 10, 11, 90, 140, 232
Griffiths, Phillip, 63, 85
Groff, Joe, 361–365
Gross, Benedict, 87, 123, 295, 297

Gross, Charlie, 66, 68, 76
Grothendieck, Alexander, 163–166, 168, 175, 184, 296
group, 303
Guatemala, 234, 334
Guernica, 40
Guerrero (Mexico), 319
Guerrero, María Gladys de Mena, 256, 265, 275
Guevara, Che, 44, 246
Guirola de Herrera, Norma Virginia, 259
Gustafson, Thane, 283
Gutman, Jeremiah, 57, 58, 60
guy work, 293

Hà Huy Khoái, 169–174, 176, 183, 184, 186, 195, 196, 198
Hạ Long Bay, 186
Hai Bà Trưng (Two Trưng Sisters), 177, 206, 207
Haiphong (Vietnam), 70
Hall, Leon, 11, 12
Hall, Sharlot, 355
Hallmark cards, 281
Hammer of Witches, 250
Handlin, Oscar, 135
Hanoi, 163, 172, 173, 175–178, 180, 181, 185–187, 189, 192, 193, 195, 197, 198, 200–202, 208, 209, 217, 221, 223, 274, 320, 336, 337
Hanoi Mathematical Institute, 172, 173, 178, 180, 185–187, 197, 200, 206, 217, 220, 222, 239, 336
Hanoi Pedagogical Institute, 165, 166, 186, 187
Hanoi Polytechnic Institute, 164, 181, 187–189
Hanoi University, 166, 187, 220
Harbison, Frederick, 76
Hardy, G. H., 297, 320
Harlem, 21
Harrington, Charles, 255
Hartwick College, 122, 195, 214, 259, 281, 288, 290, 331, 353
Harvard, 9, 11, 14, 19, 21, 22, 25, 43, 60, 66, 67, 77, 85, 87, 93, 108, 123, 125, 126, 128–132, 134, 135, 137, 138, 141, 184, 242, 273, 280, 281, 283, 286, 299, 306, 331, 367
    class of 1969, 137, 138, 365

Harvard Crimson, 19, 26, 28, 45, 46, 69, 124, 125, 133–136, 282, 283
Harvard Educational Review, 66
Harvard-Radcliffe Black Student Association (BSA), 130
hash functions, 328
Hasse-Witt matrix, 86
Hasse's Theorem, 315
Havana, 244, 245, 248, 320, 335
Hawaii, 357
Head Start, 66
Hebrew Immigrant Aid Society (HIAS), 110–112
height of points, 317
Helfgott
    Harald, 274
    Michel, 273, 274, 332
helicopter repair, 51, 53–55, 299
Hellman, Martin, 302, 304
Henderson the Rain King, 178
Hendrix, Jimi, 18, 40, 56
Hernández, Victoria, 248
Herrnstein, Richard, 66–68, 70, 73, 81, 82
Hezekiah, righteous King, 328, 329
Hibner
    Mary, 64, 80, 85, 265
    Michael, 80
Hicks, Louise Day, 124
High-Tech Heretic, 310
Highway 61 Revisited, 40
Hindu (newspaper), 209
Hinton, Jim, 77
Hinton, William, 26
Hironaka, Heisuke, 134
Hiroshima, 26
history of
    Arizona, 353–355, 359–361, 364, 366, 369, 374
    England, 123
    forensic medicine, 249
    mathematics, 23, 92, 280, 363
    Nicaragua, 230, 231, 235
    Pacific Northwest, 229
    Peru, 273
    Russia, 89, 102, 122, 278, 283
    science, 88, 91, 186, 278, 280
    Soviet Union, 29, 50
    Vietnam, 209, 221
    women, 206, 207, 236, 280
History of Science Society, 91

Hitler, Adolf, 190, 373
HMQV, 323
Hồ Chí Minh, 9, 173, 203, 209
  Trail, 223
Hòa Bình Hotel, 176
Hoàng Tụy, 180, 181, 188, 190, 197, 198, 217, 222, 223
Hoàng Xuân Hãn, 165
Hoàng Xuân Sính, 165, 168, 186, 189, 193, 195, 196, 219, 220, 224
Hoffman, Abbie, 50
Hoffmann, Stanley, 33
Hofstadter, Richard, 88
Hohokam, 360, 368–372, 374
Holcombe, Sherman, 124, 125, 127, 128, 130, 134
Home Depot, racism at, 232
Honda, T., 87
Honduras, 234, 237, 253
Honorary Doctorate (Ann), 238
Hopkins, Tom, 18, 39–42
Hopmann, Cornelius, 239
Hotel Santa Rosa, 340
How Buildings Learn, 362
Hudson Valley, 5
Humboldt School in Lima, 272, 338
hummingbird (time perception), 309
Humphrey, Hubert, 23
Humpty Dumpty, 287
hunger strike, 79
Huntington, Samuel, 22, 284, 285, 287, 288
Hurray for Me School, 339
hurricanes, 249
Huỳnh Mùi, 186, 220
Hyland, Richard, 29
hyperelliptic curves, 313, 315, 363

I.Q., 66, 69, 82
Iberoamerican Mathematical Olympiads, 274
IBM, 299, 300, 322
ice cream, 125, 296, 331, 332
ideal membership, 304
identical-twin studies, 69
idol worship, 328
Inca, 273, 333
independentista, 267
index-calculus algorithms, 298, 300, 310, 316
India, 1, 2, 4–6, 107, 121, 208, 209, 214, 233, 302, 309, 333, 350

Indian Institute of Technology, 209
Indiana Jones, 369, 371, 372
Indiana University, 16, 102
Indiantown Gap, 56, 57, 59
Indochinese Community Party, 171
Industrial Workers of the World (Wobblies), 229
Institute for Advanced Study (Princeton), 92, 189, 279, 296
Institute for Defense Analysis (IDA), 49, 73, 76, 134
Institute of Mathematical Sciences (Chennai), 208
Institute of the History and Philosophy of Science (Moscow), 121
insularity of Americans, 103, 231
intellectual property, 306
Interflug, 178, 179
International Association for Cryptologic Research (IACR), 307, 309, 327
International Congress of Mathematicians
  Helsinki (1978), 178, 180
  Moscow (1966), 23
  Vancouver (1974), 47, 168, 315
International Congress of the History and Philosophy of Science, 189
international law, 169, 287
International Mathematical Olympiads, 219, 220, 248, 256, 273, 274
International Research and Exchanges Board (IREX), 95, 116, 198, 279
International Studies Association, 219
Internet Engineering Task Force (IETF), 324
interruptions in American math classes, 343, 344
Iran, 179
Iran-Contra scandal, 233
Iraq War, 47, 62, 229, 241, 320, 373
Isaac, Ephraim, 178, 179, 280
Isis, 91
Israel, 108, 112, 178, 179, 243, 280, 282, 327, 328
Iwasawa, Kenkichi, 63
Ixtapa (Mexico), 319
Iztapalapa (Mexico), 319

Jackson, Allyn, 343
Jackson, Ethel, 352
Jacobson, Michael, 295, 318
Jagger, Mick, 242

Jakobson, Roman, 16
Japan, 342
Jehovah's Witnesses, 262
Jensen, Arthur, 66, 68
Jerome (Arizona), 366
Johnson, Lyndon, 9, 287
jointly authored papers, 88, 274, 295
Joplin, Janis, 18, 56
Journal of Cryptology, 313, 327

Kabul, 107
Kac
   Boris, 109, 110, 113
   Haika Bimuvna, 109, 113
   Jessica, 110, 111
   Misha, 109, 113
   Natasha, 109, 110
   Victor, 109, 113
Kac, Mark, 3, 4, 17, 49, 57, 63
Kahin, George McT., 9, 26, 163
Kamin, Leon, 26, 67–70, 81, 82
Katz, Nicholas, 85–87, 105
Kauai, 357
Kaufmann-Buhler, Walter, 118, 119
Kazhdan, Dima, 108, 109
Kazin, Mike, 42
Keeley, Lawrence, 373
Kennedy
   Edward, 110, 131
   John F., 9, 287
Kennedy School of Government, 128, 130, 131, 137
Kent State University, 49
Kerala (India), 208
Khâm Thiên Street, 177
Khlebnikov, V. V., 16
Khmer Rouge, 202, 209, 340
Khodzhibaeva, Sanabar, 204
Khodzhiev, 106
kid crypto, 305, 332
kindergarten, 1, 339
King County, 357
King, Martin Luther, Jr., 10, 11, 13, 28
Kings, 328
Kingston, Ontario, 139, 348
Kinzer, Stephen, 230
Kirillov, A. A., 95, 103, 106, 300
Kissinger, Henry, 22, 178, 195, 216
Klonsky, Mike, 43, 45
Klose, Kevin, 105

knapsack, 304
Koblitz
   Alvin, 7, 23
   Ann Hibner, 1, 35, 46, 63–66, 69, 71, 72, 74, 76, 78, 80–82, 85, 86, 88–97, 99, 101, 103, 104, 107–109, 111–114, 116–124, 128, 132–134, 136–141, 169–179, 181–189, 191–193, 195–197, 199, 200, 202–222, 224, 227–229, 231, 233, 236–240, 243, 245–249, 254–257, 259, 260, 263–265, 267–272, 275, 276, 278–280, 284, 287–292, 294–296, 306, 307, 310, 313, 314, 316, 319, 326, 332–334, 337–339, 342, 347, 352, 353, 355–360, 362, 364
   Donald, 7, 129
   Ellen, 7, 21, 55, 60, 90, 117, 118, 129, 279
   Minnie, 1, 2, 4, 5, 7, 8, 10, 23, 55, 69, 90
   Robert, 1, 3, 5–9, 23, 58
Koechlin, Pat, 71
Kominz, Larry, 27
Komitet Gosudarstvennoi Bezopasnosti (KGB), 104, 108, 109, 191, 199, 301
Kovalevskaia Fund, 189, 192, 193, 195, 201, 210, 212, 214, 215, 217, 224, 225, 227, 232, 234, 237–239, 243–246, 253, 256, 259, 262–264, 266, 267, 269–271, 273–276, 314, 331, 338, 341, 348, 365, 366
   Newsletter, 192, 196, 219, 224, 252, 259, 261, 292
Kovalevskaia Prize, 195
   in Cuba, 248, 249
   in El Salvador, 254, 255, 257, 259–261, 265, 268–270, 276
   in Nicaragua, 232, 236, 239, 244
   in Vietnam, 193, 195, 196, 207, 208, 214–217, 219, 221, 224, 225
Kovalevskaia, Sofia, 93, 118, 121, 186, 188, 225, 237, 257, 280, 293
Kovalevskii
   M. M., 89, 280
   V. O., 89, 118
Kramer, Heinrich, 250
Krawczyk, Hugo, 322, 323
Kuhn, Thomas, 92
Kuwait, 302
Kuyyakanon-Brandt, Ruankeo, 202

Ladd, Everett, 282
Ladinsky, Judith, 182, 189, 191, 192, 197

Lagos sisters, 340
　Becky, 341
　Edith, 340, 341
Lake Malawi, 271
Lake Ontario, 353
Lake Powell, 353
Landau, Susan, 291
Lang, Serge, 23, 57, 134, 281–285
Lao Động, 196
Laos, 9, 105, 107, 171, 213
Lay of Igor's Campaign, 15
Lê Dũng Tráng, 168, 169, 181
Lê Mai, 200
Lê Văn Thiêm, 168, 171, 173, 176, 177, 180, 181, 223
least publishable unit (LPU), 308
Lebesgue measure, 63
LeBlanc, Steven, 370, 372, 373
Legal Aid Society, 28
legal profession, 58, 129
Lemus de Bendix, Concha, 234, 237, 264, 265, 270
Lenin Library, 199
Lenin, V. I., 26, 29, 55
Leningrad, 95, 101, 105, 119, 229
Lenstra, Hendrik, Jr., 100, 299, 306, 311
Leontief, Wassily, 33
Levenson, Jim, 27, 28
Levi-Strauss, Claude, 373
Lewis, John, 9, 26, 163
Liberal Arts Professor, 344
liberation theology, 235, 253, 258
Lichty, Dan, 49, 50, 63, 65, 71, 78
Life magazine, 13, 28, 32
Lilongwe (Malawi), 262
Linares, Leonor, 264
Linder, Benjamin, 230, 241, 243
linear algebra, 305
Linnik, Ju. V., 17
Lipset, Seymour Martin, 281–283
literacy, 166, 175, 211, 219, 230, 249, 345, 348
Little Red Riding Hood, 111
London Mathematical Society, 175
Long Binh Jail, 56
Lopez y Lopez, Joaquín, 258
Lopez, Arnando, 258
Lopez-Casas, Eugenia, 255
Lorch, Lee, 246
Lord's Prayer, 233
Los Angeles, 104, 296

Louisiana State University, 134
loyalty oaths, 53
Luce, Don, 203
lumpenproletariat, 44, 56
Lương Thê Vinh, 176
Lynch, Nicolás, 340

Mabiletsa, Deborah, 132, 141
machismo, 250, 252, 360
machismo-leninismo, 244
Machu Picchu, 271
Machuca, Catalina, 269
Mad Hatter, 325
Mahoney, Michael, 91, 92
Malaspina, Uldarico, 273, 274
Malawi, 262, 333
Malayalam, 208
Maldonado, Felix, 273, 338, 339
male bonding, 360, 368, 371
Malecón (Havana), 245
Managua, 230–237, 239, 243–245
Mandela, Nelson, 141
Manin, Yu. I., 17, 88, 96, 97, 99, 103, 114, 121, 169, 170, 184, 300
Manson Family, 44
Manuel, Juan, 319
Mao Tse-tung, 26, 39, 41, 229
marathon, 100, 101
Marcuse, Herbert, 25
Maricopa County, 357
marksmanship, 51
Martín-Baró, Ignacio, 253, 254, 258, 259
Marx, sexism of, 250
Massachusetts Institute of Technology (MIT), 25, 109, 297
massacre at Central American University, 235, 258, 259
Master and Margarita, 116
MasterCard, 322
Mata Hari, 247
math education, 233, 248, 254, 268, 271, 305, 306, 310, 331, 333, 334, 336, 337, 339, 342–345, 350, 352
Math Institute Guesthouse, 187
math wars, 343
Mathematical Impressions, 117
mathematical induction, 86, 178
Mathematical Intelligencer, 117, 181, 198, 222, 286, 310
Mathematical Reviews, 295

Mathematical Sciences Research Institute (MSRI), 240, 313
Mathematics As Propaganda, 284, 285, 326
Mathematics Tomorrow, 284
Mauna Kea, 357
Mayakovsky, V. V., 16
Mayer (Arizona), 366–368
Mayer, Arno, 76
Mayer, Joe, 368
Mazur, Barry, 123, 134, 299
McCarthy period, 26, 35, 67
McCormick, Richard, 270, 293, 314, 346–349
  sex scandal, 293, 349
McDermott, Jim, 229, 261
McKean, Aldyn, 81
McKinnon, Ian, 320
McNamara, Robert, 22, 66
Meany Middle School, 345, 351
Medellín Catholic bishops' conference (1968), 253
Medical Committee for Indochina, 78
Mehta, Hansa, 5
Mekong River, 107, 211
Men and Masculinities, 360, 368
Menezes, Alfred, 295, 302, 308, 313, 316, 318, 320, 322, 325–327
Menezes-Qu-Vanstone (MQV) key agreement, 322
Mercer County Workhouse, 79
Merkle, Ralph, 304
Merkle-Hellman system, 304
Messing
  Rita, 296
  William, 296
Messing, Ellen, 35
Mexican Mathematical Society, 365
Mexico, 234, 244, 319, 333, 350
Mexico City, 319
Miami, 245–247, 273, 368
Microsoft, 365
Middlebury College, 96
Military Court of Appeals, 61
Miller, Frank, 97
Miller, Haynes, 134
Miller, Victor, 299, 300, 320
minimum dominating subset, 332
mining, 121, 355–360, 364, 366, 367
Minot, Reid, 40
missile shields, 323
modular arithmetic, 86, 304

modular forms, 186, 297, 298
Molina, Misaella, 255
Mongolia, 212
Monterrey Pop, 40
Montes, Segundo, 258
Monty Python, 210
Moon, Warren, 229
Moore, R. L., 3
Moorer, Adm. Thomas H., 70, 72
Morais, Herbert, 26
more one than anyone, 272
Moreno, Juan Ramón, 258
Moscow, 95, 97, 104, 300
Moscow Mathematical Society, 121, 300
Moscow State University, 95, 101
Moses, 328
Mother Jones, 102
Mothers of Heroes and Martyrs, 238
Mozambique, 210, 213
Muddy City, 352
Mullin, Joe, 137, 138
multiple-choice exams, 51, 55
Murray, Charles, 70
Murray, Emmett, 230
Museum of Northern Arizona, 360, 370
Museum of the Revolution (Managua), 252
Mustafin, Galim, 96, 105

NAACP, 228
  Legal Defense and Education Fund, 128
Nagasaki, 26
napalm, 21, 22, 76
Narváez, Silvia, 232
Nassau Hall, 65, 73, 74, 102, 134
Nasybullin, Anas, 96, 105, 169
National Academy of Sciences (NAS), 95, 98, 199, 285–288
National Autonomous University of Mexico (UNAM), 237, 318
National Autonomous University of Nicaragua (UNAN), 232, 239, 244, 256
National Catholic Reporter, 238
National Education Association (NEA), 69, 90
National Engineering University (Nicaragua), 239
National Football League, 229, 365
National Liberation Front (Vietnam), 44, 81, 195, 206

National Park Service, 370, 374
National Public Radio, 105
National Science Foundation (NSF), 3, 49, 200
National Security Agency (NSA), 301, 312, 319, 320, 323, 329
Native American walls, 359, 361, 362, 364
Navajo, 273
Naveda, Ruth, 338
Nazca lines, 271
Nazi Holocaust, 10, 22, 23, 115, 242
Nelson, Edwin, 63
Neruda, Pablo, 40
Nevada, 366
New Left, 24
New Left Notes, 42
new math, 233
New Mexico, 353, 355
New Republic, 286, 288
New York Times, 32, 69, 81, 111, 128, 138, 219, 230, 236, 259, 285, 288, 365
Newmeyer, John, 18, 39
Newsweek, 170, 293
Ngô Bá Thành, 204
Ngô Việt Trung, 211
Nguyễn Cao Kỳ, 190
Nguyễn Có Thạch, 189
Nguyễn Đình Ngọc, 186, 193, 195
Nguyễn Đình Trí, 181
Nguyễn Đình Xuyên, 172, 174
Nguyễn Hoàn, 167
Nguyễn Thị Bình, 195, 216–218, 221, 224
Nguyễn Thị Định, 206, 207
Nguyễn Thúc Hào, 223
Nguyễn Văn Đạo, 181
Nguyễn Xiển, 223
Nhân Dân, 201
Nicaragua, 210, 212, 230–241, 244, 247, 251–256, 263, 264, 270, 340
Nixon, Richard, 23, 70, 76
no-man's lands, 372
Nobel Peace Prize, 141
Nội Bài airport, 172
nondisclosure agreement, 302
Nong Khai, 107
Nongxa, Loyiso, 333
Norman Invasion, 359
North Korea, 45
North Pass, 369
Noticias Univisión, 231

Nteta, Chris, 132
nuclear
   test bans, 300, 301
   weapons, 26
number field sieve, 307, 311, 322
number theory, 14, 17, 22, 85, 92, 98, 100, 121, 123, 139, 169, 174, 208, 248, 297, 301, 302, 305, 314, 315, 320, 331, 363
Numbers (from the Bible, not the TV show), 328
Nurenberg, Robby, 49, 50

Odlyzko, Andrew, 299, 315
Office of Undergraduate Education, 346, 347
Offner, Carl, 22, 28, 36, 135
Ohio State University, 10
Old Testament, 328
Oliver, Bob, 117
One Column Pagoda, 173
Oort, Franz, 100
Optimal Asymmetric Encryption Padding (OAEP), 322
Ortega, Daniel, 237, 252
Orthodoxy (Jewish), 108
Osama Bin Laden, 190
Owens, Jim, 228, 229
O'Connor, Katherine, 15

p-adic
   analysis, 174, 184, 297
   formula for Gauss sums, 87, 123, 174, 297
   metric, 87
   numbers, 86, 87, 117
Pacaycasa (Peru), 333
Pacific Northwest, 94, 227, 229, 230, 243, 270, 353
Pacific Ocean, number of liters in, 337
Pagan (Burma), 107
Paine Hall sit-in, 24, 45, 46
Paracas (Peru), 271
Parent-Teacher Association (PTA), 227
Paris peace talks, 70, 195, 217
Pasternak, Boris, 15, 54, 60
Pat Koechlin principle, 71, 131
patents, 306
Pavlenko, Tanya, 119
Peace Brigades International, 270, 365
Peace Corps, 60, 138
Pentagon, 28, 57, 60, 70, 72, 76, 188, 371

perfect code, 305, 332
  cryptosystem, 305, 339
Perfume Mountain, 197
Perry Mesa (Arizona), 371, 372, 374
Persens, Jan, 141
Peru, 239, 271–274, 305, 332, 333, 338, 340, 342, 352, 368
petroglyphs, 369
Phạm Văn Đồng, 185, 203, 204, 227, 235
Phan Đình Diệu, 181
Phelps, Bob, 344
Phi Beta Kappa, 19
Phnom Penh, 209, 210, 212, 213
Phoenix, 353, 354, 357, 358, 360, 361, 364, 369, 371, 372, 374
Piatetski-Shapiro, I., 111, 112
Picacho
  Mountains, 354, 355, 360, 361, 372
  Point, 369
Pilkington, Greg, 33
Pinal County, 355, 357, 372
Pipes, Richard, 92
Plateau (journal), 360
Plato, 249
Plaza de Armas (Havana), 248, 335
Plovdiv (Bulgaria), 120
poets, 4, 15, 16, 40, 60, 112, 132, 223–225, 235, 244, 313
Poipu, 357
Pol Pot, 186, 202, 209, 211–213, 340
Poland, Professor, 368
police brutality, 23, 31, 32, 77
Political Order in Changing Societies, 284
Polly Cracker, 305
Polo, Marco, 359
Polytechnical Institute of Zihuatanejo, 319
Pomerance, Carl, 22, 25, 31
Pontifical Catholic University in Lima, 273
Posada Carriles, Luis, 190
Post-Modern Cryptography, 327
Pravda, 112
preferential option for the poor, 253
Prescott, 355, 359, 361, 364, 366
Presidential Palace (Hanoi), 203, 205
primality testing, 274, 336
primatology, 374
Princeton, 2, 27, 36, 48, 49, 63, 65, 66, 69, 70, 72, 73, 76–78, 80, 82, 85, 86, 88–93, 102, 104, 117, 134, 169, 184, 242, 274, 285, 296, 299

graduate studies in math, 48, 57, 63
Shopping Center, 85
Privacy on the Line, 291
ProfScam, 288, 289
Progressive Labor Party (PLP), 23–26, 28, 38, 44, 48, 50–55, 57, 59, 65, 66, 124, 136
Project Phoenix, 75
proof, connotations of, 323
protocol
  equals alcohol, 210
  in cryptography, 321
provable security, 320, 322, 323, 325, 326, 329
Provisional Revolutionary Government (Vietnam), 195
pseudorandom bits, 324
pseudoscience, 69, 88, 251, 286
Public Cryptography Study Group, 301
public key cryptography, 297–300, 302, 305, 307
public key validation, 323
Publishers Weekly, 288
Puerto Rico, 264, 266, 267, 333
Puget Sound, 94, 181, 187, 353
  Consumer Cooperative (PCC), 228
Purcell, Edward M., 14
Pusey, Nathan, 31, 33, 35, 36, 40
Pushkin, A. S., 15, 16
Putnam, Hilary, 35, 37, 68, 136
Pythagorean Theorem, 333

Qualifying Exam, 63, 64, 134
quarterbacks, 174, 229, 365
Quechua, 273, 274, 333, 340
Queen's University, 348

racism, 11, 23, 26, 58, 67–69, 73, 77, 80, 82, 83, 88, 89, 93, 98, 112, 124, 136, 139, 171, 228, 229, 232, 257, 273, 288, 292, 345
Radio Liberty, 14, 105
Ragin
  Debby, 117, 128
  Luther M., Jr., 117, 128–130, 133, 134, 137–141
Raiders of the Lost Ark, 369
Ram, N., 209
Ramos
  Cecilia, 258
  Julia Elba, 258

Ramparts magazine, 21
random curves, 312
random number generator, 324
random oracles, 328
rattlesnakes, 367
Ravenna-Bryant Community Association, 228
Reagan, Ronald, 119, 141, 210, 229, 234, 240, 253, 301
Red Napoleon, 188, 214
Red Queen, 325
Redford, Robert, 307
reduction, 321, 324
Reed College, 123
Reed, Lou, 40, 64
Reflections on Divestment of Stock, 133
refuseniks, 109–111
reproductive rights, 249, 254, 263, 265
Reserve Officer Training Corps (ROTC), 22, 24, 25, 27, 30, 54, 65, 73, 75, 140
Revolutionary Women's Association of Kampuchea (RWAK), 209, 213
Revolutionary Youth Movement (RYM), 25, 43–45, 49, 50, 56, 65
Reynolds, Diamond Joe, 368
Ribet, Ken, 363
Rice, Glen, 373
Riemann Hypothesis, 322
Rivest, Ron, 297, 313
Robertson, Gerald, Jr., 360, 370, 371
Robshaw, Matt, 314
rock music, 18, 64, 242
Rockefeller Foundation, 196
Rockefeller, David, 36
Rocky and Bullwinkle, 109
Rogaway, Philip, 322
Rohrlich, David, 123, 134, 295
Romero
  Edgar, 232, 235, 239, 243, 244, 255
  Frances, 233, 239
  Marco, 232, 239
Romero, Archbishop Oscar, 230, 255, 258, 259
Rony, Vera, 10, 57, 90
Roosevelt
  Eleanor, 4–6
  Franklin D., 4, 5
Rosamond, Fran, 306
Ross, Dorothy, 91

Rossiter, Margaret, 200
rotations of a circle, 303
Royer, Mayor Charles, 230
RSA cryptography, 297–299, 305, 310, 312–315, 317, 318, 322, 331
Rudd, Mark, 43
Rudenstine, Neil, 66
rump session, 302, 306
Rutgers University, 349
Rutgers University Press, 119

Saddam Hussein, 373
Sadosky, Cora, 338
Safeway, 354
Saidakhmat, 96, 105
Saigon, 56, 170, 193, 203, 206
Saint Mary-of-the-Woods College, 238
Salt River, 353
Salvadoran Communist Party, 253
Salvadoran Embassy in
  Managua, 255
  San José, 255
Salvadoran Institute for the Investigation, Training and Development of Women, 259
Salvadoran Women Doctors' Association, 262, 264, 266, 268, 270
Samarkand, 105, 106
San Diego State University, 3
San Juan Islands, 94
San Marcos National University (Lima), 271, 273, 340
San Miguel (El Salvador), 268, 269
San Vicente (El Salvador), 268, 269
Sandinista National Liberation Front (FSLN), 233–235, 239, 249, 252, 340
Sandinistas, 212, 230–232, 234, 239, 240, 247, 252, 264
Santa Ana (El Salvador), 268
Santa Ana Register, 191, 199
Santa Barbara, 301, 310
Santa Rosa (hotel in Ayacucho), 340
Santoro, Eugene, 257
Sartre, Jean-Paul, 19
Sarun, Ang, 202, 214
Satoh, T., 314
Savak, 179
Scarsdale, 2
Scarsdale High School, 4, 10, 14

Schneier
  Bruce, 341
  Karen, 341
Schopenhauer, Arthur, 250
Science and Politics of I.Q., 69
Science magazine, 288
Science, Women and Revolution in Russia, 122
Scientific American, 297
sea lions, 94, 271
Seattle, 94, 184, 227–229, 241, 245, 353
  racial climate in, 228
Seattle Times, 230, 258, 349
Seier, Edith, 274
self-witnessing problems, 274, 305
Semaev, Igor, 301
Semana Santa (Holy Week), 231
Sen, Hun, 202, 209–211
Sengo, Gonçalve, 213
serpent
  bronze, 328
  fiery, 328
sex and drugs and rock 'n roll, 64, 355
Sex and Herbs and Birth Control, 249, 355
sexism, 83, 88, 89, 91, 96, 99, 100, 118, 188, 204, 205, 218, 243, 244, 249–252, 254, 279, 289, 290, 292
sexual harassment, 239, 279
Shafarevich, I. R., 14, 17, 97, 98, 120, 300
Shamir, Adi, 297, 304
Sharlot Hall Museum, 355, 359, 366, 367
Shelley, Percy Bysshe, 4
Shining Path, 340
Shkorpilovtsy (Bulgaria), 120
Shockley, William, 98
Shoup, Victor, 322
Siberia, 119
Signs (journal), 290, 295
Silicon Snake Oil, 309
Silverman, Bob, 315
Silverman, Joe, 316–318, 324
Simmons, Gus, 300
Simon, Herbert, 285, 287, 326
Sino-Soviet split, 23, 102
Sivanna, Ros, 202
Sky and the Stars, 208, 217
Smale, Steve, 23
small but strong, 319
Small, Art, 59
Smart, Nigel, 314

Smash the Bosses' Armed Forces, 52, 54
Smith, Mark, 130, 131, 137, 138, 141
Sneakers, 307
snorkeling, 271, 319, 334, 357
Snow, C. P., 136, 277, 278, 295, 308
socialism in one country, 120
sociobiology, 374
solar power, 364
Solinas, Jerry, 312, 320
Solis, Joaquín, 237, 240
solitary confinement, 56, 57, 79, 281
Somoza Debayle, Anastasio, 234, 249, 252
Soncco, Daniel, 274
Sonora desert, 354
Sontag, Susan, 40
South Africa, 129–135, 137–141, 268, 277, 333, 342, 365
South African
  Congress of Trade Unions, 134
  Council of Churches, 132
  Digest, 133
  Student Organization, 129
South Korea, 275
Southern Africa Solidarity Committee (SASC), 130–132, 137, 140, 141
Southern Christian Leadership Conference (SCLC), 10, 12
Soviet
  Academy of Sciences, 95, 115, 119, 199
  educational system, 17, 18, 105
  Jews, 96, 98, 108, 110–112, 114
  mistreatment of mathematicians, 102
  officials, 116, 199
  socialism, 102, 106, 107, 120
  Union, 14, 16, 17, 95, 96, 98, 99, 101–111, 113–115, 117–121, 198, 218, 227, 248, 299–301, 315
Soweto, 129
Spanish language, 213, 231, 232, 234, 235, 333, 334, 352
Spencer, Herbert, 118
Sprenger, Jakob, 250
Springer-Verlag, 118, 119, 283, 284, 298, 307
Sputnik, 3, 14
spy vs. spy, 329
St. Joseph Church (Seattle), 259
Stalin, 50, 120
Stanford, 129, 279, 281
State Committee for Science and Technology (Vietnam), 188, 214, 224

State University of New York at Binghamton, 279, 281
Stauder, Jack, 35
Stein, Andreas, 295, 318
Stein, Stanley, 76
Steiner, Daniel, 127
Steklov Institute, 116, 315
Stepanov, Sergei, 315
Sternberg, Shlomo, 112, 281, 283
Stigler, James, 342, 344
Stillman, Beatrice, 118
Stoffel, Craig, 363
Stoll, Cliff, 309
strategic
    hamlets, 254, 287
    planning (guy work), 294
Strauss, Ira, 67
strikes
    1936-1937 auto workers, 26
    General Dynamics, 38, 55
    Harvard students, 31–33, 38, 81, 137
Strong, Anna Louise, 229
Struik
    Dirk, 338
    Rebekka, 338
    Ruth Ramler, 338
Student Non-violent Coordinating Committee (SNCC), 10, 12
student power, 32, 33
student ratings, 289–292, 349, 350
student writing, 4, 35, 81, 345–347, 351
Students for a Democratic Society (SDS), 19, 22, 23, 25–27, 29, 31, 32, 35–37, 40, 42, 43, 48, 49, 65, 80, 81, 134–137, 140
    national convention (1969), 42, 43, 45, 80
subset sum problem, 304
Sullivan
    Leon, 129, 138, 139
    Principles, 129, 139
Sumner, William Graham, 89
Sun Microsystems, 291
Supreme Court (U.S.), 61
surfing, 306, 357
sweat equity, 367
Sykes, Charles, 288

Tạ Quang Bửu, 164
TACA airlines, 236
Taejon Expo, 275
Taj Mahal, 107
take back the night, 358
Tanzania, 302
Tapia-Recillas, Horacio, 319
Taranovski, Kiril, 16
Tardieu, Ambroise, 249
Tashkent, 105, 106, 198
Tate, John, 14, 17, 21, 123, 134, 282
Tây Hồ (West Lake), 174
Tbilisi, 105, 106
terrorism, 54, 120, 272
Teske, Edlyn, 295, 318
Test of English as a Foreign Language (TOEFL), 192
Tet Offensive, 188
Thái Bình province, 205, 211, 213
Thailand, 105, 107, 201, 202, 209, 211
Thắng Lợi, 174
Thăng Long University, 219, 220
Third International Math and Science Study (TIMSS), 342
Thoreau, Henry David, 58
tiger cages, 203
tightness gap, 324
Tijerino, Doris, 251, 264
Tip Top (Arizona), 368
Tiutchev, F. I., 15
Tố Hữu, 223, 224
Todorov
    Andrei, 120, 178
    Betty, 120, 178
Toledo, Rhyna Antonieta, 255
Tolstoy, L. N., 15
Toro, Tatiana, 291, 292, 348
torus, 117, 303
Tourist Town, 331, 332
trapdoor one-way function, 298
Trenton County Jail, 79
Trương Mỹ Hoa, 206, 207, 217
Tsongas, Paul, 138
Tucson, 354, 369, 372
Turkey, 315
turn the guns around, 48, 52
Tutu, Bishop Desmond, 132, 141
two cultures, 277, 278, 294, 295, 308
Two Cultures and the Scientific Revolution, 277

Ulugh Bek, 106
Uncle Sam Wants You, 102
Underground Railroad, 47

United Nations Human Rights Commission, 5
University Action Group (UAG), 65, 66, 69, 71, 72, 74, 81, 82
University Hall, 13, 27, 28, 30, 134, 135
University of Baroda, 1, 5, 6
University of British Columbia, 117
University of California at Berkeley, 240, 306
University of California at Irvine, 220, 295
University of Chicago, 253
University of Dayton, 279, 281
University of El Salvador (UES), 235, 253–257, 259–262, 264, 265, 267, 269, 270, 275
    military occupations of, 253, 257, 260, 261
    School of Medicine, 260–262
    School of Pharmacology, 256, 275
    Women's Commission, 254, 262, 267, 269
University of Essen, 301, 363
University of Havana, 245, 248
University of Idaho, 332
University of Illinois, 373
University of Iowa, 82
University of Michigan, 274
University of Minnesota, 93, 296
University of Puerto Rico, 264
University of Talca, 94
University of the Western Cape, 141, 333
University of Toronto, 48, 169, 279, 280
University of Victoria, 304, 332
University of Washington, 35, 45, 81, 123, 181, 230, 234, 240–242, 244, 259, 261, 270, 277, 291, 297, 314, 331, 344, 346, 348
    racism at, 228, 229, 292
    sex scandal at, 293, 349
University of Waterloo, 295, 302, 318, 319
    sabbatical year, 248, 291, 295, 314, 316–318, 353
University of Western Kentucky, 281
University of Wisconsin, 131, 135, 189
Ursinus College, 281
USA Today, 259
Uzbekistan, 106, 107

Valley of the Sun, 353
Văn Miếu (Temple of Literature), 176
Vanegas, Nora, 254, 263, 267, 269
Vanstone, Scott, 302, 307, 309, 314, 316, 353
Varna (Bulgaria), 120
Velvet Underground, 40
venceremos brigades, 246
Veterans Administration, 62
Victory Day, 170
    marathon, 100
Vides, Irma Lucía, 268
Việt Kiều, 267, 268
Việt Minh, 9, 222, 223
Viet Nam News, 218
Vietnam, 9, 44, 45, 81, 105, 122, 163, 168, 169, 171, 172, 175, 176, 181–183, 185, 187–189, 191, 192, 195, 197, 198, 200–203, 206, 208, 209, 211, 214, 216, 218, 219, 221, 227, 232, 245, 256, 267, 271, 275, 309, 333, 336–338, 342
    National Liberation Front, 44, 81, 195, 206
    War, 9, 21, 22, 26, 27, 31, 33, 38, 40, 47, 52, 54, 56, 65, 70, 73, 75, 80, 81, 112, 134, 163, 170, 175, 177, 185, 186, 188, 190, 192, 193, 195, 201, 203, 206, 207, 224, 240, 254, 287, 288, 360, 371, 374
    Women's Museum, 206, 208, 214
    Women's Union, 189, 195–197, 200, 201, 203, 205–207, 209, 214, 215, 217–219, 224, 239
Vietnamese Embassy in
    Bangkok, 200
    Moscow, 172
Vietnamese Mission to the U.N., 185, 190
Vietnamese refugees, 183, 185, 189–192
Vindication of the Rights of Woman, 250
Vinogradov, I. M., 315
Viola, Lynne, 279
Virginia City (Nevada), 366
Visa, 322
visaless zone, 199
Vishik, Misha, 97
Vivas, Liz, 274
Võ Hồng Anh, 201, 214
Võ Nguyên Giáp, 188, 201, 214–216
Voyage of the Beagle, 94
Vương Ngọc Châu, 173, 174, 186, 198

Walker, William, 257, 258, 262
Wall Street Journal, 33
Wallace, Alfred Russell, 89
Walls, Chaka, 44
Wan, Daqing, 220

Warfield, Bob, 345
Warhol, Andy, 18, 40, 41, 64
Wari, 273, 333
Washington Middle School, 344, 352
Washington Post, 105, 288
Wasiolek, Edward, 14
We Won't Go statement, 22, 47
Weekly World News, 305
Weierstrass, Karl, 225
Weil Conjectures, 98
Weil, Andre, 92, 315
Weinberg, Paul, 59
Weintraub, Steve, 134
Weitz
    Carol, 117, 118
    Eric, 93, 117, 118
Weizmann Institute, 114, 327
Wellesley College, 279
Wenger, Steve, 54, 55, 58, 59
White Knight, 325
Who Are We? The Challenge to America's National Identity, 288
Wilcox, David R., 360, 370, 371
Wiles, Andrew, 123, 135, 363
Will, George, 286
Williams, Hosea, 11–13
Wired, 314
witch hunts, 26, 250, 286
Wobblies, 229
Wollstonecraft, Mary, 250
Wolpert, Julian, 285
women and science conferences in
    Costa Rica, 239, 255
    El Salvador, 254, 255
    Hanoi, 196, 200, 201, 205, 212, 214, 235
    Lima, 274
    Managua, 232, 234–236, 238, 243, 249

women's movement, 44, 96, 106
women's studies, 80, 122, 267, 271, 278, 295, 358, 360
Woods, Donald, 132, 133, 141
work-in, 49
Worker-Student Alliance (WSA), 24–27, 32, 37–40, 42–44, 49, 65, 66, 90
Workers Defense League, 10
World Conference on Women in Beijing, 217, 262
World War I, 48
World War II, 1, 4, 7, 23, 29, 48, 101, 115, 120, 242
Wright, Frank Lloyd, 362
Writers' Union, 114

xedni calculus, 316–318, 324
xenophobia, 232
Xhosa, 333, 334

Yale, 19, 60, 90, 129, 274, 306
Yavapai County, 357, 359, 366, 367
Yavapai Courier, 355, 366
Yavapai Indians, 359
Yippies, 50
York University, 246
Young Communist League (Komsomol), 101, 102
Young Lords, 44
Young Republicans, 67
Youth Culture faction, 24, 25, 42
Yugoslavia, 257
Yushkevich, A. P., 121

zeta-function, 87
Zihuatanejo (Mexico), 319
Zimbabwe, 333, 335
Zoshchenko, M. M., 15

Printing: Krips bv, Meppel, The Netherlands
Binding: Stürtz, Würzburg, Germany